THE GAME OF PROBABILITY

Cultural Memory
in
the
Present

Mieke Bal and Hent de Vries, Editors

THE GAME OF PROBABILITY
Literature and Calculation from Pascal to Kleist

Rüdiger Campe

Translated by Ellwood H. Wiggins, Jr.

STANFORD UNIVERSITY PRESS
STANFORD, CALIFORNIA

Stanford University Press
Stanford, California

English translation © 2012 by the Board of Trustees of the Leland Stanford Junior University. All rights reserved.

The Game of Probability: Literature and Calculation from Pascal to Kleist was originally published in German under the title *Spiel der Wahrscheinlichkeit. Literatur und Berechnung zwischen Pascal und Kleist* © Wallstein Verlag, Göttingen 2002.

No part of this book may be reproduced or transmitted in any form or by any means, electronic or mechanical, including photocopying and recording, or in any information storage or retrieval system without the prior written permission of Stanford University Press.

Printed in the United States of America on acid-free, archival-quality paper

Library of Congress Cataloging-in-Publication Data

Campe, Rüdiger, author.
  [Spiel der Wahrscheinlichkeit. English]
  The game of probability : literature and calculation from Pascal to Kleist / Rüdiger Campe ; translated by Ellwood H. Wiggins, Jr.
    pages cm. — (Cultural memory in the present)
  "Originally published in German under the title Spiel der Wahrscheinlichkeit."
  Includes bibliographical references.
  ISBN 978-0-8047-6864-1 (cloth : alk. paper)
  ISBN 978-0-8047-6865-8 (pbk. : alk. paper)
    1. European literature—17th century—History and criticism. 2. European literature—18th century—History and criticism. 3. Probability in literature. I. Title. II. Series: Cultural memory in the present.
PN744.C3613 2012
809'.93384—dc23

2012010489

Contents

Introduction — 1

PART I. GAMES FOR EXAMPLE: MODELING PROBABILITY

1 Theology and the Law: Dice in the Air — 15

2 Numbers and Calculation in Context: The Game of Decision—Pascal — 37

3 Writing the Calculation of Chances: Justice and Fair Game—Christiaan Huygens — 73

4 Probability, a Postscript to the Theory of Chance: Logic and Contractual Law—Arnauld, Leibniz, Pufendorf — 97

5 Probability Applied: Ancient Topoi and the Theory of Games of Chance—Jacob Bernoulli — 118

6 Continued Proclamations: The Law of *logica probabilium*—Leibniz — 147

7 Defoe's *Robinson Crusoe*, or, The Improbability of Survival — 172

PART II. VERISIMILITUDE SPELLED OUT: THE APPEARANCE OF TRUTH

8 Numbers and Tables in Narration: Jurists and Clergymen and Their Bureaucratic Hobbies — 195

9 Novels and Tables: Defoe's *A Journal of the Plague Year* and Schnabel's *Die Insel Felsenburg* — 220

10 The Theory of Probability and the Form of the Novel:
    Daniel Bernoulli on Utility Value, the Anthropology
    of Risk, and Gellert's Epistolary Fiction   248

11 "Improbable Probability": The Theory of the Novel
    and Its Trope—Fielding's *Tom Jones* and Wieland's
    Agathon   273

12 The Appearance of Truth: Logic, Aesthetics,
    and Experimentation—Lambert   305

13 "Probable" or "Plausible": Mathematical Formula
    Versus Philosophical Discourse—Kant   338

14 Kleist's "Improbable Veracities," or, A Romantic Ending   369

    Conclusion   391

    Notes   399
    Bibliography   465

Introduction

The term "probability" implies the notion of reality. In Europe from antiquity to the early modern era, at least, this was long held to be the case. Speaking of probability—of what approaches or appears similar to truth—meant making a statement or assumption very different in its stance toward the actual world from the truth of science and philosophy. The standards of reliability and mutual understanding that politicians adhered to in their debates and people used in everyday life, as well as those attributed to poetry and the rhetorical arts, were not supposed to meet the criteria of truth. Instead, probability was thought to provide a kind of reliability and mutual understanding that did not owe its existence to criteria for true knowledge and efforts to define and establish them. With probability, orators and writers presupposed the existence of a reality that philosophy and science had to take into account. With its universal claims and rational imperatives, science had to prove and justify itself before reality. "Probability," then, was the password for access to this real world that set up camp around the fortified city of the sciences.

This claim admittedly presents a partial view of what probability meant in antiquity and the persistence of this meaning into early modernity. It was not reality itself, but rather a watchword that opened the door to reality. With the term "probability," one assumed reality and made use of this assumption. Only under the aegis of probability could reality become a subject of knowledge, whether in philosophy, science, or history. When lawyers produced probable arguments before the court, when political orators in the agora called upon probability, or when scientists disputed first principles among themselves and had recourse to probability in order to do so, they were all availing themselves of highly developed and, as they thought, highly efficient techniques. Probability was a figure—a technical gesture—that made use of reality. Hence a large arsenal of technicalities and formalizations formed around probability in the service of an access to reality. Forms of sophistic art, including the precepts of rhetoric as well as the rules of dialectical disputation, contributed to this stockpile of techniques.[1] They constituted the first and exemplary arsenal of old European *technē*.

Probability, since the time of the ancient philosophers and rhetoricians, had been the term with which reality eluded or opposed science; it was the figure with which reality could be impugned, summoned to court, or paraded on the political stage. The term "probability" thus had a somewhat tentative aspect that distanced it from science and scientific rigor, as well as a more aggressive streak directed at practical efficiency. This ambivalence did not generate difficulties, however, as long as both aspects were united in their common distinction from the true knowledge of science. The reality to which one retreated with the term "probability," or that one maneuvered to access with its help, was one and the same. The two aspects of reality, the reality domesticated by science and the one that eluded its control, remained indistinguishable in that they both remained outside of knowledge. This one reality was the source of the efficacy of rhetorical *technē* precisely because it distanced itself from philosophy and science.

This foundational relation between knowledge and its object was transformed at a decisive turning point toward the end of the seventeenth century; and the consequences of this far-reaching event were worked out, discussed, and employed throughout the following century. The mathematical theory of probability took over the role of the old probability and it did so, significantly, in scientific terms. At approximately the same time, the new realistic novel made probability key to its form of presentation, a form in which reality now could appear to readers as an integral whole. Around 1660, and definitely by 1700, the rhetoric of probability, which had been at the periphery of true scientific knowledge, became the newest and most precious conquest of the scientific urge to expand and of the aesthetic capacity to represent.[2] This "probabilistic revolution," as it has been called in reference to mathematical probability theory,³ was not merely a revolution that changed our idea of society and nature. Indeed, with statistics, the doctrine of chances, and the form of the modern novel, it brought our modern ideas of society and nature into existence in the first place.

Both literary history and the history of science have so far either claimed that the probability theory of mathematicians around 1700 had nothing to do with that of poets and logicians, or they have left this relation in the dark.⁴ The initial claim of this book, however, is that mathematical probability settled and made itself at home on the terrain of the old European probability of orators and logicians; it was bred and nourished by the inherited terminology of reality. The new probability of mathematics would have been impossible without the assistance and innovative ideas of specialists in public speech and logic. We can even claim that rhetoricians and philosophers were the first to show that the mathematical model for games of chance expressed the structure of what we mean when we speak of probability, whether we are aware of it or not. And poets began somewhat later—in the first decades of the eighteenth century—to apply the old European terminology of probability in such a way that probability became the semblance or appearance [*Schein*] of truth. They invented the modern novel. This inextricable relationship between what would later be divided into the natural sciences and the humanities made eighteenth-century probability theory revolutionary. And it was indeed a revolution, even if it remained hidden in the terminological history of knowledge—an undeclared turnabout before Kant's Copernican and France's social revolution.

The implications and consequences of this thesis can only be grasped if we recognize that even the old European "probability" was already a technical term or figure for reality. It referred to reality in contrast to science. In the sense of rhetoric as a *technē*, probability was the technical way to lay claim to reality. Without the term "probability," finally, no reality could be made valid. "Term" in all these cases refers to an elaborated use of words and expressions that, like the rhetorical figure, is located between everyday speech and scientific concept. Though terms and figures are indeed formalized and ordered by schemata, they are organized in that way only according to their use. Terms and figures are either sorted wordings or operative concepts depending on their employment in discourse. This view of terms, and of the term "probability" in particular, differs from Hans Blumenberg's account of the introduction of probability theory.[5] In other respects, the present book's description of this process is quite similar to Blumenberg's, because he also insists on locating the roots of probability in ancient logic or rhetoric. But for him, the mathematization of probability designates a sudden departure from the old European metaphorical notion of probability. Metaphorical probability, which can be grasped in verisimilitude, or the appearance of truth [*Wahrscheinlichkeit* or *Schein der Wahrheit*], is in Blumenberg's view embedded in our lifeworld; mathematization then transfers this element of the lifeworld into the fixed and conclusive logic and terminology of science.[6] The transition from rhetorical-dialectical probability into the probability of the mathematical theory of games of chance is for Blumenberg the paradigmatic case of what he calls the process of terminologization, the evolution of a metaphorical expression into a defined concept. If, however, rhetorical verisimilitude and mathematical probability are understood as relating to each other as a metaphor to a concept, then the essential interrelation between literature and science would be obscured from the outset. The terminological links between literary verisimilitude and mathematical probability that can be reconstructed in the birth of probability theory around 1700 would inevitably lose their crucial meaning. In short, Blumenberg's view no longer allows us to see that the metaphorical gestures had already been technical terms of reality, and that even a mathematical probability theory strove at first to account for the same probability or verisimilitude that poets and orators referred to in their figures and arguments. This finally means that the historical and systematic relations between the scientific and the literary events of the emergence of probability theory and the modern novel would be lost. Blumenberg's "terminologization of probability" and the "concept of reality" he analyzes in his brilliant essay are in fact, contrary to Blumenberg's own views, consonant events in the history of knowledge.[7] "Terminologization of metaphorical probability" in this investigation, therefore, does not mean, as it does in Blumenberg, a single event in the history of knowledge. Probability instead terminologizes reality right from the beginning.

This view does not diminish what has been vaunted as the revolutionary character of modern probability in mathematics and the novel around 1700. On the contrary, only from the point of view of probability's terminological residue can we truly appreciate the overthrow of foundational understanding represented by probability theory and the concept of reality manifested in the modern novel. In other words, we are embarking on an investigation into the reversal of the relation between science and literature, the reference, that is to say, that orients both science and literature toward "reality," and, with it, toward what is supposed to be their ultimate concern and object. From now on this relation becomes paradoxical: the novel and games of chance invent models of reality, while reality is still regarded as opposed to the knowledge of science and as removed from its grasp. A paradox emerges that must be solved either by assuming a reality that is reasonable per se (Hegel) or by admitting various levels of access to reality for probability (modern constructivism). The probabilistic revolution took both paths.

A great many characters from a variety of different fields and with very different intentions participated in the probabilistic revolution. Mathematicians wished to solve a specific problem in arithmetic: they wanted to understand how advantage and disadvantage were distributed in games of chance. Moral theologians wanted to banish the dangers of gambling and make sure that chance should stand at God's behest alone. For them, this pastoral interest overlapped with their basic dogmatic interest in miracles, in the possibility of God's intervention in the lawful course of nature and history. Jurists were interested in certain forms of contract that have unforeseeable events as preconditions for their fulfillment. Natural scientists worked on developing a rubric of conditions under which experiments either succeed or fail. Their theory of experimentation was in turn connected to the question of rhetoricians and logicians about the relation between an event and its circumstances, a theme that was also crucial to the poetological investigation of literary works. The early insurance industry and the legal constitution of corporations with collective risks relied on the calculation of events that by definition could not be controlled by the parties involved. Mathematicians interested in economics were concerned with the nature of a risk that can or cannot be underwritten, and hence with an assessment of what exactly taking a risk entails. In this way they defined what might in today's terms be called risk management; and such risk taking was notably different from the then-outdated former attitude toward the future, pious adaptation to the contingency of the world by meditation and spiritual exercises. This method of evaluating risks would later be developed into the concept of value, a basic element of general economic theory. Civil servants wanted to be able to draw conclusions about the future development of the strength and productive efficiency of the state from data about population development. In this endeavor, they were very close to the attempt of academic historians and philosophers to develop a subject called "statistics." This subject—in contrast to history, which reports singular events, and to political philosophy, which contemplates the ideal form of the sovereign state in strict universality—was supposed to deal with the condition of a particular state at a particular time, an object of study midway between singularity and generality. In this attempt, they saw themselves relegated to the area of probability as Aristotle had delineated it in the *Poetics*. Their considerations hence fed on the same sources that novelists and critics drew upon. Both writers and critics had to create a new theory for this genre, because there had been no such thing as the novel, or even a place for it, in traditional poetics. According to all the criteria by which poetics recognized literary forms, the new prose novel appeared formless: it failed to follow any metric patterns and employed a rather artless diction. In order to develop a theory of the novel's form, authors and critics had recourse to the probability of its subject matter and narrative methods. For this reason, the novel required its own entirely new theory, one distinct from the treatments of rhetoric and poetics that had comprised all possible knowledge about literature up to that point.[8]

All or at least most of these questions and themes, intentions and economic necessities will be at issue in this book. In one important respect, in one certain connection and moment, each and every one of them influenced or was impacted by the probabilistic revolution around and after 1700. The book will not, however, attempt merely to sketch the panorama of fields in which probability played a role between 1660 and 1800. Nor are we concerned with establishing a hermeneutic hierarchy with certain themes as the focus of interest, and others as their (social, economic, religious, etc.) contexts. In all the various fields, the work on probability has to do with technical terms about and access to reality. We are concerned with a history of knowledge that vaulted probability into a form of knowledge, a definition of social relations, as well as the basic principle of the modern novel. This does not mean that science, society, and art should be considered as end-products of the primary operation of some Hegelian spirit. But it does mean that they first emerged as the entities we call society, knowledge, and art within the various theoretical practices this book focuses on.

Following Michel Foucault, work in the history of knowledge implies an interest in the transformations and processes by which the pre-theoretical gestures in which knowledge is incorporated are converted into the theoretical knowledge of science.[9] Probability around 1700 is no arbitrary case for such a history of knowledge. Even from the outset, the term "probability" designated the very relations that comprise such a history, the relations between pre-theoretical practices and knowledge based on theoretical assumptions—whether such relations implied distance and opposition or convergence and assimilation. The transformation of traditional probability into the probabilistic theories of games of chance, on the one hand, and into the novel, on the other, therefore bears particular significance for a Foucauldian history of knowledge; the crucial implications of such a history are at stake in this study.

Two major threads can be discerned in the transformation of the old European probability into probability theory around 1700. The two lines often overlap and permit conclusions to be drawn from one about the other, even if their respective courses of development remain largely independent. The first thread can be called the interpretation of gaming theory as probability. It is concerned with the terminological work that read the mathematical doctrine of chances back into the tradition of logical and rhetorical probability and its semantic ramifications. Alongside mathematicians and logicians, jurists, rhetoricians, and poets were involved in this process. In order to show how games could be employed as a model for the probability of a certain event, the event first had to be analyzed according to its relation to and dependence on specific circumstances. It was, however, rhetoricians and historians, the experts of probable narration, who had the technical means for such an inquiry at their disposal. Structure and meaning are at issue: what is the relation of the pure structure of games to the meaning that we attribute to this structure, and how does this meaning constitute the truth of the structure, a truth that mathematical science alone could not deal with? This line of inquiry is the subject of the first part of this book. The other thread, which the second part of the book discusses, can be called the phenomenologization of probability. In this line, the aesthetic concept of the appearance of truth is worked out and extracted from probability theory and its reinterpretation of verisimilitude [the German word *Wahrscheinlichkeit* means both probability and verisimilitude—Trans.]. This concept, which Baumgarten and the pre-Kantian aestheticians formulated, admittedly did not hit directly upon the question of probability theory until later in the eighteenth century. The first explicit encounter of verisimilitude with mathematical probability theory took place in the 1764 publication of *Neues Organon* (New Organon), a book by the mathematician, philosopher, theoretician of experimentation, and aesthetician Johann Heinrich Lambert. Lambert makes it clear that, in order to be subject to mathematical and statistical probability, reality has to be transformed into an autonomous play [*Spiel*] of operational symbols that follow the logic of games [*Spiel*] itself, that is, of games of chance. For Lambert, probability—*Wahrscheinlichkeit* in the strict quantifiable sense of probability theory—coincides in fact with verisimilitude—*Wahrscheinlichkeit* as the appearance of truth. *Wahrscheinlichkeit* is thus Lambert's key concept not only for probability theory but also for the *Schein*, the appearance of truth in our perception (which he was the first to call "phenomenology"), and finally even for the *Schein* or appearance of aesthetic theory (which for him still means a theory of sensual perception in the pre-Kantian sense). The novel, on its own and in itself, had developed precisely such a theory since the first decades of the eighteenth century. In doing so, the theory of the novel concentrated on the figure of the improbability of probability. Theorists posed the question of what conditions would make a certain form of probability, that is, a certain world referred to in a novelistic narration, count as either probable or improbable. In this second thread of thought about probability, therefore, we are concerned with the question of *Schein*—of aesthetic semblance and phenomenological appearance—or the figure of observation. The kind of observation novelists deal with defines calculable probability for the world as observed from the standpoint of a preceding, first world in which the observation takes place, which remains mired in a probability incalculable for the observers themselves.[10]

This book is concerned with determining the point of intersection between the scientific and the aesthetic lines in the probabilistic project of the eighteenth century. Since Leibniz, German academic philosophy has had a title for this sweeping enterprise: *logica probabilium*, the logic of probability. Many things could be collected together under this heading: the logic of arguments in the world of everyday life and the theory of games of chance; the juristic sophistications of conditional contracts and the application of statistical data; the probability of the event in narration and the staging of probability as the appearance of truth. All of these disparate issues were part and parcel of Leibniz's long-standing project of *logica probabilium*.

Between 1660 and 1800, probability constituted itself as a measurement of scientific knowledge and as a theory of aesthetic semblance. The present investigation pursues the history of this constitutive process in two different ways. In a first type of critical procedure, it traces the development and interaction of key concepts in the most relevant texts. Two things turn out to be essentially related in this mode of investigation: the interpretation of theories of gaming chance as the theory of probability, and the phenomenologization of verisimilitude. They nevertheless form two separate threads of development: one is cognitive and the other relates to representation. Our investigation into the constitutive history of knowledge makes allowance for this disparity by paying attention both to the cognitive and to the representational aspects of probability and even to the possibility that the two sides could be fundamentally disconnected from each other. In a second type of investigation, this book will make note of historical processes and literary endeavors that belong only to the periphery of this central constitutive process of probability as an epistemological and an aesthetic theory.[11] Such marginalia or side effects will sometimes consist in the empirical conditions, sometimes in the cultural results of the new probability. One such case is the pre-probabilistic semantics of games: how did games of chance, surrounded as they were by police interdictions and suspicions about their infringement on God's providential knowledge, become the model of a mathematical theory (part I, chapter 1)? Another question looks into the history of discourse where statistics had its genesis and traces it up to its probabilistic interpretation: where do the data come from that enter into the probabilistic calculation, and who are the agents of this knowledge (part II, chapter 8)? Literary citations and the literary appropriation of probability and statistics also play a more marginal role in the wide sweep of this history: under what circumstances and with what consequences do the formulae of chance or games of contingent events appear in literary texts (part I, chapter 7; part II, chapter 14)?[12] On the sidelines of the formative process of probability, therefore, the respectively independent stories of social semantics, the sciences, and literature play out their own acts. In other words, the histories of semantics, science, and literature have their own respective strands of meaning that remain outside and independent of the central core of the constitutive history. These individual histories are annexed to the role that they play within the history of the constitutive process of probability theory, but they cannot necessarily be reduced to this role. Discursive and literary histories of mathematical probability, conversely, also have their own legitimate place and meaning. They are individual narratives and can be summed up in full neither by understanding games as the structure of what probability means nor by the development of probability toward the appearance of truth. These histories do not coincide simply and congruently with the categorical roles they play in the scientific composition of mathematical probability and the aesthetic composition of probability in modern novels. In short, the central process of constituting probability—*Wahrscheinlichkeit*—should be accorded as much respect as the autonomy of narratives in science, literature, and culture that intersect with the central process.

The meaning of probability for modernity arises in the practice of its implementation.[13] François Ewald speaks in his book on the welfare state, *L'État providence* (1986), about the use of probability "on the entirety of those objects that are relevant to 'humanities and political sciences'" in the eighteenth century. Moreover, he refers to a "proper 'colonization' of all fields of the hard sciences" in the nineteenth century. In one particular case, Ewald adds, probabilistic reason even construed the very object on which it was employed: the "sociological scheme" did not exist until probabilistic interpretations of statistical data sets were possible. According to

Ewald, the statistical probability of existence and human behavior superseded the traditional juridical processes in modern history. In early modern theories, individuals would necessarily have been introduced into society by contract, or, in the practice of administration, they had to be bound to a state as loyal subjects through command and obedience. The political and administrative problems of *risk*, however, which could no longer be answered in the legally regulated dichotomy of individual and society or state, required theorists to search for solutions along fundamentally other paths. At the beginning of the nineteenth century, Ewald states, they were solved "elsewhere, so to say in a different world, in another philosophy, that makes possible the overcoming of the antagonism between individual and state: in *probabilistic reason*."[14]

The case for the formative role of probability in modernity can hardly be expressed more succinctly than with a summary of Ewald's argument. According to his account, the probabilistic revolution reached a decisive turning point first in the humanities and only then in the hard sciences. In doing so, it defined society in terms of a sociological scheme that was no longer conceived from the point of view of the law of the juridical system, but rather from the entirely different law of large numbers. By juxtaposing statistical and juridical law, the turning point thus brought about a whole series of further distinctions: the notions of justice and of just or paternalistic responsibility for the welfare of subjects, for instance, have been considered as distinct from an actuarial conception and regulation of society since the nineteenth century, Ewald asserts. According to Jürgen Link, "normativity," which could be expressed in semantic concepts and was in accord with legal structures, was opposed to "normalism," a mechanism for defining and propagating norms of behavior according to statistically extrapolated random distributions.[15] In short, an order of knowledge under the word of a command or the text of a contract stands in contrast to an order in numerical figures under the law of large numbers. Taking Ewald's remarks even further as an example of a broader type of argumentation, we might conclude that the turning point in the basic conception of social cohesion came about through and by probability as the mode of thinking that emerged from the elsewhere of the history of ideas itself. The "other philosophy," to which probability thinking amounted in Ewald's account, remained neutral and bland, a mere "other" from the standpoint of the history of ideas—as the older history of science in fact saw it. Only with its practical applications, when it entered into the light of social history, according to Ewald, did this *other philosophy* yield decisive changes in the concept and politics of knowledge. But the theory of probability was already enmeshed in practical application in its beginnings in the history of science, which this book will attempt to relate and analyze. Interpretation of games as probability already involved the application of mathematical structure to the semantic field of probability. This first application cannot be entirely separated in its essence from the further applications. It did not make any sense for Jacob Bernoulli in his *Ars conjectandi* (published posthumously in 1713) to apply the doctrine of chances first to any undetermined concept of probable things in order then to search out the problematic areas that could be treated as questions of probability. If, in fact, reality can be modeled in games and in the improbable probability of the novel, then the history of the modern application of probability theory, and even its very possibility, are already contained in the first application. And the paradox of the first application, conversely, is again contained in all possible further applications. If that first application involves the paradox of binding the probability of events to be evaluated to the prior probability or improbability of the modeling game of chance, then this contradiction will persist throughout the history of probability and its further applications. With that paradox, the interpretation of gaming theory as probability theory will continue to be tied to the phenomenologization of probability and the aesthetic semblance of verisimilitude, and this interpretation will be repeated implicitly in every instance of the application of probability even today.

PART I

GAMES FOR EXAMPLE

Modeling Probability

1

Theology and the Law

DICE IN THE AIR

A Prehistory of Probability: How Probability and Verisimilitude Came to Be Linked with Games of Chance

Jacob Bernoulli's *Ars conjectandi* (The Art of Conjecturing), which founded modern probability theory with its 1713 publication, includes a first and brief history of mathematical probability. According to the Basel mathematician, those who invented games of chance had, unbeknownst to themselves, already invented ways to measure probability. Bernoulli's point is that the idea of measuring probability seems implied in the invention of games of chance. If we look more closely, however, we see two different phases in his brief account. In a first step, Bernoulli's inventors prepare contrivances to engineer equally likely chance events (e.g., the hardware of regular-sided cubes—dice). In a second step, they come up with symbols for determining winners and losers (e.g., the software of the numbers carved on the cubes' sides). The two phases thus encompass the construction of strict contingency: first, the chance event; second, the additional strategic context in which chances appear as probabilities. How are the two phases connected to each other? In his redefinition of probability theory in the first decades of the twentieth century, Richard von Mises (1883–1953) developed the notion of "dice in the air" to illustrate how, in order to be subject to measurement, probability has to be radically removed from any everyday intuition of probability. The binary logic of events that are equally likely to happen or not to happen exists only in the "collectivity" of repeatable throws, and hence as if "in the air."[1] For Bernoulli, however, the invention of dice is an ingenious act that detects and implements strict binary contingency and its measure within our everyday world. The second step, the strategic weighing of probability, seems to follow immediately from the first step of creating dice. The fact that the mathematical formula for probability had been discovered by inventing dice to produce chance events remained unnoticed, or at least unmentioned. Beyond the triumph of cracking pure chance—of being able to calibrate and measure probabilities—Bernoulli and his contemporaries do not seem even to consider how the everyday, the literary, or the logical meaning of the probable might weigh in on the use of dice as a measuring device.

Today's histories of probability have not solved the problem.[2] In what sense can something that has no relation to anything beyond itself enter into a historical account—how can "dice in the air" emerge in the course of a story to be told? This book suggests that this very problem gives the literary history of probability—the history of probability and verisimilitude in rhetoric and poetics—a chance to intervene in the history of epistemological probability.[3] Or to put it even more boldly: because of this problem, a literary history of probability has to supplement the history of probability in science. How did probability come to be associated with mathematical gaming theory? This inquiry, which is fundamental for understanding the emergence of probability, will lead back to probability in logic, rhetoric, and literary works. But only when an answer to this question can be provided does it make sense to reverse it: how did the emergence of mathematical probability influence the verisimilitude of the poets? This question then leads directly to the center of the traditional concept of poetry and its transformation into an understanding of literature as concerned with reality, an understanding that came about with the modern novel. In order to look into the effects of science in literary works, we must be able to account for the use of the poetological and logical category of probability within the history of mathematics. The two investigations work to complement each other, and differ only in respect to where we choose to begin the inquiry. They are both part of a history of writing that encompasses numbers and letters, the space of alphanumeric notation.

A cursory view of more recent histories of probability reveals the urgency of an investigation that traces the relationship between gambling calculations and their interpretation as a theory of probability and verisimilitude. In his reconstruction of the history of mathematics, Ivo Schneider clearly distinguishes the theory of chance in games of chance from the theory of probability.[4] But this differentiation is merely a historical remark; the meaning of probability does not play a major role in Schneider's account. Ian Hacking and Lorraine Daston, however, in making the interpretation of calculation as probability the foundational moment of their comprehensive histories of probability, remain much closer to the eighteenth century and its way of arguing. But precisely because they are concerned with the mathematical and the semantic history of probability at the same time, they tend to overlook the historical difference between Pascal's wager and the interpretation of probability by Christiaan Huygens (1629–1695) and Jacob Bernoulli. Though the semantics of probability is finally receiving scholars' attention, the event of its inscription in the text of mathematical probability has nevertheless still been neglected. The persistence of this disregard is evidenced even by Hacking's and Daston's accounts, which are historical narratives in terms of discourses or ideas, based on a specific and unvoiced predicament for any history of science. The dilemma arises from the silent assumption that each and every semantic understanding of the mathematical term probability is a misunderstanding that, as such, affects and hampers its efficacy. If, as von Mises and others contended in the early twentieth century, a mathematical term such as probability is defined by its purely operational function within the calculation of probability—a development known as the axiomatic definition of mathematical concepts—then every paraphrase of such a term in ordinary language, every attempt to express its meaning in anything other than the strict formalism of an algorithm, is in conflict with its mathematical status. The axiomatic sense of a term is fundamentally linked to a lack of meaning according to a hermeneutical understanding. Even with Hacking and Daston, the axiomatic definition of probability remains the hidden telos of the history of mathematical probability.

Hacking, for his part, ties the origin of probability theory to the difficulties inherent in the very concept.[5] Probability discussions in the early twentieth century had led to a controversy between the epistemic view (probability as a predicate of the observer's judgment) and the frequency view (probability as a predicate of the occurrence of certain phenomena). For Hacking, this opposition of interpretations forms the very criterion for the historical emergence of probability. In his view, probability emerges wherever the two interpretations and their mutual conflict can be shown to be in effect simultaneously. Daston, however, does not follow this ingenious formula. Instead, she frames the object of her studies historically as "classical probability." For her, probability as defined by Bernoulli represents the characteristic judgment of the Enlightenment, which incorporates reason in an ideal manner. Classical probability for Daston was doomed eventually to be surpassed by history simply because it is a semantic concept, a concept involving meaning, to begin with. For Daston, no less than for other historians of science, the ultimate understanding of probability in mathematics is the one developed in the twentieth century, which, however, is an axiomatic definition that strictly excludes any semantic understanding of the term.

Probability and verisimilitude thus do not simply constitute a case of what the philosopher and historian of scientific metaphors and concepts Hans Blumenberg (1920–1996) has called terminologization or mathematization.[6] Terminologization for Blumenberg refers to a word's fixation as a perfectly defined term within the grammar of scientific expressions, particularly in mathematics. To become a term in this sense means to leave the realm of metaphors, interpretation, and ordinary language behind. Probability and verisimilitude instead pose the far more fundamental question of the relationship between scientific and nonscientific, precise and provisional, explicit and implicit knowledge: between true knowledge and its mere semblance. We must formulate our specific questions with careful emphasis: How did games of hazard come to function as the paradigm for the theory of probability and the calculation of chance in the first place? What allowed and what motivated this claim, and what did it entail? The important issue is no longer whether philosophers and mathematicians in the late seventeenth and early eighteenth centuries were on the right track, or were dealt a lucky hand, as they began to speak of probability within the theory of games of chance. Instead, presuming games of chance to be the unique example for conceptualizing probability in mathematical terms turns out to represent a simple gambit in the history of probability.

At this time, the notion of probability began to be the object of a computational method—without, however, actually meaning anything other than what probability had always meant in logic, rhetoric, and literature. This moment—the ingenious discovery or bold declaration—was nevertheless a decisive stroke in a world of thinking and writing not yet characterized by the distinctions we know today, the distinctions between hard and soft science, between explaining and understanding.

An investigation into this foundational assumption necessarily means observing the involvement of topical probability and rhetorical-literary verisimilitude. These are categories that the question of the origin of probability theory zeroes in on, not ones it should take as given starting points.[7] The task then is no longer to build a bridge from poetological terms and literary works to the idea or application of mathematical probability. Instead, the texture of probability theory and its semantic interpretation lead us back to literary works and their characteristic figuration of verisimilitude. The tension between term and metaphor already inheres in the moment when probability theory is defined and founded as the doctrine of chances and gaming. We can only gain access to this fundamental dichotomy between the calculation of chance and its interpretation as probability when we recognize the interpretation of probability as the attempt to provide mathematical probability theory with a semantic foundation. This foundation, then, is a foundation after the fact of the genuinely mathematical method.

In light of the belatedness of its foundation, a zone of conflict and paradox emerges within probability theory. On the one hand, we find Hacking's systematic paradox: for Hacking, the two opposing interpretations, the subjective-epistemic theory and the objective-frequency theory, can only establish probability when they coexist. On the other hand, we are again confronted with a paradox in Daston's historiographical argument. Probability theory in this case is only set in motion by the interpretation of probability as an act of reasonable human judgment, an interpretation that is, however, inadequate from the beginning, because all interpretation will prove to be incommensurate with the purely mathematical term. Finally, along with Blumenberg, we may characterize verisimilitude as essentially metaphorical and the calculation of chance in probability theory as scientific terminologization. In this case, modern probability becomes a terminally inconclusive back-and-forth between terminologization and re-metaphorization. Rhetorical and mathematical devices line up alongside each other and provide heterogeneous models for probability within probability theory.

This book contends that the invention of the modern novel, the novel of the eighteenth and nineteenth centuries, has to be seen through the lens of probability theory and, in particular, of statistics. The phantasmagorias of social normalcy and social laws based on probabilistic thinking[8] indeed play major roles in the fabric of the modern novel. But before any attempt can be made to elucidate this relationship between the novel and probability theory, we must return to the interpretation of games of chance as a model of probability. We have to look even further back to the meaning of games before its interpretation within probability theory. In the pre-probabilistic understanding of gaming, the mutual relation between literary meaning and mathematical theory is negotiated in its most basic form. Even the relation between a semantic concept of the law and the law of large numbers finds its fundamental determination at this early stage. Today, we tend to see the nineteenth century's statistical norms and legalities as an overcoming of the semantic norm or an avoidance of the juridical concept of law.[9] The semantics of games, however, which provided the origins of mathematical modeling, were themselves grounded in norms and laws of even older ancestry. The mathematical foundation of the theory already presupposed the fair game, *lusus justus*, before any norms could be deduced from its statistical application to morality and society. The very interpretation of gamblers' reckoning as the calculation of probability, which has been the subject of so much scrutiny, removed this older law of games, the law within probability theory, from our view.

A Brief Survey of the Semantics of Gaming: Theological and Juridical Politics

The semantic history of "game" provides clues for its use as a model example in probability theory, and this history is deeply connected to the politics of gaming in the early modern era. The church's fiats, the laws of the political sovereign, and police interventions against games of chance all worked together slowly and circuitously to carve games out as a sphere of calculable chance. An extensive literature of commentaries and tractates that separated the immanent world of chance from God's intervention in this world was inspired by the condemnation of games of chance by theologians and Christian jurists.[10] A phenomenology of human behavior vis-à-vis chance events emerged as distinct from the Christian history of redemption and the divine sphere of miracles and providence. Jurists, especially Calvinist ones, picked up the debate in early modern times. They increasingly turned their focus away from the individual chance occurrence and concentrated on the contractual frame in which an event can first arise, or not arise, as the chance that decides the outcome of a game as a win or a loss. Concentrating on the rules of the game helped to neutralize the theological implications of the singular chance event.

In the sixteenth century, theological and moral themes found expression together in a fascinating allegory that relied on games of chance, not only for its colorful imagery, but also for its nature as allegory itself. The moral-theological elements can probably be traced back to the third-century *Liber de aleatoribus*, a pastoral pontifical decree that has come down to us misattributed to the Carthaginian Bishop Cyprian. The devil, as the oft-quoted text claims, is always present at the dice table. The game's diabolical nature reveals itself through the vices and crimes that follow more or less closely on its heels: the players' extreme passions, the squandering of fortunes, murder, and whoring.[11] The fifth book of Saint Augustine's *City of God* is often cited for the dogmatic elements of the trope. In this book, the bishop of Carthage devotes himself to the struggle against the allegory of Fortuna and, by extension, the entire pagan pantheon. In condemning the divinatory arts, especially astrology, for their diabolical character, the very core of Christian dogma comes into play.[12] The equilibrium between free will and divine power to decide the salvation of the individual and the whole world is eventually at stake in the game and the predictability of chance. In bulls and papal edicts of the sixteenth and early seventeenth centuries, therefore, indictments of divinatory writings and prohibitions of secular lot drawings always appear in near proximity to demands for executive measures against games of chance.

Two typical, early baroque power politicians in particular conjoined moral theology and dogma to pinpoint the nature of games and gambling in administrative and executive terms: Pope Urban VIII, with his 1631 bull against astrologers; and the French king Louis XIII, with his ordinance against gambling addiction. As the pope and king both saw it, every throw of the dice was an incursion into the prophetic competency of God in things to come, and simultaneously on the mystery of his sovereign power to bring order to the world. The player's attempt, within the rules of the game, to control chance was "diabolical." According to the understanding of Aristotelian logic, which scholastic interpreters followed in their admonishments and prohibitions, chance has the logical quality of the contingency of singular events in the future. The propositional logic in Aristotle's *Peri hermēneias* (On Interpretation) had shown that predictions about *contingentia singularia futura*, about events in the future, are neither true nor false. This exclusion from the realm of ordinary true-or-false propositions could be interpreted in two different ways: on the one hand, they could be understood as a clue to a field beyond decidability between true and false. The other and more often preferred solution, however, was simply to concede that, in judgments on future events, man cannot lay claim to truth or falsity. In hindsight, one proposition always proves to be true and its negation to be false. In any case, the *contingentia futura singularia* are for God to know and for Him alone. Any curtailment of His sovereignty in the area of contingency would deprive the Last Judgment of its absolute mandate. Any breach in the logic of future events would open up possibilities for humans to demand an account of God's soteriological decisions. The entire theological coherence between sin and the promise of salvation would be endangered: man would become a debtor able to calculate his outstanding liabilities and prospects. All further divine decrees pertaining to the regulation of contingency follow from this soteriological core: God must be recognized as the highest and the only authority over the imperilment and the salvation of individual souls and over the providential plan for grace or damnation of the world.

Certainly, the devil tempts souls, first of all with the glamour of winning, fraud, and cheating. Authors in the Patristic tradition, however, often pile on additional polemics referring to gaming as performance and representation: as theatrics, jugglery, and circus tricks. Games, which after all involve playing with chance, also always have this additional relation to play, and to plays.[13] Play and plays, however, introduce into gaming a sphere of events that constitute performance of mere outer appearance, and worse, they end up as idolatrous show. Thinkers up through Kant had written about performance in these terms,[14] and Tertullian's brilliant diatribe "De spectaculis" influenced an author as late as Nietzsche with respect to his notion of *Schein* (appearance, but also semblance and even show).[15] The kind of semblance that Tertullian and Saint John Chrysostom, for example, saw in the *spectaculum* of the theater and other conjuring acts can by no means be adequately understood as a mere perversion of aesthetic play and mimesis. Rather, the offending spectacle had its source in the theologoumenon of bedazzlement and of idolatry. Play and the game orbit together around the abyss of contingency.[16]

Authors around 1600 still made use of Patristic turns of phrase to express their fiery admonitions: gaming and play invite sin and crime—gluttony and harlotry, theft and murder; and they fulfill themselves in the subtle moves of treachery with which the players try to gain the upper hand over one another. The thousand delinquencies form a colorful wrapping inscribed with the sinner's deeper ambition to break up God's monopoly on contingency. For the keepers of Christian order, the game is thus the prime allegory of transgression against the law: the many misdemeanors and crimes that surround the gambling table mirror the intimate attempt to infiltrate the sanctuary of the law and its Christian perfection in grace. A multitude of moral iniquities follows from the fundamental attempt to break the law.

The proximity of games of chance to the idea of tempting God (Deut. 6:16), who alone recognizes the *contingentia futura*, was hence at the center of both metaphysical and law enforcement's concern about gaming. The relation between God's intervention and games of chance, however, also provided the clues to give a new direction to the discussion of gaming from the sixteenth century on and finally made the game's neutral stance as an example in models of probability theory since the late seventeenth century conceivable. Post-Tridentine moral theology continued to discuss the balance of sins attendant on gaming, and particularly on games of chance, in a manner almost unchanged since the arguments of the church fathers. Playing dice and card games "in which luck predominates over exertion is a mortal sin, all the more so when one plays with avarice and burning desire for victory": such were assurances that came once more at the beginning of the seventeenth century, still accompanied by all manner of impressive quotations from Saints Thomas Aquinas, Bonaventura, Isidore of Seville, and Cyprian, the prefect of the papal sacristy.[17] But the casuistry of games of chance at this point had already taken a new direction, as we see in the works of Puritan and Protestant, especially Calvinist, authors. The standard sixteenth-century book on gaming was published in Calvinist Geneva in 1574 by Lambert Daneau with the significantly double title *Briève remonstrance sur les jeux de sort ou de hazard* (Brief Remonstration on Games of Lots and Luck); the London Puritan clergyman Thomas Gataker titled his 1619 book on the subject *Of the Nature and Use of Lots*.[18] Even beyond the confessional links, however, Calvin's doctrine of providence, based on Augustine's *City of God*, as expounded in the famous eighth chapter of the 1539 edition of *Institution de la Religion Chrestienne,* served as the conceptual basis for such tracts on lots, games of chance, and the nature of chance itself.

Calvin declines to differentiate between general and particular providence (*providentia generalis* and *specialis*), between the providentially ordered course of the world and the chance that is active in specific moments and individual lives. His exegesis of Proverbs 16:33, "Man throws the lot, but it falls as the Lord wills," reveals the point of this refusal: "Who, likewise, does not leave lots to the blindness of fortune? Yet the Lord leaves them not, but claims the disposal of them himself."[19] Calvin argues for the immanent physical lawfulness of the created world; divine providence, the ordering power behind this law, remains an ineffable mystery. The underlying order only becomes manifest, however, if the question, and indeed the struggle, arises of who has final authority and title to the sovereign decision. The Lord not only has control over how a lot is thrown or drawn, Calvin explains, but also over the very moment of contingency itself. He even arranges for the alternatives of winning or losing. God therefore comes onto the scene in the aspect of a decision: his intervention comes essentially in terms of a ruling or judgment on the chance of a lot to win or lose. Acknowledged as sovereign over the decision, His lordship is also then indirectly involved in the physical events of drawing lots and the laws of nature that determine those events. Decision takes precedence over physical mechanics, in which it ultimately becomes manifest. To put it in even more general terms, God's universal physical and moral providence (the destiny in the *contingentia futura* of the world) is incorporated into His juridical predestination over souls (the predestination of salvation).[20] In this way, a fundamental new concept of games of chance becomes possible that cannot be traced back to the Aristotelian analysis of the chance event. Aristotle's *automaton* referred to an event within the course of nature that lacked causal explanation. For an incident to count as *tychē*, individuals had to make the additional judgment that such a chance event happened contrary to their expectation and desire. *Tychē* was thus an *automaton* with special features for human minds. For Calvin, however, the chance event is essentially not an occurrence within the course of nature but the result of God's decision. As such it underlies the alternatives of yes or no. The lot corresponds to the theological and ontological nature of the chance event insofar as the binary logic of winning or losing echoes that of the affirmative or negative decision. But despite this consonance between lot and chance event, there is no way for humans to know when and how God in fact does or does not reveal his decision.

Calvin develops a second argument on the chance event that presupposes the first one but that now concerns human perception. For Calvin, however, the point is not, as it was with Aristotle's *tychē*, that individuals relate certain events to their expectations or desires. Instead, he emphasizes the fact that humans have to cope with something like chance events in the course of nature, even if theologically and ontologically there is none. In the long run, this argument will open the possibility for games to model a kind of probability to reckon with. In order to support the human spirit in its weakness, Calvin allows us to employ the following distinction: "Notwithstanding the ordination of all things by the certain purpose and direction of God, yet to us they are fortuitous."[21] In the aspect of the decision, particularly, of course, the judgment about salvation, the Lord demands the acknowledgment of His unqualified competency. With respect to human knowledge and action, however, chance is a reality for men. In absolute acknowledgment, God demands the concession of His unlimited providence over the physical event in light of the decision by lot. In the concession to knowledge in practical life, what the lot reveals or how the dice fall becomes a question of chance. Behind that concession, however, the assumption of God's providential control of the event remains untouched. The world's obedience to physical laws becomes a neutral zone between God's sovereign decision and the actions of men under the pressure of decision making in this world.

Thus Calvin can let chance have its way for humans before the backdrop of the unconditional attribution of every event to God's final and sovereign command. By doing so, the teacher of predestination opens a space for worldly calculation. This bracketing of human knowledge is Calvin's second important move with regard to chance; and it makes calculation in games of chance possible. Precisely because we know that chance-for-us is in fact no chance at all, it can become the object of practical wisdom and finally even of rigorous calculation. Calvin marks out a sphere of calculation—and, especially for the *contingentia singularia futura*, a sphere of chance event—that at once upholds all of the church's theological and ontological prohibitions and, nevertheless, makes them moot in an "as-if" scenario.

Initially, however, discussions of gaming concentrated on Calvin's first consideration: the correlation between chance and decision. They hence reintroduced some old Renaissance practices of interpretation and investigation, practices that at first glance seem odd enough. But a closer look reveals the interest they could have for an understanding of hazard and chance: according to Calvin, the theological discussion of gaming is at first a question about lots. Lots can be classified with Lambert Daneau into holy or serious secular lots, both of which must be strictly distinguished from the frivolity of chance in games of hazard. For such holy or "serious" lots, then, the connection of chance and decision is presumed to be established. Lots that are not played with luck in a mere game should either reveal or determine decisions. The holy lot consults the will of God. Colleges and societies commit themselves to serious secular lots in questions for which there are no criteria of judgment, or for which they do not wish to use such criteria. Such cases occur, for example, in estate distributions or elections.

So far, the realm of holy or serious secular lots is still hermetically sealed off from straight games of chance, the frivolous chance events. In games of hazard, decisions are linked to the operations of chance only on a structural level of win or loss and therefore are not strong and visible enough to lend legitimacy to the reckoning of chances. With holy and serious lots, by contrast, only singular events and visible and previously formulated decisions are always bound by the lot. Nevertheless, French and—especially—English authors of the late fifteenth and early sixteenth centuries based their arguments for the establishment of lots on the logic of the ritualistic or legal process that, as such, enabled chance to become authoritative in decision making.[22] In doing so, these theologians and jurists were not concerned with invoking mythical practices, but they nonetheless revealed their peculiar institutional logic.

In anticipation of the future epistemic role of games of chance, it is possible to distinguish and characterize two aspects of such an institutional logic. With lots, holy or secular, the logic of the institution places the quasi-experimental setting of the distribution of chances at the disposal of a religious questioning or of a procedure of election. Conversely, the setting of a game of chance, even under the most frivolous circumstances, owes its logical and technical possibility to something like an institution that establishes the symbolic dichotomy of yes or no, win or loss, in the first place. Holy and serious secular lots are bound together by the power and the technology of an act that mounts and posits the setting for the binarism of the occurrence or nonoccurrence of an event. A symbolic order of criteria for judgment is thereby installed in the physical world. However much the Calvinist polemics against the obscenity of games of hazard and the idolatries of semblance remind us of the moral theologians of the fifteenth and sixteenth centuries, a fundamental change in perspective divides them from Lambert Daneau and his followers. For the latter, the player's sin always finally consists in his attempt to encroach upon God's authority over chance, but the theologians in Calvin's retinue disclose the institutional logic of the Arcanum. God's decision is still the subject of the institution's inquiries. Through the discovery of institutional logic, however, the sovereign God for Calvin and his followers occupies an increasingly structural place, the place of making the first and final decision; this and every other sovereign can always thereby become just the name or the function of the sovereign. This makes Calvin and the authors of Calvinist pamphlets the first theoreticians of modern contingency politics.

Calvin's clarification of the connection between chance and decision was certainly theologically more important than his linking of providence and works, but perhaps it was even more consequential in sociohistorical terms. Precisely because no wisdom and no metaphysics can arbitrate between the divine order of nature and chance-for-us, the immanent world's quasi-acknowledgment of chance follows directly from its theological negation. Gataker, for instance, distinguishes in this context between "chance" (casualty, adventure) and "contingency." Chance or adventure is an event that appears to be without a cause to us but assembles a dramatic cast of signs about itself demonstrating its strange and striking nature. Contingency, the intrinsically causeless event, on the other hand, remains hidden in the immanent world and only emerges clearly when subjected to institutional inquiries. In adventurous chance, we easily think that we are acting in a world in which God constantly operates by special interventions and miracles. This active divine involvement may, but need not, be the case. More thorough examination would reveal most chance events of this sort to result from a chain of contingent causes. Even the special providence of God much prefers to work through such chains of chance than through miracles. We must be on our guard against the theatrical danger of dissembling and illusions lurking in these marvels and wonders. We would behave more correctly theologically and more cleverly in practice if we looked away from such apparent chances to the nonnecessary causal chains of contingency. The critique of chance as adventure leads to its minimization and to attribution of chance and nonchance to the theater of intentions. Contingency, on the other hand, is fundamentally hidden in the cosmological providence of the natural world and its coherence, which in its entirety remains inaccessible to us. Contingency also comes to light, of course, in precisely circumscribed spaces of institutional logic, whether in the random choice of scriptural passages in the vein of the holy lot or in the determination of political leaders by drawing short and long straws.

The game of chance, which in medieval and Renaissance teaching had been the focus of collusion between idolatrous spectacle and encroachment on God's competence in contingency, becomes a more marginal issue in Calvinist doctrine, even if admonitions and warnings remain valid. On the one hand, the game of chance is closely related to the—holy or serious—lot that points toward God's unique decision. On the other hand, it is often connected to adventurous chances, from which we can and should abstract and enjoy a maximum of order in the world.

The image of the game did not fundamentally change until, in the seventeenth century, casuists from various Catholic orders as well as Calvinists and other Protestants developed a new approach.

First of all, this literature forged an anthropological doctrine based on the casuistry of prohibitions and transgressions, a doctrine that justified the distraction of gaming and play from serious life. A harmonic rhythm of exertion and relaxation, tension and release was bestowed on man in his very creation. In paradise, man followed this natural rhythm, alternating between rest and work in the Garden of Eden without a contradiction ever developing between the two. Original Sin and the subsequent Fall formed the opposition that now exists between seriousness and play, between vital necessity and diversion.

Play and games are justified to the extent that they compensate for the cares and hardship of the work imposed on us by the Fall.[23] Though amusement remains susceptible to sin and frivolity, distracting games nevertheless find relative justification in their contrast with care and work.[24] Moreover, the compensation reminds us of the lost rhythm of rest and activity in the Garden of Eden. Not many further contributions to this theological anthropology would appear until Schiller's *Briefe über die ästhetische Erziehung* (Letters on Aesthetic Education) in 1794 and Karl Marx's Parisian writings. In the course of development of this Christian anthropology, the old distinction between games of skill and games of chance is also qualified. Seigneur du Tremblay, for instance, makes the following argument in his *Conversations morales sur les jeux et les divertissemens*: a pure game of skill would not be a game at all, and there could be no alternation of winners and losers if the ability of the players alone determined the outcome without further qualification. On the other hand, it is the minute physical inequalities and the subtle psychological tricks that allow experienced players to apply their routine in dice games, which seem to be pure games of chance. Skill and luck are two poles of a continuum.

This theological anthropology now joins up with Calvin's second argument, his concession to epistemology and the practicalities of life. The explanation of divine will and the intervention of God in the laws of nature can now be rigorously distinguished from the chance of gaming, which only counts as chance *for us*. Traditionally, gaming had been problematic since the endless strings of chance events in games lacked the clear institutional logic of the holy lot that always resulted in just one event occurring or not occurring. With Calvin's concept of chance for us, this same feature of games of chance now helps avoid the danger of encroaching on God's competence in contingency. Protestant authors such as Christoph Wittich and Jean La Placette, as well as the Catholic theologian Jean-Baptiste Thiers, present the case thus.[25] In La Placette's 1697 tractate on gaming, the intervention of God in the course of nature—the miracle—no longer has any relation at all to the cast of the die. The miracle is the highly visible form in which God can undertake an intervention in the course of the world and the law of nature in accord with *providentia dei specialis*. Incidentally, there are also divine interventions that do not appear as miracles. For this reason, picking a lot at random from an urn, for instance, does not necessarily reveal a position taken by God. "Imagine that I am alone in a room, and dice are lying on the table. Now if I should throw them without thinking anything in particular, where pray tell is the voluntary decision revealed by God?"[26] The lot is a decision by chance only for him who has provided certain rules of procedure for inquiries. It can coincide with a divine *providentia specialis*—or not. Particular providence and *revelatio immediata*, the two miraculous interventions of God, both have their place in a divine politics. They have no more and no less to do with such chance events as games of hazard or lot drawings than with any other occurrences in this world. Games of hazard and decisions by lot are arrangements of events that count as chance for the players and those who cast the lots.

Hence it is possible to bring Calvin's first argument, the constitutive coupling of institution and decision, to bear on gaming, and particularly on games of chance. This invocation of institutional arbitration occurs no longer so much in respect to a theological as it does now to a juridical anthropology.

The Protestant theologian La Placette, for example, stresses that games represent a very particular type of contract. The gains won in a game of chance are no less legitimate than all earnings resulting from any legal business contract made for profit. The only condition is that the parties involved do not smuggle any secret clauses into the contract—the rules of the game—and do not cheat in the course of its transaction—the game itself. As with every contract, the implied contract of the game is valid regardless of what moral theology has to say of what the contracts are about, that is, to gaming and winning.[27]

It is no coincidence that La Placette's tractate on gaming falls right between his treatises on mental reservations and interest on the one hand (situations of mutual relationship and contract), and those on the right to self-defense and the arousal of offense on the other (situations of asymmetrical relations). No matter how much the game threatens to infringe on God's law and police directives, it is equally justifiable and litigable as a quasi-spontaneous form of contract. The game is justified as a contract because it is just and hence fair play. This is the second form in which games made their appearance in modernity with a positive meaning: as sphere of the contracting partners' morally indifferent sincerity and reliability.

This view naturally crops up among juridical authors, as in J. A. van der Muelen's *Forum conscientiae* (Forum of Conscience), for instance, and Jean Barbeyrac's 1709 *Traité du jeu, où l'on examine les principales questions de Droit naturel et de Morale qui ont du rapport à cette Matière* (Treatise on the Game, Examining the Principal Questions of Natural Law and Morality That Relate to This Subject).[28] Barbeyrac, the son of a Protestant refugee family, was a professor of fine arts in Berlin's Collège Français before going on to teach public law in Groningen and Lausanne. He gained a reputation as an authority on the Saxon jurist and political philosopher Samuel Pufendorf among the numerous Huguenot exiles in Prussia through his commentaries and translations, transmitting, explaining, and propagating the modern doctrine of natural right for readers of French in Germany.[29] The juridification of gaming comes most clearly to light, of course, in Pufendorf's *De jure gentium et naturae* (1672). Barbeyrac had already commented extensively on the pertinent passages in his translation of this work before it subsequently supplied the essential thesis for his later treatise.

The chapter on "contracts in which chance plays a role" in *De jure gentium et naturae* places games of chance in a significant constellation in two senses. First of all, the chance Pufendorf focuses on is not an event in the merely natural state of the world, no simple physical occurrence that then might be processed and preserved in the order of justice and society. Chance in Pufendorf is rather a construct or something instituted by justice, just as obligation and validity are settlements of justice. The theoretician of justice can only analyze chance from the point of view of its framing conditions, which means from the definition of its possible occurrence. The question as to what chance is in itself pales in comparison to the investigation into under what suppositions and in what framework chance occurs as a moment of decision in a given situation. The metaphysical essence of a singular case of good or bad luck is everted into the settlement of a framework in which it can transpire and make a distinction. In Pufendorf's analysis, this framework is a "wager" that "is a reciprocal and conditional promise or bargain where chance plays a role, because whether or not an event or a thing exists does not depend on the players' efforts."[30] Pufendorf is alluding to a complex form of Roman contract law. Conditional contracts are agreements in which both parties bind their obligation to a future event that can occur or not occur and on which they have no influence. With this form of contract, jurists had something at their fingertips that would not be recognized as a logical structure until the development of probability theory. As such, however, it would have a promising and long-lasting career ahead of it: the logic of if-then constructions. Pufendorf thereupon immediately comes to speak of games. Games, after all, can be seen as mutual agreements in which every party takes equal risks and has the same chances—if one understands them according to the pattern of conditional contracts. Barbeyrac, already in the notes to his translation, indicates that Pufendorf shortly thereafter, when he speaks about the justice of games and the connection between fair play and probability, seems to be referring to a section of Pierre Nicole and Antoine Arnauld's book *La logique, ou l'art de penser* (Logic, or The Art of Thinking), produced at the Jansenist Port-Royal monastery near Paris.[31] And this passage, we should hasten to add, is the very section that in turn alludes to Pascal's wager argument and mathematical theory of games of chance. This constellation of citations, as will be made clear in later chapters of this book, lays the groundwork for the mathematization of chance, and prepares for its connection to the semantics of verisimilitude. For the time being, however, we should make careful note of one simple fact: the analysis of games of chance in terms of contract theory goes back to the first attempts to comprehend games mathematically, just as much as conditional contracts provide them with a logical structure in hindsight. As long as we look at chances only as this or that occurrence, we are not able to subject them to procedures of computation. Calculation becomes possible only with artificial framing. The framework of a logical if-then-space that only a legal act—a contract—can bring into existence is required.

Pufendorf places the wager theory and the game theory between two significant aspects of political science. On the one hand, we find the question of how one sovereign state should act among other sovereign states; on the other hand, Pufendorf asks how the state should care for its populace. These are the two great areas in which "contracts in which chance plays a role"[32] should conspicuously regulate conditions. The first area is the law of war. In a "public and orderly war,"[33] a silent agreement must be assumed according to which the opposing parties have dispensed with a peaceful accord and have bound their respectively just causes to the chance settlement—the *fortuna*—of combat. The peace treaty that ends the war retrospectively determines precisely the actuality of such a conditional contract.[34] The second area in which chance plays a role in law is the contract that forms the basis for several forms of insurance. Pufendorf is thinking in particular of maritime insurance. This contract presupposes bilateral contingency as a condition: the insurer must not already know that the ship he insures has reached port safely; and the insured must not already know that his ship has been sunk in a storm.[35]

At the moment when the mathematical theory of gaming and the conceptual validity of probability converge under the demands of the law, the *fortuna* in the ancien régime's military actions and the rationality of insurance are pervaded by one and the same logic. It is the Janus face of the modern state: outwardly, the state displays its mask of sovereignty in the politics of *fortuna*; inwardly, facing society, the contingency-related politics of insurance shows its pastoral countenance of care and welfare. The dates between the first appearance of Barbeyrac's book in 1709 and its second edition in 1737 mark out quite precisely the time frame for the initial reception of Jacob Bernoulli's foundational text of modern probability theory, the *Ars conjectandi* (1713). In his book, Barbeyrac again summarizes what the work of jurists had made of the theology of games of chance in the course of the seventeenth century. Once more he shows how to remove games from the domain of divine intervention, characterizing it as constructed by contract law. The theological notion of the institution, which for Calvin had marked the lot as a coupling of contingency and decision precisely in opposition to games of chance, shifted around the beginning of the eighteenth century to an as-if contingency within games of chance and its own institutional logic, the logic of natural law. Just as the theological concept of the institution reveals the fabrication of the singular chance event, the juridical logic of the institution similarly reveals the fabrication of serial contingency. Discussions of gaming between Calvin and Pufendorf did not merely neutralize the game's frivolity, thus setting it free to serve as an example for mathematical theory. More important, contingency's status as an epistemic object apparently became conditional on the institutional and contractual form of games. In the exemplarity of games, which probability theory over the course of the eighteenth and nineteenth centuries tended to forget but could not wipe out, the connection of (natural) law and mathematics persisted as a valid precondition in all probability theory and all statistics based on it throughout the nineteenth and twentieth centuries.

The Game Interpreted as Probability

Histories of science have, with greater or lesser deviations, provided two types of narratives for the development of probability theory. Both date from the initial decades of the eighteenth century and became current in the wakes of Pierre-Simon de Laplace and the Belgian statistician Adolphe Quetelet. The first account deals with authors and inventions. According to this story, probability theory came about due to an incidental problem that was posed to the great religious thinker and mathematician (later labeled a mathematical dilettante) Blaise Pascal, which he discussed in an extensive correspondence with the mathematician Pierre de Fermat. In this account, Pascal takes on a kind of authorial function for the problem, despite the historical facts that he neither was the problem's inventor (which distinction belongs to Chevalier de Méré)[36] nor formulated a complete mathematical theory in response to it (which Fermat managed in any case earlier than Pascal). As the story continues, the discussion between Pascal and Fermat very quickly brought about a textbook-like mathematical elaboration (by Christiaan Huygens) and led (with Jacob Bernoulli's *Ars conjectandi*) to the systematic inclusion of methods of higher analysis. If Pascal is the author of the problem of probability, then Bernoulli is the author of the concept that then initiated the history of probability proper. The great authors of the later eighteenth century (d'Alembert, Lambert, and Laplace) then respond in turn to Bernoulli's framing of the problem. According to the second type of story about the development of probability theory, urgent practical problems arose that the new method of calculation was best suited to solve: the calculation of compound interests or insurance policies; the average life expectancies that could form the basis for the rating of pensions and life insurance; and the growth of the populations of nations. Finally, the juridical push for a calculation of judgments and decisions also contributed to the new-felt need: How highly should a statement that reaches us via a chain of several witnesses be valued? And is there a measure for the tenability of judgments according to the strength of various indications?

We are not concerned in this book, however, with a decision between an account of the development of the concept (for the theory of gaming) and a pragmatic history of its use (with respect to probability in daily life). Instead, we are interested in a reading of the relevant texts in which both accounts intersect *in the interpretation of games as the prime example of probability*. First of all, therefore, we must examine the two texts about gaming theory before this theory is interpreted as a model for probability—with Pascal's text, which offers the calculation in a literary context, and Huygens's, which presents the example of games in the vein of a mathematical textbook. Only then can we turn our attention to the interpretation of probability. This interpretation, which must be distinguished from the previous doctrine of chances, led to Jacob Bernoulli and his foundational work in probability theory. Bernoulli, then, inaugurated the use of the doctrine of chances in games and launched the actual history of probability theory.

On the one hand, treating games as the paradigm of probability justifies including Pascal and Huygens in the history of probability. On the other hand, their texts on the factor of chance in games, though composed before the interpretation of games as the model for probability, have a consistency of their own: they demonstrate how the mathematical structure of games can be seen as paradigmatic in itself and thus as an epistemic object in its own right. In doing so, Pascal and Huygens, in different ways, follow up on the earlier semantic history of games. The important innovation in their thinking is the notion of *fair play*, the quasi-juridical definition and positing of a structure that lends itself only to mathematical formalism. This institutional logic of games leads to and supports its later interpretation as exemplifying probability. The internal logic of games and their constitutive structure in the doctrine of chance then becomes more invisible with each interpretation of probability imposed on the mere structure of chance in games. Our first job, therefore, is to understand the constitutive role of this antecedent internal logic of games before the advent of probability. Only by doing so will we be able to recognize the event of interpretation and its power to shape the history of probability.

In a reversal of traditional views, we can claim that the mathematical structure of games was expropriated, as it were, through the sense of probability, a sense that only added the fresh meaning of a Husserlian lifeworld to the mathematical structure of chance. This expropriation in the late seventeenth century was a precondition for the mathematical structure of calculating chance to settle and spread in the understanding of probability, which in its turn has been a formula for reality in aesthetics and in the novel since the eighteenth century. The history of knowledge and its interest in the foundational understanding of games as the primary example of probability therefore neither reports on the taming of chance through calculation nor bemoans the defeat of an original practical orientation with respect to probability through its mathematicization. Instead, literature and calculation, sense and structure became elementary building blocks in the process that made the calculation of games into the primary example for the living sense of probability.

2

Numbers and Calculation in Context

THE GAME OF DECISION—PASCAL

Aleae geometria

In neither Pascal's nor Huygens's writings on the calculation of chance in gaming does the term "probability" play a major role. The word does not appear even once in Pascal—neither in his correspondence with Fermat nor in the *Adresse à l'Académie Parisienne*, which recounts his mathematical and scientific works up to 1654, nor even in the fragment on the wager. In Huygens's tractate on games of chance, the term turns up only once, and that occurrence is in the Latin translation of the Dutch original. There is no indication that either Pascal or Huygens regarded probability as the question they set out to solve by calculating chances.

Ian Hacking inscribes Pascal's wager into the history of probability by the detour of a reference to modern decision theory and its use for and interpretation of probability. In doing so, he leaves two things implicit in his view of Pascal: according to Hacking, Pascal was well aware that he had invented a general theory of decision making but decided not to come out and say so; and, although the literary success of the *Pensées* familiarized many readers with the wager's application to Christian faith, and, more broadly, Christian Apologetics, the enigmatic quality of Pascal's text refuses to spell out how gaming and betting came to refer to such a daunting conundrum.[1]

Understanding Pascal's wager in terms of decision theory apparently makes sense only from the point of view of later phases in the history of probability, when, from the early twentieth century on, people began to discover that decision theory was part and parcel of probability's own domain. What can and must be shown, however, is that Pascal does not simply postulate the exemplary nature of games for a general concept of probability that he otherwise chooses not to mention by name. He rather proposes to read gaming and betting as self-exemplifying processes of the calculation of chance. Making a specific decision, as we shall see, is in fact a pertinent moment in these processes.

In the letters that Pascal and Fermat exchanged on the problem of calculating chance in mid-1654, nothing indicates an attempt to give any further semantic interpretation to the issue of gaming or of putting the method of calculation to any further conceptual use. At issue in the letters is the purely mathematical problem of distribution: how is the sum of winnings agreed upon by the players in a dice game to be divided up if the game is interrupted before the regular ending? A typical debate between the two men addresses the situation in which four wins in individual rounds are required for a player to secure the final victory. They then pose the question: what share of the winnings are the different players entitled to if the normal condition for victory has yet to be fulfilled? If, for example, the game is broken off at a point when one player (A) would still need two wins and another player (B) would need three to claim a victory? Fermat resorts to the then usual method of combinatorics to develop a solution. He first inquires after the highest possible number of rounds that could fall before achieving victory. For the present example, the answer would be four rounds (in the case that player B wins a round: two rounds are left for B to achieve victory and two for A; in the case that player A wins: one round is still required for A's victory and three for B's). Then, assuming the longest-lasting possible game (here four rounds), he notes down all the combinations of the possible outcomes from this point on. Pascal on the other hand employs more of a calculating strategy than an exhaustive combinatorial process. He begins by asking about the possible paths the game could still take from the moment of its premature cessation until a proper completion. If A were to need to win one and B two rounds, that would mean that if A wins, the game is over; and if B wins, then A and B each need to win one more round. Then Pascal goes on to calculate the distribution of winnings backwards from the game's final result: at the point when A and B still need to win one round, both are entitled to half of the total stakes. At the juncture in the previous example, A can either reach this condition of an even split (if he loses), or decide the game in his own favor (if he wins). At this point, therefore, he is entitled to half of the halved winnings plus half of the remaining half (= 3/4). Expressed in a formula, this becomes: $S = L + (C - L) / 2$, where S indicates the just share of the favored distributed winnings; L the risk of loss, and C the chance of winning the next game. Pascal calls this process the *méthode générale*. The general method involves calculation in two senses: it is both a process of mathematical reckoning and an analysis of the tactical situation. Pascal sticks to his general method even after acknowledging the validity of Fermat's combinatorics (and he apparently continued to insist on it even later when confronted with Huygens's chance theorem).[2]

In developing the *méthode générale,* Pascal did not, therefore, attempt to forge an understanding of games as an example by which any further questions and cases in life might be modeled. On the other hand, however, he did integrate games of hazard into the rich and loaded context of the wager on the chance of salvation. The famous wager passage, which was not part of the bundle Pascal classified, but has been included in the *Pensées* ever since the Port-Royal edition of 1660, takes up gaming and calculation for unexpected purposes.[3] Gaming and calculation appear in it as a type of situation and action that are supposed to be equal to the question of faith, a question without any measure or limit in human knowledge. This move by Pascal significantly does not introduce the universally flexible status of the example that would be applicable to all the various social and moral dimensions of life, as we shall find later on with Huygens and Bernoulli. Instead, the game for Pascal has the incomparable exemplarity that applies to one instance alone, though this single application is valid for life as a whole.

In his exchange with Fermat, Pascal neither shows any interest in a debate on probability nor engages in any way in an epistemological discussion on calculation and its meaning. Whereas later students of probability would prefer to devote a distinct part of their understanding of the mathematical problem to gaming and the calculation of chance, Pascal simply perceives it as a question and even a paradox in the relation between the method, calculation, and its object, chance. We find this approach expressed in a paragraph from Pascal's *Adresse à l'Académie Parisienne* (1654). In terms of argument and time, these lines represent a hinge between the mathematical texts on the division problem and the wager. The term Pascal uses in the *Adresse* for the general method of the division problem is the mathematics of the dice, *aleae geometria,* a phrase that we may render metonymically as the mathematics of dice games or allegorically as the calculation of chance. This is, as Pascal notes, a *stupendus titulus,* since it brings to our attention how this calculation "connects the demonstrations of *mathesis* to the uncertainty of chance, thus reconciling entities that appear to be of opposite natures."[4] This peace and reconciliation, however, are only accomplished by a use of force. Pascal pointedly notices that reason makes use of the geometrical method in a field where no stable experience has been established beforehand. For this reason "this question has heretofore been wandering aimlessly around."[5] Pascal for his own part now wanders within a few sentences through the Latin vocabulary of chance, speaking of *anceps fortuna, ambiguae sortis eventus,* and *fortuita contingentia,* as if even the process of naming were affected by uncertainty. "Now, however, what so far had rebelled against experience, could no longer escape the realm of reason."[6] In its rebellion against experience and reason, chance constitutes the profane counterpart to metaphysics and religious belief. The physics of the die, the seriality of the dice game, and the metaphysics of chance all come together under the aegis of *stupendus titulus.* Only when relating the physical object, the seriality of the game, and the singular event of chance to one another does the calculation, which corresponds to no certain experience, achieve validity. The announcement of the *aleae geometria* thus makes clear how the transition from mathematical texts by way of the division problem to the wager is possible. Here, we see a first level of meaning for the calculation emerge. As yet, however, there is no indication that the calculation might stand for any kind of semantics of probability. Instead, an understanding of the mathematical procedure lies in the paradoxical stance of calculation itself: by forcing the epitome of the nonrational—chance—to become rational, calculation mirrors, as if in a farce, the problematic place of faith in the realm of knowledge.

So we must turn our attention once again to Pascal's text on the wager. Not that after three centuries of interpretations anything has gone overlooked regarding the context of the *Pensées*, which mainly revolves around a justification of faith. The topical embedding of *aleae geometria* in this notion of apologetics has also been painstakingly described. For both of these projects, we need only consult the authoritative commentaries by Henri Gouhier and Jean Mesnard. Patricia Topliss has shown moreover that the motif of the wager on salvation stands in a long tradition of Christian apologists, stretching far back.[7] Instead, this study concerns itself with a reading that takes into account how the mathematical problem emerges in Pascal's text. The philosophical and literary interpreters see in games and the calculation of chance only an illustration of the possibilities and limits of rationality in the realm of faith. Hacking's scientific interpretation, on the other hand, manages to reconstruct the mathematical structure of the text quite ingeniously, but at the same time destroys the conditions under which the mathematical element actually makes its entrance in the text.[8] This is precisely the project at hand in the following pages: to examine the precise point at which and the gesture by which the mathematical structure, without any previous semantic legitimation, intervenes in the construction of meaning, which is concerned with understanding the place of faith outside of understanding.

In the course of our reading, a bipartite architecture of the text on the wager will emerge that, as such, echoes Pascal's bipartite essay on method, *De l'esprit géométrique et de l'Art de persuader* (Of the Geometrical Spirit and the Art of Persuasion). This structure has hitherto been largely overlooked by interpreters in its relevance for the wager. Put briefly, *De l'esprit géométrique* deals with demonstrative method, rhetorical persuasion, and their mutual relationship. In the Aristotelian tradition of epistemology, the realms of philosophical and geometrical knowledge and of persuasion that operate through semantic effects are strictly divided. In the unity of the essay, however, which consists of two parts or even two essays, demonstrative argumentation and rhetorical persuasion seem to intersect in a kind of argumentative-rhetorical evidence.[9] If such an intersecting point of evidence can in fact be alleged and made plausible, however, then the Aristotelian division of knowledge into demonstrative *scientia* and into *historiae* that use rhetorical topoi is quite fundamentally called into question. The text on the wager poses this same question of the relation between method and persuasion. It does so, however, not in a topical manner, but through its architecture. With nothing other than the gesture of mathematical structure entering a text purporting to be an apology of faith—an entrance that, as we shall see, is in fact twofold—the intrinsic division and intersection of cognition and persuasion come even more clearly to the light of day. By staging the emergence of numbers and formulas in the framework of words and meaning, the text on the wager demonstrates that of which the essay on method was speaking.

The following pages reconstruct the architecture of Pascal's text on faith and the calculation in games of chance to show how he fundamentally rethinks the relationship between demonstrative knowledge and the rhetorical production of meaning. He accomplishes this reinvention in a setting where readers after Pascal often thought they recognized the first exemplification of a theory of mathematical probability.

The Fragment on the Wager: Part I, The Inner Limit of "natural light"

Pascal's *pensée* on the wager changes course twice, albeit in two different ways. The wording of the first transition (I–II) obviously implies the division into a previous and a subsequent section: "Let us now speak according to our natural lights" (*Parlons maintenant selon les lumières naturelles*).[10] This injunction creates a distinction between two levels of meaning. The meaning of the fragment discussing the question of God is determined by differentiating it into a "natural" and apparently a "non-natural," speculative, sense. On one side of the divide, we find a discussion about the infinite as number, and on the other, an argument about the wager and the calculation of chance in gaming. This division begs the question about the meaning of introducing number and calculation into the text.

The second transition makes a split between current, ongoing discourse and words that are already present and at our disposal, a discourse referred to by anaphora: "This is conclusive [*cela est demonstratif*], and if men are capable of any truth this is it.—'I confess, I admit it [*je le confesse, je l'avoue*].'"[11] At stake in this transition is the status of a *pensée* that leads to mathematical deliberations in the context of a possible proof of God's existence. This transition thus subdivides the text into a primary level and a meta-level referring back to and commenting on the first level. On the primary level numbers and words are placed side by side; the meta-speech—the discourse referring back to that configuration of words and numbers—probes its validity in the form of a disputation on how to use numbers for making sense.

The first division, the dividing line of meaning marked by the "now" of "natural light" requires some commentary to start with. The coexistence of numbers and letters in the text is first crystallized here in the initial part of the text. Pascal thus sums up the old tradition of considering numbers to have meaning like words and regarding calculation as a procedure that bears no meaning. According to this tradition, arithmetic consists, on the one hand, of the *notatio* of numbers that can be traced back to the meaning of the corresponding notions, and, on the other hand, of the *numeratio* of calculative procedures without such meaning.[12] In the course of the section that opens with the "now" of natural reason, Pascal sketches a new configuration of number (or meaning) and calculation (or purely operational procedure) that will endure through much of the eighteenth century. With a single stroke, this transition invites discussion and revision of the old division between *notatio* and *numeratio*.

The limiting mark of insight's *lumières naturelles*—the limit of sense and meaning—does not simply separate a first speculative part of the text somehow from a mathematical part that would only then be bound to natural reason. It is much more the case that the first part leads up to the moment of speculation, the "non-natural," while the second part gets its impetus from assuming the opposite position of the *lumières naturelles*.

How does the natural light function as a dividing mark of sense and meaning in this text? Pascal speaks of the *lumen naturale* in respect to the "simple things" and "fundamental names" that underlie reasonable arguments,[13] though reason cannot analyze these names and things any further. It is in the natural light that we find the references that reason cannot overtake and the vocabulary of reasonable discourse and judgment in our everyday world, as well as in science, a vocabulary that cannot be further analyzed. This limitation holds true, for example, for the four basic geometric terms of space, time, number, and motion. This, however, is precisely where the first part of the fragment begins. Its points of departure are the basic geometric terms: "Our soul is cast into the body where it finds number, time, dimensions; it reasons about these things and calls them natural, or necessary, and can believe nothing else."[14] This is exactly the quintessence of the *lumen naturale*: the "thinking over" (*raisonner*) and "naming" (*appeller*) of the soul (the phases of its reasonable use of judgment) under the condition of its natural situation ("being-thrown-into-the-world") on the dimensions of this same situation (on the geometrical dimensions that the soul cannot avoid accepting as such, in view of which it "cannot believe anything else"). But we also recognize that, with these words, the actions and the situation of natural reason are merely described or named. The text forms a judgment on natural judging. In a constative mode of speaking, the text leads us from its starting point of the natural situation and behavior of the soul on to the rational description of faith as a particular act of judgment. This first part of propositional statements ends in determining the nature of faith. After the caesura of insight, of the "Let us now speak according to our natural lights," a different way of speaking begins to take hold. By drawing conclusions, arguing, and dialoging, the reverse mode of discourse makes itself heard. The question now at issue concerns what it means to declare faith, to accept an unprovable proposition; what does it mean to involve oneself in the game of accepting or denying faith? Here the text no longer speaks about natural reason. It shows reason acting under its natural situation of "being-thrown." It lets us see the soul "thinking" and "naming" and carrying out reason's operations in the situation when it cannot help but believe certain things. While the first part runs its course from "having-to-believe-certain-things" (*ne pouvoir croire autre chose*) to religious faith (*foi*), in the second part natural reason rather works its way forward within that mode of "having-to-believe-certain-things" (*croire*) to the act of confessing the faith of Christendom (*foi*). Thus both parts, before and after the cut, deal with "speculative" as well as "natural" basic reasons and given facts. More precisely, they both deal with the basis of what is naturally given, and that, as such, presents itself only to speculation, that is to say, with natural light, which provides the basis for insight but is inaccessible for it. Both parts addressing the speculative basis of the natural, to make use of Louis Marin's pun, have to do with sense and the senses. In taking up the given condition accessible to the senses, they establish the domain of sense and meaning, which in turn is inaccessible to sensual perception.

In order to spell out at this point what "Let us now speak according to our natural lights" means, or rather what it does as a division mark between the two parts, we should introduce the difference between a propositional statement (one constatively ascertained) and carrying a statement out (performance).[15] In terms of this distinction, the first part discusses in propositional statements how the logical core of "faith" is contained in the simple things and words that natural reason "must believe." Reason develops this logical core toward a proper confession of faith—but only if we remain strictly in the mode of making ascertaining statements and do not actually venture to make a confession. The second part, on the contrary, displays reason in the practical mode of argumentative disputation with the "speculative" and "natural" contents of the first part. Here we see natural discourse on its way to ascertaining a statement of faith—but this is done only by performatively accepting given modes of performing acts of speaking. To put it pointedly, "Let us now speak according to our natural lights," means: "Let us now talk" (instead of clinging to propositional statements), that is: "Let us switch the text to the mode of discourse corresponding to the *lumen naturale*" (to the execution of reasonable discourse instead of the description of reasonable contents and the constative ascertainment of reasonable matters). With this, Pascal also relates ascertaining statements about numbers and other geometrical magnitudes in the first part to the execution of chance calculation in a wager. "Let us now speak according to our natural lights," is a stage direction and the first utterance following the stage direction at the same time. It does what it says and says what it does. In speaking of making a statement, it is the urgent call to perform it.

As yet, however, we have not really taken the numbers into account. Only the introduction of numbers and operations of calculation into the fragmentary text can fully justify and articulate the division of the text at this inner limit, the limit of its meaning.

In both parts of the text, number or calculation contributes to the discourse on God and on morality. They do so, however, in different ways and by different means. The establishment of logic and the logic of institutional establishment are at stake at the line between *we* and *us*, at the words *lumen naturale* in the exhortation we direct to ourselves ("let us speak").

Continuation of Part I

"Infinitude" as Number and Concept: The Analogy Between Numerics and Verbal Meaning.

The first part of the text, as regarded from the point of view of the dividing line at "Let us now speak," employs the mode of constative ascertaining. It introduces the mathematically infinite and its relation to the number One in order to articulate the abyss of disproportion (L199; 72) between God and finitude: "Unity added to infinity does not increase it at all, any more than a foot added to an infinite measurement: the finite is annihilated in the presence of the infinite and becomes pure nothingness. So it is with our mind before God, with our justice before divine justice."[16]

This passage—along with many similar passages we are familiar with from the *Pensées*—insists upon a metaphorical analogy between mathematics and the semantics of morality and religion. The analogy relies on a paradox of measurement in a geometrical as well as in a moral and religious sense: the infinities in both cases result from measurement and at the same time disturb its operation. Precisely through their self-sabotage in the case of infinity, the measurements in geometry and in morality, in mathematics and in religion, function in strict analogy to one another. The point of the analogy is therefore not to claim any kind of similarity between the moral and the physical world, but rather to indicate the structural analogy in the paradox of measurement that manifests itself in the face of the infinite in both realms.

Already within geometry, the various kinds of geometrical measurement are strictly connected to each other only through the fact that they all allow respective modes of infinity to emerge, which then evade those selfsame types of measurement. Part I of *The Geometrical Spirit and the Art of Persuasion* exhibits the classical formula for this general rule. In this tract, Pascal introduces the notion of the two infinities. The two infinities characterize what the geometrical magnitudes of motion, number, space, and time have in common; they thereby guarantee the measurability of the natural world in the first place. The "two infinities" are constituted by the infinite approach toward the maximum and the minimum that is possible for each existing measurement.[17] All kinds of measurable dimensions are characterized as measurable relations by the fact that any large magnitude can be increased (multiplied, accelerated) and any small one can be reduced (divided, decelerated), according to the method of continuous measurement derived from Descartes's concept of the representation of nature (as a final indivisible magnitude an atom can thus not be posited by any geometrical term). This fact defines Cartesian nature as continuously mensurable, that is, as the object of representation.[18]

A path now leads from the analogy of measurements within the geometrical representation of nature to that between geometry and morality. Since the paradox of infinities always reveals itself anew and surprisingly to the observant eye as the unexpected structure of the analogy in all things mensurable, the two infinities, as the one common trait in all respects of creation, are objects not only of cognitive, but also of moral and aesthetic perception.[19] This is why fragment L199 can claim outright that both infinities in nature are the "image" that of itself and its creator nature engraved into all things: "When we know better, we understand that, since nature has engraved her own image and that of her author on all things [*ayant gravé son image et celle de son auteur dans toutes choses*], they almost all share her double infinity [*double infinité*]."[20]

> The structure of infinite bounds, both great and small, that is transferred analogously and that transfers itself anew with every insertion of further measurement, itself therefore has the status of the similar, of the portrait, or (to borrow from Antoine Arnauld and Pierre Nicole's 1662 *Logique*) of the natural sign. And this very movement of turning itself into a picture and portrait, which belongs to all measurement and reveals itself in particular in and as geometry, leads as if by itself to a moral and aesthetic proposition, to wonder and the sublime.

But those who clearly perceive these truths will be able to admire the grandeur and power of nature in this double infinity [*double infinité*] that surrounds us on all sides, and to learn by this marvellous consideration to know themselves, in regarding themselves thus placed between infinitude and a negation of extension, between an infinitude and a negation of number, between an infinitude and a negation of movement, between infinitude and a negation of time.[21]

Perceiving the two infinities in geometry provokes a state of wonder, first, because it is immeasurable in itself and recurs with all geometrical magnitudes, thereby guaranteeing the unity of measurement and its possible application.[22] When this circumstance is further joined with the formula of self-knowledge, the view of infinity becomes an opportunity for moral reflection in a clearly figurative, and at first sight rather artificial manner. Pascal picks right back up on this artificiality by continuing with the common metaphor of the *juste prix*, the just value and fair price, and derives self-knowledge back to the problem of measurement and proportion. "From which we may learn to estimate ourselves at our true value [*s'estimer à son juste prix*], and to form reflections which will be worth more than all the rest of geometry itself."[23] The *juste prix*, the just value that self-knowledge can develop in and of itself, is in turn a kind of a fair price, the price man has to pay by recognizing the *disproportion* of his own finitude in view of the infinities of creation. Proportionality and appropriateness are terms of cognition: only in areas to which man stands in a certain relation and that are appropriate to him can he establish relations and make measurements. He is only able to grasp, appraise, and recognize relations among things to which he already has a relation.[24] The disproportion of finitude compared to God has the same relation to this law of proportion in man's understanding as, in number theory, the two infinities have to the law of countability for the discrete series of numbers.

The cardinal phrase for the fragment's first part can be summed up as the "proportion of disproportion." It conveys both the geometrical and moral-religious themes of the text. If we think here of Aristotle, who defines metaphor by analogous proportions, we would have to speak of the metaphor of the hyperbolic, and thus of the proportion of disproportion. But it seems as if, at least at one point, a disproportion finds its way into this proportion of the disproportioned. It is as if it were affected by the very lack of proportion it seeks to place into the web of proportions: "There is not so great a disproportion between our justice and God's as between unity and infinity [*entre l'unité et l'infini*]."[25]

Even if the mathematical problem of the infinite complicates and strains the verbal expression of traditional analogies between numerical laws and the notions of morality and religion, we can thus far still speak of a classical use of number in the text. The relation of numbers and the geometrical proportions representable in numbers have belonged to the realm of forms and the immutable truths ever since Plato's dialogues and particularly in the Neoplatonic tradition. This understanding and use of mathematical number theory is, as we shall see, however, fundamentally different from the way Pascal employs the calculation of betting chances in the second section.

Operations with Infinite Numbers: Topoi

Before we even completely exit the register of speaking-about—the cognitively ascertaining statements in part I of Pascal's wager fragment—and enter into the register of the performance of calculation in part II, however, this difference already resurfaces within the first part. It is precisely in the theory of infinite numbers that the concept of *notatio*, the traditional notion of writing and the meaning of numerals, proves to be insufficient. Even for the sake of pure number theory, certain aspects of *numeratio*, the operation for calculating, have to be taken into account from the outset. In this respect, Pascal's real point in the first part has not yet been mentioned: the observation of—and wonder at—the analogy of geometrical infinity with ontological and theological infinity embodies an element of inadequacy, which is bound to unravel the analogy.

According to Pascal, infinity is characterized by its accessibility to the defined operations of geometry, on the one hand, and, on the other, by a certain indeterminacy, which has a particular reason in number theory. Pascal focuses, in this context, on a discrepancy he detects between an intuition in number theory and the failure to give this intuition adequate expression. We are no longer dealing with number as an example and case of measurement, nor with the urge to transfer such characteristics of numbers into the moral and aesthetic ideas of measuring. Instead, we are confronted with a purely mathematical operation that is possible or impossible with regard to infinite numbers. This problem will then transform—through a violent but exact transition—into a question of morality and religion.

In order to grasp the problem in question, we should start from this point: the arithmetical operations of addition and subtraction, as well as the corresponding physical processes of increase and decrease, acceleration and deceleration, let us know that infinity exists. We can indicate the operations in geometrical systems of measurement, and particularly in the succession of numbers, through which the infinite comes into being. According to Pascal's *Réflexions sur la géométrie en générale*, infinity is an operational term referring to other terms given by natural light, such as number, time, space, addition, increase, acceleration, and so on. The assumption of the infinite follows from succession in numbers and measurement. The text complex of the wager leaves the question open as to whether "infinite" refers to the operation itself or to its result, that is, whether an infinite continuation of arithmetical operations is intended or the limits zero and infinity are meant. This ambiguity is Pascal's hidden point, however: the value of the infinite magnitude simply cannot be made to stand out from the infinite continuation of mathematical operations. Or to put it in epistemological terms, the knowledge we can attain about the infinite magnitude cannot be clearly separated from merely operating with calculation.[26]

The major paradox toward which Pascal has been heading with regard to the infinite follows from this hidden point. There is in fact one question that can neither be answered nor even posed in the case of infinity, though for every other number we do so without any hesitation or doubt: is it odd or even? Infinity is a number that can be produced by mathematical laws and in the realm of natural reason, yet it nevertheless cannot be judged according to the rules of the realm in which it falls by derivation or determination.

The claim that every number has to be either even or odd is an old topos, which follows its own history and tradition. The formula originated in ancient mathematics and turns up repeatedly in Aristotle's *Topica*.[27] It represents mathematics in the ancient doctrine of dialectical judgments and argumentation. The argument that every number must be "even or odd" is, one might say, the topical probability of mathematics and the point where its exact knowledge becomes persuasive outside itself. In this role, the topos has made its career in occidental logic and philosophy: Boethius makes it the model *maxima propositio*, or indubitable proposition: "for those propositions that are undoubted are generally the principles of demonstration for those propositions that are uncertain. Propositions of this sort are 'Every number is either even or odd' and 'If equals are subtracted from equals, equals remain,' and others whose truth is known and unquestioned."[28]

The humanist Pierre de La Ramée adduces the "odd or even" topos in order to exemplify *axiomata catholica*, phrases supposed to serve as organizing principles of individual sciences: "Man is a rational animal. A number is either even or odd. A wolf is born to howl."[29] Pascal himself reckons a number's being even or odd to be an example for a nominal definition.[30] Nominal definitions, however, may only use defined or (in Pascal's words) self-evident terms. The topos "even-or-odd" thus comes to serve in Pascal's work as evidence for self-imposing truth pure and simple.[31] With this topos, Pascal introduces the paradox that infinity is defined by operations of calculation but lacks the nature of a number. This introduction takes place still within the first part, the part within which the theory of number and meaning engraves on things the portrait of nature and its creator: "We know [*nous connaissons*] that the infinite exists without knowing its nature."[32] In this awkward crook between the existence and the essence of infinity, the disproportion of proportion finally comes into effect in Pascal's text on the wager.

Pascal cites the fact that infinity can be produced by continuous multiplication or division in order to indicate that our knowledge of the infinite exists. The fact that the question of whether infinity can be divided by two does not yield an answer is, according to Pascal, an indication of our failure to recognize its essence. He thus invites an interpretation that, again, opens up findings in mathematics for consideration in the spheres of morality and religion. This time, however, only the structure of the paradox is transferred. Thus an image emerges in the realm of morality that offers no imaginary qualities.

This transfer leads us to the paradox of faith in the following way: in terms of a cross-classification between "knowledge" and "ignorance," on the one hand, and "existence" and "nature," on the other, the argument of the "infinite," which is to be gained from number theory and the history of topoi, ends up in a middle position between our knowledge of things finite and our ignorance when it comes to God. We know both the existence and the nature of our own finitude. God is inaccessible to our knowledge again with regard to both His existence and essence. "Infinity" and the knowledge of the infinite, however, combine knowledge and ignorance: we know that there is such a thing as mathematical infinity (we can produce it through defined operations); but we know nothing about its nature (we cannot decide about its nature as a number). Infinity thus appears as a hybrid of knowledge and ignorance. Faith, too, is just such a hybrid. The same argument and its combinatorics can also be seen as a stepladder of reason, mediating between knowledge and ignorance. In this sense we might say that by citing the topos of infinite numbers, we set what is perfectly known to us into relation with what is perfectly unknown. This topos then in itself would prove to be the known-unknown that straddles this boundary. "Infinity" would turn out to carry nonevidence as a rupture of knowledge in itself.[33]

Considered according to its quality as a kind of knowledge given to us before Judgment Day, the faith of a Christian comes to be characterized by the same hybrid that marks the argumentative power of the "infinite" number. By faith (*foi*) we know (*connaître*) of God's being (*existence*). But we shall get to know (*connaître*) God's nature only through the glory (*gloire*) of salvation after Judgment Day's act of grace.[34] The epistemological structure of infinity as the connection of knowing-the-existence and ignoring-the-nature repeats and in fact manifests itself in the peculiar soteriological position of the Christian in this world, since she already knows God's existence in faith without knowing His nature. Pascal confirms infinitude as a model of faith in this structural sense with the first insertion of an *I*, which occurs immediately before he reaches the borderline of *lumen naturale* between the first and second part: "Now I have already proved that it is quite possible to know that something exists without knowing its nature" (*Or j'ai déjà montré qu'on peut bien connaître l'existence d'une chose, sans connaître sa nature*).[35] The argument repeats its own structure: the relation between knowledge and ignorance in mere number theory corresponds to the relation between knowledge and ignorance in the soteriological quality of faith in this world. We are dealing here with the transfer of knowledge (in geometry and number theory) to a realm of human ignorance (faith), with the analogy depending, however, on the relation between knowledge and ignorance in both mathematical knowledge and faith.

At this point, the fragment's first part breaks off with the signal phrase, "Let us now speak according to our natural lights." The introduction of a notion of operative proceduralism that characterizes the determination of the number infinity and the performative manner in which the argument is used for purposes of religious persuasion have indeed led to a new stage in the understanding and the semantics of the infinite number. But this point is only truly realized in the mode of theory, the realm of ascertaining declarations. This is so since the transfer of mathematics to the semantics of faith, the moment of operative procedure and the performance of persuasion, is built on the paradox of the nonrelation between the theoretical and the evident within theory.[36]

Part II of the Fragment: Performative Speech and the Operative Procedure of Calculation

In the first part of the fragment on the wager, the speaking subject is consistently called "we" (*nous*) or "one" (*on*). "We/one," up until the caesura of the "Let us now speak," really means "everyone" or "all of us." It is either the implicit justification of an ordinary language that authorizes its statements by the universal norms of what is evident in everyday life, or it is a speech that becomes universal through the authority of an institution—the eternally one and all-embracing church, even if in fact it should be limited to the territory of a monastery just outside the gates of Paris.

In the second part, following the breaking point of "Let us now speak," the ambiguity of the attempt to produce unambiguousness is dismantled. In place of the universal point of view incorporated in the "we/one," three distinct voices now make themselves heard. The first party is the proponent of the disputation (who designates himself with the first-person plural, though "we" and "one" still resurface for a rhetorical unity between "we" and "you"). Joining this voice, we now hear the addressed second party, the respondent of a skeptical non-Christianity (who himself says neither "we" nor "you," but who refers to himself—and nothing but himself—with verb forms of the first-person singular). Finally, this duet grows to a chorus: we have "they" (*eux qui, ceux qui*), the Christians of the "we"/"one" of the first part, who have now become third person plural and who are the ones about whom the proponent and respondent are speaking. What "they" say (and have said in the first part) is now a citation in the current dispute. The analogous metaphorics of measurement and the topos of infinity are absent discourses and discourses of absences; discourses, to be precise, in the Foucauldian sense of the word. And rightly so. For, whatever images and rhetorical tricks are employed, no one speaking in a disputation "according to natural insight" can rely on propositional representation alone. The pragmatics of the situation in which speaking takes place—the conversation or disputation—in part II constitutes the everyday normative or institutional default frame the parties must face, and in which their arguments prevail or fail.

Immediately before the "Let us now speak," a first person, "I," summarizes and concludes the monologue of part I: "Now I have already proved. . . . "[37] The "I" thus gives his signature of approval to the monologue of the first part, before the roles for the disputation in the second part are even distributed. Several readings of this gesture are possible: the signatory of the first part may be nominating himself with the "Let us now speak" as the proponent of a disputation; he may fall back into the position of institutional administration and minute-keeping for the disputation; the monologue in the first part might in truth be an introductory speech already within the frame of the disputation; or, finally, the disputation could merely be a figure of speech within the monologue, a *dialogismos*. In any case, the brief entrance and immediate disappearance of the "I" on the boundary line of sense and meaning simultaneously designates a similar boundary for the possibilities of speech, defining the institutional logic of the fragment.[38] The analogous and metaphorical, the Platonic and Aristotelian qualities of magnitude, number, and infinity in the monological statements of the first part always already make implicit reference to the authority of an institution of normative evidence. They bear witness to and activate a decision about instituting measures, which they thus stipulate has already been made. Now, in the second part, an institution grants permission to speak to the interlocutors. They must search for a decision according to the rules of the institutional establishment of their parts in the order of the discourse.

This is where calculation and its operational procedures come directly into play. We are no longer dealing with a situation in which number theory is only complemented by operations of computation. The first-person plural now introduces the calculation of games of chance as an example by the very act of performing the calculation. Its rhetorical nature has a status entirely different from that of the infinite number in the first part. It is no longer a matter of analogy, transfer, and illustration, but rather of modeling a situation in which action must be taken. Calculation, the operation of reckoning, makes its entrance in a performative style and ambience. The difference between number and calculation, by which the example of the game distinguishes itself, is the same as the difference between stated propositions and the open display of performances. It is this difference that designates the unity of the speech act.

(a) Introducing the example of the game.—After the call to speak "according to our natural lights," a passage follows that can be understood as determining the conditions for the dialogue to come. The conditions are these: because God has no unities or extending parts and is unlimited, "we" are without any proportion and measure in relation to Him, and neither His existence nor His nature is open to "us" to know. Therefore "we" are not the ones who can make a statement about God's being. This logic seems to repeat exactly the measurement and number theory with which the "I" had signed off on the first part: "Now I have already proved that it is quite possible to know that something exists without knowing its nature." But there is a fundamental difference between the two versions. In the monologue of part I, the "I" repeats and gives its signature of approval to the expression of the proposition of the possible separation between knowledge of existence and nature. In the present case in part II, this expression is treated as a performative contradiction and contained as if in brackets. The claim to know the existence of something about whose nature nothing is to be known is now applied to the implementation of the claim itself: though the structure of this claim can be recognized and put up for debate, one cannot assign it any place of its own in the order of things. The utterance cannot be carried out under the regime of the "Let us now speak": "We are therefore incapable of knowing either what He is or whether He is. That being so, who would dare to attempt an answer to the question? Certainly not we, who bear no relation to Him."[39] This performative turn amounts to disclosing the question of the speech act. Through this turn and disclosure, the separation of "we/one" takes place as described above—the separation into the voices currently speaking in the debate, who cannot implement any speech acts that include performative contradictions, and the absent "they" of the theologians and geometricians who simply draw on the foundation of institutions for the right to use discourses implicitly based on such paradoxes.

The first and basic act of the first person consists in bringing the second person to accept being in a situation of decision and under limitations of time and information: the second person must answer. The tempo now picks up. The distribution of performative roles and the introduction of the example of the game of chance can hardly be isolated from one another in the text on the wager. In the argument of the game put forward by the first person in the disputation, the "I" repeats and hones the rules of the disputation that force us to make a judgment. We can also read the passage the other way around: in this case, it is the entrance into the game, that is, the entrance into the impossibility of not playing the game, that first stamps the entrance into the rule of disputation and the duty to discursive judgment onto the performative situation of the speakers. From the outset, the performance of speech and calculation of chances seem about to overtake each other in turn. Everything in the second part depends on seeing that this rash meshing of the rules of speech and the rules of calculation is not to be blamed on the fragmentary condition of the wager text, nor, much less has anything to do with Pascal's negligence at this point.[40]

It is particularly helpful to read the fragment again from the viewpoint of an architecture of text and number. This perspective allows us to recognize that speaking and calculating here coincide for one single moment in an absolute act of simultaneously performing (the speech act) and implementing (the operational procedure of calculation)—an act radically different from the mere coexistence of number theory and propositional statement in the first part, where they only relate to each other through analogy and metaphor. We also should not compromise this moment of the coincidence of speech and calculation by associating it with the logic of the probable, which remains foreign to Pascal. Pascal's point is precisely that the regional problem of mathematics, the calculation of games of chance, relates to the semantically loaded situation of man's obligation to decide about God's existence. Pascal does not take recourse to some intermediary step where the structure of the calculation of chance would be interpreted in general as a model for understanding cases of uncertainty. The particular task of mathematics makes its entrance immediately and specifically in the apologetics disputation. In doing so, it becomes clear that only this one singular discursive situation, the disputation over God's existence, is comprehended by the calculation of chance. Games become the example for this moment in life alone, though indeed their consequences then affect each and every individual's entire life.

We should identify two phases in the coincidence of the performance of speech and the operation of calculation. Phase one: the second person has made the objection that renouncing argumentation as championed by the faithful may be consistent with his position, but within the institution of the disputation, it still would not amount to any effective behavior. To this the first person replies: "let us say: 'Either God is or he is not.' But to which view shall we be inclined? Reason cannot decide this question. Infinite chaos separates us. At the far end of this infinite distance a coin is being spun which will come down heads or tails. How will you wager?"[41]

In this initial phase, the first and foremost condition for calculation in games of chance—the isolation of a strictly contingent event or matter—is connected to the discursive situation. It is this and only this absolute lack of information that creates the artificial situation of a game of chance for life: only with respect to God's existence can it be said categorically in life that something can be or not be, take place or not take place. A wish to say anything more than this means having to calculate. If we take up the calculation of chance at this point, and only at this point, we are taking up the rules of a discourse over the existence of God in general.[42] Phase two: the other voice accepts that whoever makes a choice in this situation must choose and thereby play Yes or No. He does not accept, however, that he must choose and play in the first place. But this is precisely what the first person assumes in answering this argument: "Yes, but you must wager. There is no choice, you are already committed" (*mais il faut parier. Cela n'est pas voluntaire, vous êtes embarqués*).[43]

We are not faced just with recognizing that the situation of the decision about God's existence is a game of chance, but rather with accepting the necessity to make that decision—to express it—and play the game—to calculate. Once again the game of chance is not brought up as an example to model situations of uncertainty and pure probability in general; instead, it reveals the regional mathematical problem to be the immediately singular case of this decision. In this context we should recall the second principle of chance calculation, the division problem in the event of a game's interruption. Pascal's *méthode générale* and Fermat's combinatorics of possible events are both solutions for this second principle of what in general can be analyzed and calculated as chance in a game—a principle that had already been grasped in the sixteenth century by Niccolò Fontana Tartaglia and Girolamo Cardano. This second principle again proves to have a singular correspondence to man's decision about the existence of God. It is no longer so much the case that we must enter into the disputation—which is a game of chance—because the question demanding a decision is so important. Instead, we are already playing the game of chance—which is a disputation—because this game and this disputation can never be played to an end in this finite life. It will always and in principle be an interrupted match. The dice will never fall—they are always already and always still in the air. In the methodological sense of the suspended final round, the question of the existence of God is even first of all and above all a game of chance that can only be verbalized and reenacted in a discursive disputation. This second aspect obtrudes itself in Pascal's general method more clearly than in Fermat's combinatorial solution. As the letters to Fermat and the corresponding chapter in the treatises on the arithmetical triangle show,[44] Pascal deals with the method of the interrupted game directly with the strategic moment of the deferred decision in mind. Fermat constructs the game with the help of combinatorial tables of the possible outcomes from the first match, and he derives thence the fair distribution of stakes among the players. Pascal, in contrast, identifies the relevant opening positions necessary at the outset of the last round of the game for a final victory or defeat. The opening conditions at this moment point directly to the fair distribution of stakes throughout the game. They determine in reverse the positions in each preceding round in their relation to the attainment of the relevant starting positions for the—structurally absent—final round. Pascal's calculation is based on a strategic consideration from the point of view of a final match that never takes place.

It thus becomes obvious that the fragment on the wager hinges on the distinction between two different situations regarding number and calculation: one of them is the situation in which number theory and operational procedures are kept well separated from each other, as is the case in the world of theologians and geometers in the first part. The other consists in a process of calculation that within itself involves the theoretical view of numbers and their interpretation. This is the case in the process that develops in the calculating performance between proponent and respondent in the second part. The fragment thus also displays in its overall structure both of the possible connections between numbers and discursive speech, between the two possibilities to speak discursively about and in numbers. These two possibilities, however, are built into the mathematical structure. The illustrative meaning of the infinite in the topoi of the first part turns out to be connected with the hidden ambiguity between the number infinity and the infinity of mathematical operations associated with it. Already in the first part, this ambiguity between infinity as a number and a result of operations points to a concept of mathematics transcending the dichotomy between number and calculation and, with it, the dominance of the symbolism of numbers. Conversely, in calculating games of chance, the difference between number and calculation will return in the second part, although in a different way, as soon as the number infinity is inserted into the process of calculation.

(b) Calculating as speaking.—The example of the game, which the first-person plural uses in order to persuade the interlocutor in the debate, confronts the respondent with the situation of unavoidable judgment that results from insufficient information. This, however, is the very rhetorical situation in which he as disputant already finds himself.[45] The only rhetorically conveyed expression of the second part stands for exactly this scenario of finding oneself in a rhetorical situation: "Vous êtes embarqués." A persuasive and hence rhetorical use of calculation is introduced by this move. The precise coincidence of performative speech and calculation is then superseded by a mere parallel between reckoning and persuasion. The game of chance at this point actually becomes an example that the speech employs for persuasion. Finally, not only the unity, but also the parallels that made it possible in the first place will rupture according to the consequential development in this whole section. In the first part, the truth of propositions about nature and numbers could not become performative or effective; in the second part, the unity of speech and calculation exceeds and destroys itself in the very moment when it claims to have reached a truth.

Turning the game of chance into an example in the rhetorical sense is bound to a rephrasing of the meaning of the wager. The first person gives this new and allegorical interpretation only after the game of chance has imposed itself silently and, as it were, procedurally on the interlocutors in the course of their disputation: "You have two things to lose: the true and the good; and two things to stake: your reason and your will, your knowledge and your happiness; and your nature has two things to avoid: error and wretchedness. Since you must necessarily choose, your reason is no more affronted by choosing one rather than the other. That is one point cleared up. But your happiness?"[46]

The text's attention thus shifts from the methodical issues of contingency and its inherent problem of division to focus more squarely on gaming, and on the questions of stakes and speculation on winning in particular. At first sight, this seems to finalize the identity between calculation and argumentation. As soon as a person lays a stake and expects a certain sum of money as winnings, the model of the game immediately begins to move away from its unique correspondence to the question of the existence of God. The game then becomes a mere example in the manner of later probability. Nothing of the sort occurs to Pascal. Connecting stakes to speculation on the existence of God and the issue of salvation, however, brings about another problem.

The value of the stakes and winnings in this game will turn out to be quite dependent on its result. When the first-person voice of the dialogue says: "If you win, you win everything; if you lose, you lose nothing," it presupposes the interpretation of the absent party, the Christian who bets on God. In his view, God is offering eternal life as the award, and a finite life compared to infinite life equals nothing (as we learned in the first part). When, however, the nonbelievers addressed in the dialogue, who bet on the non-existence of God, then answer: "Perhaps I am risking too much," they are applying their own interpretation of stakes and winnings. A nonexistent God can offer no life at all, and so the same stake, finite life, has the value "one." The coin toss—the contingency machine—now paradoxically helps determine the quality of the stakes and the winnings retroactively.

The first-person voice now carries out a number of different calculations. Bit by bit, it thereby brings into play the notion that the result of the game retroactively affects the meaning and the value of stakes and winnings. In the end, this process will systematically have destroyed the sense of the calculation, its purely performative accomplishment. Five possibilities of a simple coin toss are reckoned out all in all: twice two variations of the two first basic models, and then an invariable third model.[47]

The first model is characterized by the chances of the coin turning up heads or tails amounting to 1 : 1, so that the odds of winning are finite. It is hence a matter of the ideal model of a coin toss. In this case, we can take into account the idea of a retroactive impact of the game's outcome on its stakes without any logical difficulties cropping up. If, heads or tails, the bets on *God exists* and *God does not exist* have the same chance, then there is also an equal chance that the God at stake has infinitely many lives to give out, or that He has none. Thus we can understand why Pascal discusses cases under the supposition of a 1 : 1 division of chances as having the party betting on God to wager 2 against 1 life, and then 3 against 1 life (and so on for n until infinity).

The second model is very difficult to understand. It is determined by supposing that the chances of winning for the party betting on God amount to one to infinity. With this supposition, we obviously are no longer looking at an ideal coin toss. The difficulty lies in the stakes or the winnings respectively. On the one hand, the same cases are named as in the first model: 2 lives against 1, 3 lives against 1. On the other hand, both the beginning and end of this passage mention "an eternity of life and happiness" (*une éternité de vie et de bonheur*) or "an infinity of infinitely happy life" (*une infinité de vie infiniment heureuse*).[48] Both of these provisions are difficult to reconcile with the calculation of chance in a coin toss. Henri Gouhier accordingly brackets the sentences dealing with an infinity of life to be won and removes them from the context.[49] But in this case, clearly no calculation would lead us logically to place our bets on God. Only if the sentences about the infinity of life are given some consideration does this second series remain feasible as a calculable series. But if an infinity of life is to be won, how can another such series as 1 : 2, 1 : 3, (etc.) apply?

The following solution is conceived in terms of the overall architectonics of the fragment. According to this interpretation, the second model is constructed on the pattern of the first one, which in its turn was characterized by reckoning with finite stakes and a finite distribution of chances. At the same time, the second model is already preparing for the third, which, as we shall see, reckons with infinity for stakes and distribution and, for reasons of calculation, is bound to transcend the realm of calculation. The second model hence appears as an internally contradictory intersection of both models. (In terms of its architecture of thought, such a structure would mirror the three instances of knowledge—human knowledge of the finite, faith in the state of finitude, and knowledge of God in the state of grace—in the first part. There, too, the middle case was a paradoxical one that combined knowledge and ignorance, though in the first part, the middle case was admittedly more clearly designated as such.) Based on the consideration of a retroactive impact from the game's outcome on its initial stakes (a consideration commentators do not deny, even if they tend to give it short shrift), we can argue as follows: nonbelievers have their share in the second model insofar as the relation 1 : ∞ is given. We can then have the series of calculations begin as in the first model by betting 2 : 1 life; 3 : 1 life; etc. The faithful, with their own outlook, lay their bets on the side of the winnings, in the opposite direction from the distribution of odds: for them, behind the two or three lives that are posited as prospective yield, there is the eternity of life that the Christian God dispenses to the saved. For Christians, then, the 2 : 1, 3 : 1 cases are in a manner provided with the added clause, "etc. with n approaching infinity." In this scenario, as long as "infinity" remains a possible reward—as long as there is even the slightest chance to win an infinity of life—then logically the game should be played even according to a model based on chances of winning at 1 : ∞. This interpretation would mean that the second model can assume two possible sets of values, according to whether players follow assumptions *a* of the nonbelievers or assumptions *b* of the faithful.[50]

The third model, however, is again quite transparent. Here the presupposed distribution of gain and loss is "x : nx = 1 : n," where n is a finite number. The stake is m (where m is a finite number), the possible gain is "infinite"; the rate of winning ∞ : m, therefore (according to part one), equals infinity. This means that the calculation commences from the perspective of the believer. The quotient for the distribution of chances may be small, but it has a finite value; the rate of winning may exhibit as large a denominator as you wish: it will have an infinite value. At this point,

Pascal, or the speaker of the first-person voice in the dialogue, comments, "That leaves no choice."[51]

What does all of this mean? According to Pascal's general method, the formula for a fair division (the distribution of the sum of the stakes equaling the possible winnings) after a game's interruption before its final round is provided by the ideal coin toss, that is, with the ratio 1 : 2, and hence: $S = L + (C - L) / 2$ (whereby, as explained above, S is the just share to which player A is entitled; L is his risk of loss in the coming game; and C is his chance of winning in the coming game). This gives rise to the following applications according to the three models discussed above:

The first model (the ideal coin toss with the ratio 1 : 2), case A (2 lives to be won): $S = 0 + (2 - 0) / 2 = 1$; case B (3 lives to be won): $S = 0 + (3-0) / 2 = 1.5$. Conclusion: the game in both cases can be calculated and turns out to be reasonable or even advantageous in accordance with Pascal. By staking one life, the distribution rule in the first case yields precisely the staked value, and in the second case it yields a greater value than the stakes.

The second model (with the ratio 1 : ∞)—first, according to interpretation *a*: case A (2 lives to be won): $S = 0 + (2 - 0) / \infty \approx 1 / \infty$; case B (3 lives to be won): $S = 0 + (3 - 0) / \infty \approx 1 / \infty$. Mathematically, both cases yield the same result. By staking one life, contrary to Pascal's claim, either stake would be infinitely disadvantageous to risk (= ∞). Secondly, according to interpretation *b*: case A (∞ lives to win): $S = 0 + (\infty - 0) / \infty$. If we assume that by "infinite" Pascal understands the same calculable value both in its sense of eternal life and as an infinite series of rounds to play out, then this is still a reasonable risk, since the fair expectation of gain is equal to the stakes. According to this interpretation, which admittedly does not accord completely with the text, the result Pascal asserts would finally arise.

The third model (with the ratio 1 / n, where n = a finite value), only specified case: (∞ lives to be won): $S = m + (\infty - m) / n \approx m + \infty / n \approx m + \infty \approx \infty$. We may begin to understand here what is meant when Pascal or the first-person voice of the disputation claims, "that leaves no choice," or, "with this, all consideration of one's due share becomes impossible." Introducing infinity as a gain makes all other components of the formula irrelevant, as long as there is no ∞ in the denominator, that is, if there is no ∞ in the chance of loss. Infinity thus "leaves no choice," because it destroys the structural requirements for a choice, or, in other words, the determination of one's due shares. In the third model the formula of probability carries out its operation (regardless of what might be inserted for m or n) as its own destruction. Calculation in this case leaves no choice because no calculation of a due share is any longer possible according to its own rules and formula.

The number of possible cases has not been exhausted in discussing the three models, but we have touched on all that seem relevant to believers and nonbelievers. The discussion of the cases functions in fact as kind of a rhetorical climax. Pascal stresses this character by claiming in the context of the last model that "no choice is left" because one's appropriate share and portion can no longer be calculated. This is the rhetorical success of the example of gaming and its calculation. At the point where calculation "destroys" itself by successfully implementing its own rules, the persuasive character, which had been hidden in the example of the game and its mathematical calculation, becomes explicit: "there is no room for hesitation, you must give everything."[52]

Once the features of insufficient information and necessary decision characteristic of the rhetorical situation have been entirely invested in the example of the game, the example and the logic of calculation seem to exhaust and even transgress themselves. An entire paragraph is now added in order to persuade nonbelievers, the dialogue's second person, in a purely discursive manner, that they have been persuaded by the logic of calculation.

Recall the text's motto for introducing the game and its formula as a persuasive example: the question of God's existence should be played as a game of yes or no "in infinite distance." This mode of play, according to the claim, is the precise prerequisite for addressing the problem of God's existence directly and uniquely by the ideal construction of a game of chance. After working through the calculation of this example, which derives from its rhetorical rephrasing, the first-person voice of the disputation takes up a clearly rhetorical persuasive technique to push for the example's acceptance, as if after the fact. (1) The logic in the example, calculation, has its value by virtue of the logic which makes gaming an example for all and even less certain problems of risk taking: "Every gambler takes a certain risk for an uncertain gain, and yet he is taking a certain finite risk for an uncertain finite gain without sinning against reason." (2) The abyss between the example and what it is meant to illustrate is mirrored by the abyss between man's stakes in the situation modeled by the example. The decision to invest in God or in this world is an absolute and incomparable choice. But things that are incommensurable as long as we think of them as choices outside of a game become the origin of proportion the moment we interpret them as stakes within the game: "There is, indeed, an infinite distance between the certainty of winning and the certainty of losing, but the proportion between the uncertainty of winning and the certainty of what is being risked is in proportion to the chances of winning or losing." Like geometrical dimensions and numbers in particular, now the calculation of chance, too, is supposed to present a proportion. This proportion obviously does not address and measure any existing dimension of nature; instead, it retroactively introduces a proportion of proportions by itself, the example of the game of chance. (3) The example of the game of chance thus makes use of something that infinity and its symbolism lacked in the first part: it applies the force of persuasion. "Thus our argument carries infinite weight [*notre proposition est dans une force infinie*], when the stakes are finite in a game where there are even chances of winning and losing and an infinite prize to be won."[53]

The persuasive force of calculation in the example is dependent on persuasion as defined by the institution of the debate, the situation in which it is proffered.

As this rhetorical reinterpretation of the example forfeits its striking identity—and even its parallelism—with persuasive performance, it becomes clear that the second part of the fragment is nothing other than an essay on persuasion. The concluding passage (which can be understood as uttered by the dialogue's second person, or as a remark outside of the dialogue) only seems to contradict this claim. Although this last sentence of the section with its opening salvo: "This is conclusive [*démonstratif*]," obviously refers to *docere*, the continuation restricts the meaning back to a more rhetorical understanding: "and if men are capable of any truth this is it."[54]

Calculation assumes the rhetorical value in part two that geometry and the theory of numbers have in part one. But while measures and numbers belong to the register of representation and meaning activated by analogy and topoi, the calculation of distribution plays its role within the process of persuasion and judgment. The relation of the numerical to the verbal parts of the text in this section is therefore different from in the first part, in which the analogy of measurement and the metaphorics of the infinite duplicate the text. Text and number here represent each other in analog transference or in the topoi of argumentation. In contrast, the second part's example of the game mirrors the rhetorical situation in a radicalized form, while reckoning on the basis of calculation is an exemplary argument in the service of rhetoric. Calculation and persuasion thus put each other to use, the one instrumentalizing the other in turn.

We begin to discern a neat architecture at this point: the first part of Pascal's text on the wager is characterized by propositions and cognitive statements with the topos of "even and odd" mediating between text and number. In the second part, the performance of argumentation is in the foreground, with calculation carrying out the argumentation in a unique manner. The first part of the fragment, composed of text and number, makes points and statements cognitively and ends up presenting nonillustrative and inevident images; the second part, composed of text and calculation, becomes self-destructive in the final consequences of its persuasive function. Within each part, therefore, the dominant character—the proposition in the first part, the performance of persuasion in the second—is combined with elements of the other. A moment of persuasion is contained in the topos of infinity in part one. But when the computation of distribution is finally implemented in the second part, we find a moment of persuasion's self-theorization, a propositional account of its performance as was characteristic for the previous part. This architecture presents us with a chiastic crossover of the opposed attributes of what we call a speech act. It is important to recall that a chiasmus is a formal union, a balance of opposite characteristics, without, however, dissolving them in the synthesis of a new unified concept. The generality of the constative proposition and the singularity of performances remain unreconciled. In the chiasmus of the two parts, the wager text thus projects and, at the same time, rules out as a unified concept the speech act of probability—a speech act that would unify the generality of number theory and the singular decision of the game of chance.

Propositional and Performative: The Discourse of Religion, or *triangle arithmétique*

In the wager fragment, no such unifying concept as probability imposes itself. The architecture of this text hence proves compatible with the pairing of Pascal's essays "On the Spirit of Geometry" and "The Art of Persuasion" and the assumption that they were meant to form one text or, if not, the obvious fact that they relate closely to one another. This does not mean, however, that we cannot find forms of representation in Pascal that do in fact join together the two chiastic halves. But they are either of a purely textual or a purely mathematical nature, and thus subject the unity only to either verbal discourse or numerical presentation; consequently, they fall short of unity again.

We find the intrinsically textual solution at the end of the wager text. Both parts we have discussed close with respectively unexpected, parallel objections. The pure representation of faith as knowledge in the first part breaks off with the indication that we should now speak "according to our natural lights." This injunction is meant to be explained by the fact that the question of topoi, of persuasion and of numerical operations, has already made a first appearance in the propositional representation. The epistemological split between existence and essence is indeed already contained in the mathematical concept of infinity, but only the verbal text could claim this to be so. On the other hand, the second, performative part breaks off by finally admitting that persuasion by means of the necessities of calculation can only be experienced as a rational state. Human emotions, however, do not necessarily comply with such a state, and they may even remain directly opposed to it. Rational persuasion is a compulsion that feeling resists. When judging persons perceive decisions to be forced upon them, their judgments are rendered impossible by the very nature of what induced the decisions in the first place: the compulsive nature of calculation. The idealization of pure decision that makes the game of chance its exemplary model robs it at the same time of the practical use of exemplarity.

The two parts are united in the institution of Holy Writ and the rituals of the church by the subsequent coda, which closes the wager text as a whole.[55] The conclusion thus points as if in code to two of the most powerful and most comprehensively discussed areas of apologetics. Scripture lies, as Pascal says, "behind the game" (*le dessous du jeu*); it is the flip side of calculation. It remains invisible when reckoning the distribution of the winnings, but is also the only thing that imparts sense to the numerical proportions in the first place. When Pascal speaks at this point of "Scripture and the rest," he is referring to the rules of typological hermeneutics. Below or behind the game, a space opens up for the law, and here the discourse of the law and its exegesis begin their work. But this disclosure still does not help an unbeliever who may have been intellectually convinced by the example of the game but remained unable to match intellectual with emotional persuasion. For him, since law again acts through compulsion, even if of a different nature, it runs into the same dilemma as the example of the game. In contrast to law, however, ritual is first of all another kind of game, or rather a play, but a play of as-if, a theater play. This play is demonstrated by the absent others, the believers, and the watching unbelievers can in turn imitate them in a play: "learn from those who were once bound like you and who now wager all they have. . . . follow the way by which they began. They behaved just as if they did believe, taking holy water, having masses said, and so on. That will make you believe quite naturally, and will make you more docile."[56]

The mere observation of ritual pretends to originate from some calm knowledge of the law. No persuasion seems to take place. But the imitation of those for whom faith is already secure knowledge—the pragmatics, so to speak, of their knowledge-turned-faith—implies a hidden performative force. The ritual, then, that according to appearances seems to presuppose only the knowledge of law, is a kind of return to where the text of the wager began.

The law of the Holy Writ can either be read in order to see behind the game of chance, and to realize what makes the game and our playing such games compulsory, or we can imitate the faithful, and, by doing so, make the law our own through the performance of a play, a theatrical play in this case. The one method amounts to the hermeneutics of typological references; the other is the self-rehearsing fiction of an as-if. Both paths lead us to the center of the *Pensées*. Propositional truth and performative gestures are already unified, or bound together, in scriptural hermeneutics and ritualistic acts, but this is true only of the realm of scripture and the institution. No connection with number remains.

Correspondingly, a space can be specified where number and the calculation of chance are located together in a purely mathematical manner. This is the case in the *Treatise on the Arithmetical Triangle*.[57] The mathematical, also known as the magical, triangle—Pascal adopted it from sixteenth-century speculations by the German monk and mathematician Michael Stifel—consists in generating series of numbers by constructing an elementary configuration of position and number. First, the unity 1 is arranged along ten ordered positions:

1	2	3	4	5	6	7	8	9	10
1	1	1	1	1	1	1	1	1	1

Then the following rule is applied nine times to generate nine new series of numbers: on each "position" n of a row, enter the number produced by the sum of the numbers on the "positions" 1 to n of the preceding series. This procedure results for rows 1–4 in:

1	2	3	4	5	6	7	8	9	10
1	1	1	1	1	1	1	1	1	1
1	2	3	4	5	6	7	8	9	
1	3	6	10	15	21	28	36		
1	4	10	20	35	56	84			

etc.

In his treatise, Pascal discusses the different kinds of structures contained in the generating process, which as such originates in the organization of position and number. The discussion concentrates primarily on the two structures that appear in the fragment *Infini—rien* (more commonly known as the "wager"). First, we recognize the problem of the *ordres numériques*. Pascal detects in every row of the triangle an order that in each case corresponds to one single generative rule: the first row is called the series of "simple unities"; the second that of "the natural numbers"; the third, "triangular numbers"; the fourth, "tetrahedral numbers"; and so on. It will be recalled that the particular nature of "infinity" was based on characteristics of the order of the "natural numbers." Hence, what we see represented in the arithmetical triangle are, firstly, orders as defined by the theory of numbers. Secondly, in reading the diagonal series of the triangle, Pascal recognizes the rules of combinatorics and the arithmetic series of numbers that had been produced in the calculation of chances.[58] In other words, following the diagonal lines in the triangle leads to calculation in the sense of the calculation of chances. Hence, the arithmetical triangle, in its graphic evidence, combines number and computation into one. Viewing the tableau comprises numerical order and calculation in one single form. In short, the horizontal and diagonal rows of the arithmetical triangle in this respect are the flip side of religion—a numerical space that excludes any literal ciphers of a verbal text and renders them superfluous. The texts of Pascal that belong to this enterprise are therefore also the works that are formulated *more geometrico*. The measure of measures and the calculation of the distribution of chance are both contained within the latticework of the triangle.

This is to say that in religion and in mathematical theory, respectively, a self-sufficient sphere of texts on the one hand, and of numbers on the other, has been achieved. As long as text and number, whether metaphorical or persuasive, make use of each other, the linguistic aspects of proposition and performative speaking remain as separate from each other as the mathematical aspects of number and calculation. These parallels are therefore to a certain extent superficial; they are not reflected in the center of religion or of mathematics. But whereas the texts' theological center displays at once truth and performance and the mathematical center reveals both number and computational operations, no correspondence between the alphabetical text and the numerical one holds sway anymore. The parallels in each case have been lost at the center. There is a center, but only a doubled one.

It may seem far-fetched to push the interpretation quite so far. And yet this radicalized interconnectivity is inherent to the wager text's initial point of departure both in terms of content and structure. What we dub the "doubled center" points back to the bipartite structure of the essay *Of the Geometrical Spirit and the Art of Persuasion*. The systematic unity in the representation of knowledge reveals a rupture between ascertaining statements and performative acts, between the parallelism of text and number and the rhetorical function of the example of calculation.

In his text on the wager, Pascal not only leaves behind an example for the "geometry of chance." He also offers a unique example for the verbal presentation of a mathematical problem. On the one hand this verbal rendition seeks to predetermine the problem for calculation, and on the other hand it is driven by the same mathematical problem and forced beyond itself. It seems that we can cash in on this ambivalence for the profit of science if we read the text in the way that all histories of probability do: as a text, unsure of its own actual goal, that applies computation to a problem that hitherto had only been approached by means of the topoi and of rhetoric, the problem of probability. To the modern reader, it seems that Pascal merely forgot to arrange his text around a central word that would solve its architectonic problems with a single stroke: "probable." The difficult passage that closes the wager argument in the narrower sense seems almost carefully written to avoid this precise word. Where today we expect the word "probability," the text speaks of gain and loss, putting at stake and necessity, *certitude* and *incertitude*.[59] This is the terminology still used by Leibniz in the deliberations about the division problem that he jotted down in 1678 and brought back from Paris, "*De incerti aestimatione* (Estimating the Uncertain).[60] The Basel professor of mathematics Jacob Bernoulli also happened to study theology in his youth. In contrast to Leibniz's treatment, a solution to Pascal's wager argument has survived among Bernoulli's papers, the so-called *Meditations*, which could be read as a transformation of gaming theory into a theory of probability.

Question: is it better to lead a pious or a sinful life? Assuming that eternal happiness and punishment infinitely surpass worldly enjoyments and adversities in both intension and extension: then—if we call the mortal goods a and the mortal evils b—the eternal goods equal ∞a and the eternal evils ∞b. Now the atheist can assume that it is much more probable [*probabilius*] that what is told about eternal happiness and eternal punishments is simply invented rather than true. So long as it is not more probable by an infinite magnitude (which the atheist cannot assume unless he wants to take it for entirely impossible, which he will certainly not do), then it can be shown that he is acting with extreme imprudence.

For if he claims that the probability [*verisimilitudo*] that it is not the case—be it by however great a number, yet less than infinity—surpasses the probability [*verisimilitudo*] that it is the case, then the expectation of the pious man can be calculated thus:

$$\frac{m \times b + 1 \times (\infty a + b)}{m + 1} = \frac{mb + b + \infty a}{m + 1} = b + \frac{\infty a}{m + 1}$$

And the expectation of the sinner amounts to:

$$\frac{m \times a + 1 \times (\infty b + a)}{m + 1} = \frac{ma + a + \infty b}{m + 1} = a + \frac{\infty b}{m + 1}$$

The expectation of the sinner is worse than that of the pious man; for, since m + 1 is a finite number, $\infty b / (m + 1) >$ and in fact infinitely $> b$; and $\infty a / (m + 1) >$ and in fact infinitely $> a$.[61]

Sober computation and verbal text can take on this appearance with a turn from the theological figure of the finite decision about infinity to the language of the probable; if, in other words, one switches from a situation in which one is forced into a game whose computation then destroys its own procedure to a computation of probability that is supposed to be the epitome of rationality.

3

Writing the Calculation of Chances

JUSTICE AND FAIR GAME—
CHRISTIAAN HUYGENS

The First Textbook on Games of Chance: Translations

We may look at the history of probability as the gradual translation of game calculation proper into a more generally applicable doctrine. This general doctrine, the goal and result of the translation process, is what we call the theory of probability. Christiaan Huygens's tractate about the calculation of chance in games of hazard is a prominent piece in this translation, yet it is not easy to determine its proper meaning in the process. On the one hand, Huygens made a definite contribution to shaping the story of how the calculation of chances developed by acknowledging the priority of Pascal in the introduction to his tractate. By this point in time, however, nothing relating to calculating chances had yet been made public by either Pascal or his learned partner in discussion, Fermat. Their correspondence about gambling did not appear in print until 1779. In other words, though Huygens was the first theoretician of chance actually to appear before the public, he nevertheless referred to precursors who had come before him but had remained invisible to a broader audience. On the other hand, the tractate figures as the first book in Jacob Bernoulli's *Ars conjectandi*, which appeared posthumously in 1713. Through its incorporation in Bernoulli's truly foundational publication on probability, Huygens's work on games of chance came to form the mathematical basis for what would become probability theory in the modern sense of the word.

Contrary to prevalent perceptions, it is not immediately evident where in the history of probability we should locate Huygens's *Van rekeningh in spelen van geluck* (On Calculation in Games of Chance), in which "the basic concepts" were for the first time made "publicly accessible ... in a manner explicitly stated and correctly utilized."[1] Huygens had learned of the private debate between Pascal and Fermat in 1654 and knew quite a bit about their arguments and methods. Although he evidently never met them and was clearly never able to peruse the letters or other documents of their exchange, he drafted an account of it in 1656 after his return from Paris to Leiden.[2] Unlike Leibniz and Jacob Bernoulli in the 1680s and 1690s, Huygens could not in 1656–57 have known anything about Pascal's employment of the model of gaming in his wager argument. Hence the emphasis on decision and judgment, which was later to form the basis of Bernoulli's interpretation of the game of chance as probability, harks back to thinkers before Huygens, transmitted as it was by Arnauld's version in his *Logic*. And yet it was precisely Bernoulli who, in preserving Huygens's text by incorporating it as the mathematical foundation of his own *Ars conjectandi*, appended the interpretation of probability to it. In a similar way, even Huygens's indisputable contribution to Leibniz's theory of the logic of probability would remain almost hidden; whatever Leibniz owed to Huygens, in the printed passages of his *Nouveaux Essais* in any case, he would also follow Pascalian decision theory as Arnauld had elicited it from the wager. This avoidance of Huygens's *Van rekeningh in spelen van geluck* conflicts oddly with the fact that it was the obvious and, until Laplace, the officially acclaimed work of reference on the mathematics of probability. But as important as Huygens's tract was for the development of the mathematics of probability theory, it only became indirectly part of the development of understanding the calculation of chances as the theory of probability. It is, therefore, difficult to determine exactly where Huygens stands in the process of the translation that would eventually insert probability into the mathematics of games of chance.

For Huygens, the central issue was a current but also entirely local problem discussed at the time by the most advanced mathematicians in Paris. In the terms of the day, this problem was part of what they called *mathematica mixta*, which describes a realm of problems that are neither purely mathematical nor exclusively applied to physics. With Pascal, we saw the deep separation between the mathematical calculations and the apologetic text of the wager. The brief passage mentioning the allegory of the *aleae geometria* in his letter to the Académie Parisienne merely highlights the possibility of connecting philosophical and literary themes with mathematical problems.

Huygens formulates *Van rekeningh in spelen van geluck,* the first treatise on games of chance that is mathematical throughout and constructed in an entirely systematic way, in the space that the *aleae geometria* charts between Pascal's mathematical notation and the literary solitaire of the wager. He apologizes for the triviality of the subject, gaming, but emphasizes its particular exemplarity for mathematics, a move that can be seen as inaugurating the transition from particular mathematical problems to universal theory.

Huygens's presentation of the calculation of chance shines in contrast to Pascal's purely technical discussion of the subject, notwithstanding the imposing baroque gesture of the wager. For Dirk Struik, whose overall characterization of Huygens also applies to his game theory, Huygens is a towering figure of the Dutch Golden Age.[3] *Van rekeningh in spelen van geluck* develops a mathematical theory of gaming and nothing else. As with the reflection of light in a mirror, or the acceleration of bodies on an uneven surface, the game of chance circumscribes a realm of problems that occur in the order of things and thereby provides the framework for a possible mathematical theory. Theory and case, example and calculation are situated, as if indistinguishable, within each other. The outline of a gulf that in truth is deeper than with Pascal is evident precisely here. For however blatantly Pascal's discussions with Fermat about mathematical problems may differ in terms of style and theme from his apologetic *Pensée* about the acceptance of God, both share the same constellation of concerns. Both are characterized by the risky potential of reason to act in a situation precluding its use, the situation of making a decision based on deficient information. The question of the possibility of strategically rational decisions is essentially identical in the limited everyday case and in the case of life's finiteness. Huygens steers clear of this kind of climactic drama. Considering the close relationship Huygens's father, Constantijn, had with Descartes, a circumstance that led the French philosopher, newly relocated in the Netherlands, to become frequent reading material for young Christiaan in his earliest years, we might construe a quite devious but telling genealogy for Huygens's take on probability: Huygens, we might argue, delivered a Cartesian *aleae geometria* that beat Pascal to the draw in the theme of chance. With this move, however, a new problem emerges for Huygens: in his presentation of the mathematics of chance in gaming, the dice game and the coin toss become observable phenomena to which the structure of calculation must adjust itself. With his analogy between the structure of finitude and the structure of gaming, Pascal had assumed without further ado the ideality of the casting of dice. With Huygens, this becomes a problem in need of explanation. In this regard, another model provided by Huygens's father comes into play even more clearly than Descartes. During his years in the London embassy, Constantijn Huygens had come to know and to admire the Dutch inventor and experimental physicist Cornelis Drebbel, who lived in England at the time, constructing, among other things, submarines for the Royal Navy.[4] By composing his treatise initially in the vernacular, Dutch, Huygens did so not so much draw on the model of Descartes's French publication of his *Discourse de la méthode* as he followed the standard set by the great humanist mathematician of the Netherlands Simon Stevin. When Stevin published his *Tafelen van Interest* (On Calculating Interest) in 1582 in Dutch, and then in the following year his *Problemata geometrica* in Latin, he did so according to the assumed dichotomy between pure and applied mathematics, the latter arising from technical and everyday skills.[5] At the beginning of his *De Beghinselen der Weeghconst* (The Principles of the Art of Weighing [or Statics]) of 1586, in a section titled "Uytspraeck van de Weerdichheyt der Duytsche Tael" (On the Worthiness of the Dutch Language), Stevin calls for the use of Dutch in all publicly relevant branches of science. In line with his poetic ambitions, this demand for the widespread use of the vernacular was the decisive document in the politics of language in sixteenth- and seventeenth-century Netherlands for Huygens's father, Constantijn,[6] even more so than the literary argumentation of Hendrick Spiegel in *Twespraeck van de Nederduytsche Letterkunst* (Dialogue of Dutch Literature) of 1584. Stevin's "Uytspraeck" proclaimed above all the accuracy and forcefulness, the *energeia*, of the people's language.[7] Constantijn Huygens, meanwhile, retraced the basic pattern of argumentation in the following way. Every people in the world, he explained, has a language in which the *plebs*, the children, and the women can express anything they may wish to. There is no natural deficiency of language under natural conditions. The case is different, however, with cultured men and scientists: for "in re barbarâ versantur"—they deal with barbaric things. The fact that he refers to the matter of science and of cultured literature as a *res barbara* is merely an indication that science is a matter of other languages—of Greek and Latin, and also of French and English. The vernacular thus suffers from a *penuria verborum*, a poverty in words, in scientific matters. Only the inventor of a specific thing knows the word for it. We might rightly conclude that this notion implies an Adamitic language for science and technical invention. Even if vernacular languages often lack the word for a technical invention, one's own language in principle remains the most accurate, according to Constantijn Huygens. All other languages, even the richer ones, will perforce in contrast count as barbaric: according to this argument, they are languages that fall short of the original language that one's own tongue has turned out to be. Consequently, the vernacular must find an equivalent for the exotic foreign word, because it does not have its own original one. The foreign word is still barbaric in relation to the "own" word that does not exist.[8] Stevin's linguo-politics coalesces in a culminating move of bringing figures and numbers of mathematics into play. In order to understand this, one must first become familiar with what, exactly, mathematical literature means for Stevin. For Stevin, mathematics formed by and applied in technology is the fundamental condition for pure mathematics. This assumption amounts not so much to an outright attack on the long accepted distinction between pure and applied knowledge. More important, a distinction held to be self-evident in the tradition of mathematics is made a crucial problem to be questioned. This analysis becomes clear as a focal point in Stevin's famous method of decimal notation which appeared in 1584 in Dutch as *De Thiende* (The Tenth) and in French as the fifth part of Stevin's *L'Arithmétique* (*La Disme*).[9] The brief text deals with notating numbers in a concise duplication of what that can mean. First, the work introduces a notation for decimal fractions, which is accomplished by a whole number that indicates the value of the digit and an encircled number that indicates the decimal place in modern understanding. Stevin then develops a general theory of unities of magnitudes in the widest array of areas, constructed with decimals.[10] For this reason, in the heading of his introduction, Stevin addresses himself to astronomers, surveyors, cloth workers, wine dealers, stereometers, and to minters and merchants in particular. Even before Galileo and Newton, whom Husserl would later criticize for having forgotten the foundation of mathematics and technology in the life world of primary practices, the rehabilitative grounding of the relation between pure and applied mathematics was thus made an issue for discussion. This topic, which Stevin explicates in an exemplary manner in the case of decimal notation and the standardization of measurement, is here integrated into the theme of popular language. On the one hand, the idiom of science, which happens to be Latin (and here additionally French), is to be translated into the vernacular language of the everyday. On the other hand, a fresh start of thinking and writing in the common tongue is required, a start that will subsequently need to be translated back into the Latin of academic communities and the great world. Hence we can even more pointedly characterize both versions as translations. *De Thiende* is a translation from scientific language into everyday language with its requirements for basic writing, calculating, and measuring practices. *La Disme* translates this invention of the everyday, of pre-theoretical technique, into the language of the great and cultured world. Thus we are dealing with the figure of translation and its epistemological sense: even regarding its theme, the notation of decimal calculation and of decimal units of measurement, *De Thiende / La Disme* targets nothing other than translation, the translation of scientific constructions into the everyday, and from the everyday into the idioms of erudition and learning.

This early modern program of translation implies a certain view of the relation between theory and practice. In terms of the history of science, this view manifests itself in what Husserl calls the "scientific 'formula-meaning'" (*Formel-Sinn*) in his *Crisis of the European Sciences*:

The *decisive accomplishment* which . . . makes determined, systematically ordered predictions immediately possible, going beyond the sphere of immediately experiencing intuitions . . . , is the establishment of the *actual correlation* among the *mathematical idealities*. . . . If one still has a vivid awareness of this correlation in its original meaning, then a mere thematic focus of attention on this meaning is sufficient in order to grasp the ascending orders of *intuitions* (now conceived as approximations) indicated by the functionally coordinated quantities (or, more briefly, by the formulae); or rather one can, following these indications, bring the ascending orders of intuitions vividly to mind.[11]

Huygens, we might conclude, discovers in *Van rekeningh in spelen van geluck / De ratiociniis in ludo aleae* the vivid formula-meaning (*Formel-Sinn*) at a place where Pascal had used the allegory of the *aleae geometria*. When he incorporated Huygens's treatise into his *Ars conjectandi*, Bernoulli would in turn later add the relation of "predictions" (Husserl's *Voraussicht*) and "immediately experiencing intuition" (Husserl's *unmittelbar erfahrender Anschauung*), that is, the application of probability.

By the time the young Huygens came to write his treatise, some eighty years after Stevin's treatise on decimal notation, the back and forth between the vernacular and Latin had indeed become routine. Yet something special about this practice persisted in the particular case of this tractate, which is one of the earliest important works of Christiaan Huygens. The brief text made its way to the public in two different authorizations. In 1657, Frans van Schooten published Huygens's tractate in Latin as *Ratiocinia in ludo aleae*. Van Schooten was a professor at the Leiden institute for technical arts and a private preceptor of the brothers Christiaan and Constantijn Huygens during their years at the university. Van Schooten gave the tract of his former student a spot in his *Exercitationes mathematicae*, and in this context he published it in his own translation, which took Huygens's attempts at self-translation only partially into account. Not until two years later in 1659 did the Dutch original by Huygens, *Van rekeningh in spelen van geluck*, appear. The authority of van Schooten's editorship and translation could hardly have been greater— around this same time he also published Descartes's *Géometrie* in his own Latin translation.[12] And yet, or perhaps for this very reason, there was no lack of animosity between Huygens, who had entrusted van Schooten with his Dutch manuscript, and the former teacher and patron, who now appeared as editor and translator. The altercations, which one can follow in the correspondence of 1656 and 1657, concerned nothing less than the terminology of the calculation of chance and the exemplary meaning of the game.

Issues of Translation: *kans* and *verisimilius*, Frivolous Game and Mathematical Problem

In *Van rekeningh in spelen van geluck* (1656), the Dutch original of his treatise on games of chance, Huygens, like Pascal, makes no reference to *waarschijnlijkheid*, or probability. In his first, unfinished draft of a Latin translation that same year, the word *probabilis* turns up just once. The very fact of its singular appearance shows that Huygens evidently did not at this point see it as a term of art. But he opens the door as it were for later translations of his chance theory into "probability theory."

The point of interest occurs in the second sentence of the introduction. In the first sentence, Huygens speaks of the source of fascination in the new method of calculation: namely, the claim that, precisely where everything is left to chance, a certain determination or stipulation of conditions (*seeckere bepaling*) should be possible. This takes up Pascal's letter to the Académie Parisienne, which Huygens may well have seen. The second sentence describes and amplifies this claim with an example. If one should set oneself (*neemt / sibi sumat*) the task of coming up with six on the first throw of a pair of dice, it is uncertain whether one will win or lose. In Dutch, Huygens continues: "maer hoe veel minder kans hy heeft om te winnen als om te verliesen [how much less chance one has of winning than of losing] dat is in sich selven seecker, en werdt door reeckeningh uytgevonden [this is certain in itself, and is shown by calculation]." In Huygens's own translation, this reads: "at quanto verisimilius sit perdere quam vincere, per se definitum est, calculoque invenitur."[13] *Verisimile* (not *probabile*, which would be the more usual word choice in the context of rhetorical topoi) thus appears for once to explain what constitutes the object of the theory at hand: the calculation of things subject to chance. Within the sentence, with its illustration by means of example, the word *verisimilius* now indeed already occupies the place of the conceptualization that has just been launched by introducing the game of dice as an example. For it is in games of chance that the *kans*—an event or phenomenon that Huygens says is "in sich selven seecker" [certain in itself] "per se" or "reipsa" [i.e., Latin, res ipsa loquitur, "the thing speaks for itself"]— makes its appearance. In Huygens's Dutch, *kans,* translated by Huygens and van Schooten as *expectatio*, means not only the physical throw of the dice but also what such a "fall" ("chance" from Latin *cadentia*) means for the one who plays the game of chance.[14] Thus it is a physical description and a theoretical term, semantic and numerical formula in one. *Verisimilius* does not provide any additional denotation to this complex meaning of *kans*. Since it only appears in the comparative, "probability" remains inseparably connected to the proportionality of the "more probable," that is, with the technique of tossing and with the quotient that is defined "in itself with certainty," "through itself," and "in the thing itself," in short: the *kans*. Only the transition from pure chance—about which we know nothing—to knowledge about the possibilities of winning and losing in the game first constitutes the knowledge of what is "more probable." The "more probable" creates the transition or translation. But this still does not constitute a theoretical position or an interpretation we can extract as such for this translation. The place of the interpretation is indicated but not filled out. It is the place of the self-exemplification of the example of the game, before it devolves into the terms of probability, that is, of the sphere of applying the model of gaming to everyday situations and experiencing them as such. As with Pascal's wager text, the Dutch original only knows of the opposition of certain and uncertain where Latin marks the place of probability.

"Probability" in Huygens's Latin text appeals to the familiar semantics of scholarship and rhetorical elegance. The tractate, whose original Dutch version distinguishes itself by lexical terseness and supple stylistic fluency from van Schooten's semantically more varied and syntactically stricter Latin, reflects gaming theory before the invasion of probability, that is, before the invasion of rhetorical and logical topoi and the final translation into the discourse of humanism. Huygens's own translation marks the point at which a rendering can ensue that will have been both a translation of mathematical theory back into literary sense and meaning, and a translation into the history of great proclamations and infinite applications.

But to begin with, at any rate, Huygens and van Schooten agreed that the *Ratiocinia in ludo aleae* and *Van rekeningh in spelen van geluck* offered a particular work of virtuosity in algebraic calculation. This, for contemporary readers, was its crowning achievement. Huygens's letter of dedication to van Schooten skates over the frivolous—the trivial and questionable—aspect of gaming, without denying its existence outright. This whole negotiation between the frivolity of the game and the achievements of chance calculation already begins within the rhetorical conventions of the dedication. The letter points out that his former teacher's gestures of translating the treatise and adopting it into his own work display the high value hidden in the trivial subject matter. Indeed, Huygens had first written the dedication containing this claim in Latin, the idiom of rhetorical elegance, in contrast to the original language of the treatise. Only then did he translate it into Dutch for the planned vernacular edition, a circumstance reversing the direction of the translation that was supposedly meant to demonstrate the higher value of trivial gaming.[15] The rhetorical turns that address van Schooten in the preface thus have a certain ambiguity. And this rhetorical play of the foreword goes further where it aims at the subject matter of the book, the game. The greater the apparent difficulty in determining uncertain and chance events in a rational manner, the greater the resulting wonder—"verwonderinghs waerder"—at science's success in treating them. This phrase once again highlights the aura of Pascal's allegory of the *aleae geometria*. The successful mastery of uncertainty and contingency in an ordered representation would deserve, as the son of the John Donne translator Constantijn Huygens should know,[16] the *admiratio* of baroque poetry. Science and gaming are arranged in contrast to each other, and poetic effect results from the successful figuration of this contrast. This figure is the counterpart to Pascal's use of the rules of division in his apologetics. Pascal stages the calculation of game chances in a drama where faith would actually have to play a role and should in fact emerge through the use of reason in apologetics. Huygens makes the discrepancy between the uncertainty of chance and the certainty of reason's determination visible precisely within the calculation, thus revealing the poetic effect of wonder. The game transcends its own triviality when we perceive the example hidden in it. But the example at this point still does not represent something else, as it will do later on in Bernoulli's probability theory. Instead, it composes its exemplarity by its own means and thereby differentiates itself from itself.

The calculation of the *kans* in games obviously does not yet unfold the horizon of the many uses in life that it will take on with Jacob Bernoulli in the eighteenth century, nor the aura of the great proclamation of a new logic that Leibniz will already want to extract from it in the coming decade. It is only a single, admittedly very beautiful, object in the garden of mathematical *exercitationes* or *Oeffeningen* that evokes the great theme of the Dutch baroque classicist poet Joost van den Vondel: *fortuna,* or *geval.*

The foreword's play with translation has its source in the subject matter of the treatise. In a private letter to van Schooten, Huygens admitted that in composing the tractate, he initially lacked the necessary vocabulary in Latin.[17] His effort to acquire this and translation attempts on the theme of the *kans,* the dice throw, and related gaming terms are testified to by a word list dated to the time of the treatise's genesis: "alea. sors. fortuna. casus. lusiones. deponavit. certare. sibi sumere. qui ter superior fuerit. ubi majorem numerum jecerit. senarium jacere. jactus. contendere."[18] The frivolous and the vernacular were at first one and the same. Together they made up the terms of calculation that could not easily be expressed in Latin, the language of discursive theory. Huygens's translation policy is clearly ambivalent in his correspondence with van Schooten. In the letter in which he mentioned the difficulty of the Latin formulation, Huygens enclosed his own translation samples, which, he suggested, might assist van Schooten in his job as translator.[19] From the beginning, Huygens thus attempted to keep a hand in his teacher's work of translating the trivial game into the language of theory. In the further course of the correspondence, he obviously tried to take back the translation request—fluctuating between an appeal and an offer—especially as van Schooten for his part continued to postpone the announced translation.[20] Huygens went so far as to place the publication itself in question. After van Schooten completed his translation, which often deviated from the author's samples, Huygens rarely missed an opportunity to insinuate his dissatisfaction with the teacher's translation to others.[21] He thus communicated both that van Schooten had to translate the tractate because the Latin vocabulary did not yet exist and that the translation was actually already prepared and would have been better taken care of by himself. The frivolity of the game was inevitably transferred to the discourse of theory, but the theory came to realize that it itself already had its roots in the frivolous game and its numbers.

The Economy of Games and the *fondament* of Chance

The ornamental figuration in the dedicatory letter and the mini-diplomacy of the translation are worthy of note because they already partake in the mathematical and theoretical structure of the treatise. In text and calculation the *spelen van geluck* becomes a quasi-object that evokes wonder and can lay claim to respect without seeming to have a function or relation outside of itself. The case with Pascal is quite different. On the one hand, Pascal discusses with Fermat the mathematically fascinating structure of distribution in suspended gambling rounds. On the other hand, he stages the stakes and the function of gaming in the wager text as an irrefutable model for infinite decision in finite life. Between the intellectual pride of the mathematician and the earnestness of the apologetics, a gap yawns that is a modifying component of both, an element of pride as well as of earnestness. For Huygens, in contrast, a quality hidden in the game should emerge in the mathematical and analytical translation from the game itself as if without disruption.

Huygens expresses his basic principle—his fondament—thus:

that in gambling the chance that somebody has toward something is worth as much as that with which, having it, he can arrive at the same chance by an equitable game, that is, where no loss is offered to anybody.[22] [dat in het speelen de kansse, die yemant ergens toe heeft, even soo veel weerdt is als het geen, het welck hebbende hy weder tot deselfde kansse kan geraecken met rechtmatigh spel, dat is, daer in niemandt verlies geboden werdt].[23]

In its tendency to plain similarity between *definiens* and *definiendum*, Huygens's determination of chance is just as terse as it is exact. This becomes particularly clear when we read it with an eye to the question of the exemplary validity of chances in gaming. The similarity of *definiens* and *definiendum* should not be seen as a mistake, as Lorraine Daston tends to do when she remarks, "Huygens's fundamental principle ... sounds suspiciously circular."[24] Daston considers "the path [to be] still blocked" for Huygens to support the rationality of the game of chance with the concept (and the formula) of probability. For this reason, Daston thinks that Huygens uses the "intuitive, or at least nonmathematical, notion of equity" in a circular way as the constitutive principle of gaming theory.[25] But this is a retrospective point of view with the hindsight of a history of probability.

The argument below will instead claim that the self-exemplification of the game takes its point of departure from the well-nigh circularity of its "justice" (or fairness). To do so, we must return to the task Huygens set himself. With the transformation of regular chance into chance in a "just (or equitable) game," Huygens tacitly refers to the problem of dividing the stakes in an interrupted game. Chevalier de Méré had posed this very problem to Pascal, and it had been shown again to Huygens in Paris. In contrast to Pascal's method in the wager on salvation, Huygens attempts to find the validity or function, that is, the exemplarity, of gaming theory in the semantic field of the division rule. Huygens makes the distribution of stakes when a game is interrupted or broken off into the definition of chance—the particular situation of a player—in the game. As will be shown, "in the game" for Huygens means two things at once: the conditional phrase refers firstly to the respective strategic situation of the game at a specific point in its course; and secondly to a certain quality of the game, the quality of being a true or just game.

The conspicuous structure of near-tautology in determining the *fondament* results from the initial translation that occurs in Huygens's tractate, the translation of the special mathematical problem into a general theory. The game of chance in which one finds oneself at any given moment is, by virtue of the regularity by which the analysis proceeds, already nearly identical with the epitome of the game of chance, the just game that only allows such games to be invented, organized, and participated in. The game of chance seems to translate its systematic quality as if by itself from itself.

At the core of the definition lies the calculation of chance according to a measure of value that is generally valid—a kind of ideal money, even if the word does not occur here. This measure of ideal value in the ideal just game determines the condition and the rules of any game to be played.

Gambling for money offers the simplest case, of course. Huygens determines the equivalency between chance and value (or money) in a second step through the equivalence between the odds a in game A and those b in game B. The value (or the money) now only symbolizes the medium for converting one game of chance into another. Chances are calculable to the extent that they can be traded like goods or bought and sold for money. Two goods have the same value if, for the sale of one, we receive a certain sum that buys the other. The third step then involves a question of classical money or value theory. All exchange requires a fixed measure of value in money, such as gold, for instance. This role is played here by the "'just' or fair game." And this game is for its own part determined in economical terms: it is defined as a game "in which no one's loss is intended." The *kans* amounts therefore essentially to the chances of winning or to their equivalent under the condition that all participants are granted the same odds. In a *rechtmachtigh spel*, no one can be prevented from winning by the rules of the game.

By following how, from the situation of the interrupted game, the very definition of chance produces its universal character, we also understand the economy of distributions and equivalencies surmised within a single game to lead to economy and distribution between games. The meaning of chance in gaming is defined by its equivalence to the odds we would have in the ideal, fair game in which every player may win. The semantic foundation of the concept of chance leads to a closed system of economy whereby game odds take on the position of second-order goods. The chances of a game only have a value where there is a second game chance and a convertibility of the games according to the measure of the fair-just game. For Huygens, the game does not need to turn into an example of anything other than itself, since the very prerequisite for the game—or more precisely, for the concept of chance in gaming—already implies an economy of games and odds. The situation resembles a trompe l'oeil: where we search for the definition of chance in gaming, we already find an economy of games in which chance is present.

In reflecting on the situation of the interrupted game, Pascal discovers the impossibility of the infinite decision in a finite life. From the concept of chance that emerges from the situation of distribution, Huygens develops the equivalency of nothing other than games. The exemplarity of the game in Pascal is imposed and unique, a model pattern that no one can escape acknowledging and following. In Huygens's exemplarity, the game unfolds an infinite series of games from itself. It makes itself the example of a perfect economy of games. Whereas Pascal requires measurelessness,

Huygens looks for the measure. The fair and just game, understood as the model that makes games exemplary, is the condition of Huygens's *fondament* guarding against mere circularity in definition, thereby establishing the theory and calculability of chance.

Citing Justice

At the end of the introduction, Huygens repeats his definition of chance, and this time he does so with a numerical example. Again, he points toward the equivalence with chance in the fair game: "and namely with the just game [*rechtmatigh spel*]: which should be demonstrated in the following." Indeed, the first theorem (*Voorstel*) and the first algebraic representation that follow do not present the formula of favorable and unfavorable cases as we would expect later in the probability theory of the eighteenth century. Instead, they introduce the mathematics of fair distribution, or of the justice of chances. In this respect, it might be open to debate whether Bernoulli really took Huygens's definitions and formulas seriously when he placed them at the beginning of his *Ars conjectandi*. In any case, no other theoretician of probability since Abraham de Moivre and Willem Jacob 's Gravesande has thought again of taking Huygens's formula of chance as his point of departure. Huygens's distance from the standard interpretation of mathematical gaming theory as probability theory is marked precisely by his beginning and this formula.

The phrase in question reads as follows: "When I have equal chances of having a or b, then the value for me is as (a + b) / 2."[26] We should first note how this formula, upon which Huygens builds his entire mathematical exposition, is related to the semantics of justice. The formula (a + b) / 2 is nothing other than the formula of the arithmetic mean. It is, however, in the guise of the arithmetic mean that numerical measurement—and indeed number and measurement per se—appear in the conceptual field of justice at all: this is Aristotle's formula for distributive justice. In the pertinent context of the *Nicomachean Ethics*, Aristotle speaks of contracts between equally entitled subjects as agreements for regulating gains and losses. Subsequently, with the help of the arithmetic mean, he defines the second type of justice by the balance between unlawfully acquired gains and innocently suffered losses. It is the judge's job to restore this balance based on the assumption of the parties' basic equality. The point here is the distribution of the two oppositions, legal/illegal and gain/loss. With the crisscross ordering of unlawful gain and innocent loss, Aristotle ties economy and law, value and evaluation, number and meaning together in an irresolvable knot. The knot cannot be unraveled because after it is tied, neither economy nor law, neither number nor meaning, can stand outside of this relation that institutes balance as justice. "[B]etween them [i.e., win and loss, or, in terms of the single parties concerned, advantage and disadvantage—RC] the mean is what is equal, which we say is just, so that the justice that sets things straight [which in the Western tradition has been known as "corrective" or "rectificatory justice"] would be the mean between loss and gain."[27] The first formula in *Van rekeningh* should be read in light of the Aristotelian formula of "corrective justice." Obviously, Huygens recognizes the structure of gaming with chance precisely in this equalization between unlawful gain and innocent loss. Games of chance are characterized by the economy of their convertibility. This economy of an order of converting positions in one game into positions in another game introduces the *rechtmatigh spel*, the just game, which sees winning by chance as a kind of guilty gain and a loss by chance as an innocent forfeit. Without this equation, which becomes increasingly invisible in the history of natural law, there would be no economical rationality and calculation to begin with.[28] At this decisive moment in the history of gaming theory, the pages of the *Nicomachean Ethics* function as Huygens's subtext in the precise sense of Michael Riffaterre, who uses the term "subtext" to refer to the presence of classical authors from antiquity in old European literature and education.[29]

This discovery concurs particularly well with Christiaan Huygens's course of studies. According to the schedule set by their father for the studies of Christiaan and Constantijn Jr., the scientific part of the daily instruction plan was devoted mainly to jurisprudence and mathematics. The paternal *norma studiorum* dates from May 9, 1645,[30] and the seventeen-year-old Christiaan and his brother began their studies according to it one year later. The father's pedagogical plan dictated the following classes of private instruction and university lectures: from six to eight-thirty, methodical *studium Juris*; after a brief recess, a lecture on elective juridical themes at the university; then an hour of mathematical studies, which likely meant practice in the scientific "geometrical method." Then followed a day of instruction in draftsmanship and music, physical and literary exercises, and preparations for the next day's lessons in jurisprudence.[31]

Constantijn Huygens employed Frans van Schooten, later the publisher and translator of the *Ratiocinia*, to instruct his sons in mathematics. Arnold Vinnius (1588–1657), one of the most important jurists in seventeenth-century Netherlands, gave them private lessons in law, and they also had to attend his university lectures in jurisprudence. Christiaan soon switched to study only mathematics, but the jurist remained important for his game theory.[32] The relevant area in this context is Vinnius's doctrine of contracts, which falls somewhere between Grotius and Pufendorf. In the treatise "De pactis" in his *Tractatus quatuor*, Vinnius goes into detail on the subject of contractual agreement. In particular, he discusses the fine differences between contracts legally agreed upon and popular or natural law forms of contract, focusing on cases in which legal obligation and natural law obligation contradict each other. With all of this, Vinnius elaborates on the question that for Grotius had become constitutive for natural law and its relation to Roman law: can there be a binding obligation of contractual acts—of promises and agreements—that certain defined forms of executing these pacts and promises antecede, as stipulated by Roman law?[33] Even the so-called *nuda conventio*, agreement without any legal formalities, creates legal obligations—thus does Vinnius decide in the end—when the will of the contractual parties is sufficiently articulated. "Nothing, namely, corresponds to natural justice more than holding fast to one's decisions and one's given word, and not deviating from one's agreements."[34] We may conclude that what made Huygens the author of the first systematic textbook on gaming theory was the fact that, as a good student of Vinnius, he was well equipped to recognize the fundamental and universal prerequisite of an ideal game and its rules to lie in the legal sphere of contract and the obligations of natural law. The notion of justice discloses a sphere in which the laws of chance are universal and—in contrast to Pascal's wager argument—can be valid again and again. The methodological *aequalitas* of equiprobability, required for all computation of chances, is assumed as the legal *aequum* of justice.[35]

Huygens sets up corrective justice, the arithmetical mean, as the first and foundational formula for what he, without referring to probability, still calls the "calculation of chances in games of chance." For Huygens, the fair game, ruled by the Aristotelian measure of justice that sets things straight, is the standard of all games in which the equivalency and convertibility of chances are established. Thus does the just (economical) game make possible the justice (economy) of games. The formula (a + b) / 2, which provides the basis for all further derivations of game calculation, derives from the Aristotelian subtext as actualized for the young Huygens by Vinnius and modern natural law. The formula appears as if it were just translating the last words from the preface to *Van rekeningh in spelen van geluck* where Huygens elaborates on the notion of fair and equitable games. This is not a matter of Platonic numerical symbolism, nor of speculation in number theory. Justice in its regulative variant is numerical up through the move of abstraction. For Aristotle, a disregard for the subjects forms precisely the prerequisite for contract and corrective justice.[36] In this formula, the semantics of what is meant by justice and the numerical representation of calculation are entirely exchangeable sides of the same thing. Thus, they are perfectly transparent for one another.

Pascal develops his analysis as well as the scenario of the wager directly from the problem of the interrupted or postponed game. Fermat counters this method with the combinatorial construction of all possible results of a given game. Huygens, for his part, takes a middle path with the help of the algebraic *fondament* of the fair game. His *Van rekeningh in spelen van geluck* resumes the problem of the interrupted rounds in its fourth theorem. The example is the same one that Pascal and Fermat had discussed: one player (B) has won one game; the other (A), two. What are the odds for the second player (A)? To answer the question, we must examine the possible outcomes of the game. In the first round either A wins—in which case A is the victor and the game is over; or B wins, so that both players would need one further round. Under the latter assumption, then, either A or B ends the game with a victory in the last round. Before the last game, the formula of justice gives both players half a claim on the victory prize (*en daerom elck gerecthigt zijn tot 1/2 a*).[37] In the situation of the penultimate round, A is facing one plus and one minus: either victory or half a chance of victory. Huygens points once again to the justice formula here. The just mean between the claim to the award for victory (= 1) and the claim to half of the award (1/2) comes to 3/2/2 or three-quarters of the winnings. It is certainly possible to carry out this train of thought without recourse to justice and its formula, but its foundational effect lends an additional methodological point of reference to the strategic analysis, which is reminiscent of Pascal's solution, without leading on the other hand to Fermat's combinatorial construction. Expressed mathematically, Huygens's formula of corrective justice proves to be an algebraization of the problem at hand.

Huygens has, however, departed from the foundational basis of his treatise in its narrow sense with this fourth proposition, the problem of division. This proposition connects the formula of justice in the first proposition [(a + b) / 2] with the formula of chances in the third proposition [(bp + cq) / (p + q)].[38] Huygens's methods as a whole can be most profitably viewed as being demarcated by these two basic formulas:

(1) The formula (a + b) / 2 will never play a role in a single probability theory after Huygens. But even for Huygens himself, this was not the theorem from which he deduced the intuition of his solution. There is a difference between the foundational power of justice and the quasi-intuitive power of the chances formula. In his letters to Pierre de Carcavy, written between the composition and the publication of the tract in which he tried to investigate further into Pascal's and Fermat's methods, even Huygens, too, already asserts the chances formula to be the basic one. Even for Huygens, that is, the formula of the third proposition appears as the one that every calculating player himself would draw from reflecting on games of chance. In Huygens's example, if one has two chances to acquire a third of the stakes and five chances to receive half of the stakes, then that means the same thing as if he expected a share of (2 / 3 + 5 / 2) / (2 + 5) of the stakes. At the moment of the game when the status can be thus calculated, the odds are 19 / 42.[39]

Taken as the opening theorem of a theory of gaming, the formula leads to a table of possible chances. In a simple case, all possible dice throws are divided into desired and undesired. Desired throws win the entire stake; undesired throws, nothing. Thus when a win—the desired chance—is the throw of a six with a die, then the throw of a die presents one chance of getting a six and therefore the entire stake, and five chances of getting a non-six and thus nothing. If one calls the win G, then according to the chances theorem the following is true: $(1 \times G + 6 \times 0) / (1 + 5) = (1/6) G$.

The evidence of the chances formula supports itself on the intuition of a combinatorics of possible cases. It is not obviously clear, however, from the evidence of justice, which for its part consists of evidence based on the immediate translation of semantics into computation and of computation into semantics. The chances formula formalizes the players' conduct while gambling, whereas the justice formula formalizes their entrance into the game.

We can grasp the gap between the counting of chances and the foundational formula of justice if, following Huygens, we try to read the chances theorem directly as a just game. To do so, we must first take up the following prior considerations on which Huygens's example in this context is based.[40] The chances p and q become players designated by p and q, all of whom play on their own account with equal odds and the same stake (= x). Accordingly, the total stakes amount to px + qx. This arrangement secures the equality of the game. The shares b and c become consolation premiums that the winner of the total stakes (px + qx) must pay the losers. With these prior considerations in mind, we can now understand the peculiar story that illustrates the theorem of chances in Huygens's tractate: I make a deal with the players q that, whoever of them wins pays c to me, whereas I will pay c to everyone of them, should I be the winner. And I agree in the same manner with the remaining p − 1 players (I myself being one of the p players) that whichever of them wins will pay me the sum b from the total stakes, and that should I win, I will similarly pay him b. In so doing, the p and q players become so many chances to win or lose a and b respectively, with $(pb + qc) / (p + q)$ again expressing the total stakes in relation to number of chances, or, in this cases, of players participating in the game. The chances theorem $(pb + qc) / (p + q)$—the view on the state of play from within the game—turns out to be, as one sees, exceedingly complicated if one wishes to understand it on the basis of the theorem of justice that grounds the game of chance and makes possible the entry of players into the game.

(2) There is, however, also a specific point at which the players experience the evident certainty of justice, the foundation for the calculability of games of chance while gaming. But this point occurs only in the case of simple games.

In the introduction, the argument of the arithmetic mean or Aristotle's corrective justice is connected to the account of a game meant to illustrate the convertibility of chances as measured by the fair game. The example runs as follows: one person holds seven shillings hidden in one hand, and three in the other. A second person may choose either the contents of the left or the right hand. The chance or expectation contained in the possible choice has the equivalent value of five shillings, which is the value each player may rightly claim for himself at the point of choosing. The story is told to exemplify chance or expectation, but it turns out that it has done its job only halfway. The story has shown what it means to express the expectation that one has in a game as the mean between a greater and a lesser value. It narrates a game that acts out, as it were, the Aristotelian formula of regulative justice. Justice happens here in the course of a game, but this does not yet exemplify the reconversion required by Huygens's definition of chance. The story still does not tell us what it means to set the sum resulting from the game that is justice back into the model game that should be the norm of just play from now on. What is still missing, and what the final clause of the introduction, appended to the example story, announces, is the conversion of justice as a game into the just game. The conversion alone is what first brings about foundational generality and with it the lesson of the foundation of games and their calculability.

To this end, in the first proposition, Huygens tells the story of another game. The game, as here assumed, must be fair; and it must offer the chances a and b with equal ease to each player who stakes the sum resulting from $(a + b) / 2$. First step: two players each stake the sum x, since an equality of entrance risks is the norm in the standard game of just games. Second step: it is agreed that the total winnings go to the victor and that from these winnings he gives the loser a consolation premium a. For in the standard game of just games, losers will be compensated for the loss of their stake. The share of the winner hence comes to: $b = 2x - a$. If we break this down to the stake that should be the ante of expectation from the first game, then the result reads: $x = (a + b) / 2$. This shows us that if one takes off from a game that exemplifies the Aristotelian justice of equality, one's result will correspond to a game that we can define as the just game.[41] The game of justice is translated into the justice of games.

94   GAMES FOR EXAMPLE

Huygens thus models the theorem of chances, the special mathematical case that is his semantic starting point, as the translation of one story into another. The translation proceeds in a mirror-inverted or chiastic manner. Huygens's message to Paris when he pins down Pascal and Fermat from Holland on the expectation theorem has an elegant balance: the text of the doctrine of chances is the chiasmus of justice and play, on the fulcrum of which stands the formulation and the formula of the *Nicomachean Ethics*.

Spinoza's Addendum: Contours of a Game of Chance in the Politico-theological Doctrine of Contracts

There was a follow-up act to Huygens's translation attempt. A certain Jan van der Meer, who was appointed to be the Dutch government tax collector in Leiden in 1670, had asked Spinoza about a universal proof for the calculation of games of chance. On October 1, 1666, Spinoza finally sent the answer, which had apparently been long in coming, from his peaceful retreat on the countryside.[42] Spinoza possessed a copy of van Schooten's Latin translation of Huygens's *Van rekeningh in spelen van geluck,* which had appeared nine years earlier, and his formulation of the problem and its solution clearly relate to it, although he does not mention Huygens. Now that Spinoza had taken the time and had the leisure to work on the answer, he found the question "very simple." The "proof, universally stated," assumes "that the fair gambler is one who makes his chance of winning or losing equal to that of his opponent." As he continues, Spinoza also comes across the Aristotelian formula of justice and Huygens's term of fair or just play: "but if the chances are unequal, one must stake and lay down as much more money as his chances are greater. Thus the prospects on both sides are equal, and consequently the game will be fair."[43]

What distinguishes Spinoza's wording from Huygens's is the fact that he puts the emphasis on the honest player instead of the fair game. Spinoza derives the calculable game not by the juridical-geometrical method of a system of games in which it is defined as the standard game of all calculations. Instead, the analysis of geometric proportions remains bound to the situation of the player at play.

Compared with Huygens's *Van rekeningh*, this stance constitutes a shift in perspective. The shift may also be traceable in a short text titled "Calculation of the Probable," published posthumously, in which Spinoza resumes the discussion of a number of problems posed by his predecessor.[44] The shift from Huygens's point of view is in particular mirrored in how the notion of *kans* appears in Spinoza. The High Dutch original has been lost, and we can hence only judge based on the Latin version of the *Opera omnia*, but the key word here is *sors*. *Mensura sortis* would in fact remain a common title for the theory of chance calculation until long into the eighteenth century. Its field of meanings corresponds to the concerns of the moral-theological debate, and particularly to its increasing gravity among Calvinists: *sors* can mean (1) contingency and contingent event; (2) the agreed conditions under which a certain event precisely can take place or not; and finally (3) the winnings that, by the event's taking place or not, fall to the person who agreed to the conditions. In Spinoza's Dutch *Nagelate schriften* (Posthumous Writings), in which a certain amount of later editorial reworking cannot be ruled out, however, *lot* [fate] and *geld* [money] appear alongside Huygens's term *kans*, which Spinoza uses inconsistently.[45] According to Ian Hacking's persuasive supposition, *kans* for Spinoza should be read as a word used intuitively, which only gradually becomes the "dominant concept" in the course of the argumentation as it replaces *lot* and *geld*.[46]

According to the doctrine of natural law in his *Tractatus theologico-politicus* (Theological-Political Treatise), for Spinoza the economic system of justice or equal opportunity does not come first, but rather the fair or honest players. It is the utilitarian and voluntary aspect of Spinoza's contract theory that lets the question of the foundation of fair play appear "very simple."[47] The renunciation of betrayal and deceit that is given only under certain conditions and that can always be taken back is in each case a singular decision for cooperation. Precisely because it is a purely utilitarian act, it does not require any particular explanatory reasons. Although Spinoza's intervention in the subject and the solutions he suggests have only survived in fragments, they still make clear what Huygens, by means of his stylistic force and algebraic refinement, leaves hidden. Speaking of fair play as Huygens does implies the assumption of a voluntary act, namely, that of laying down the rule of corrective justice. The translation that plays a role on so many levels in Huygens's *Van rekeningh in spelen van geluck / De ratiociniis in ludo aleae* is not self-evident. It must first aim at exemplarity and gain a possible understanding of the example of games of chance in a general theory of chance.

We may speak of two basic versions in which games of chance can become the example for a general theory and part of a semantically meaningful text without leading us to the interpretation of probability. Either—and this is Pascal's version—there is only one case in life that is a game of chance. That is the decision that affects finite life as a whole. Alternatively, games of chance produce their own exemplarity as if from within. This is then either the economy of justice (Huygens) or the respective decision to enter such an economy (Spinoza). In both cases, the economy refers to the systematic construction of a game of games. It methodologically presupposes and juridically stakes the entirety of the social order.

| 4 |

Probability, a Postscript to the Theory of Chance

LOGIC AND CONTRACTUAL LAW—ARNAULD, LEIBNIZ, PUFENDORF

Bordering on Interpretation

In Pascal's strategic calculation of chance and Huygens's algebraic theorem of chances, a structure of calculability emerged that later theorists would call a priori probability.[1] Pascal and Huygens had respectively, however, already assigned specific functions to such chance calculation. The former employed it to give structure to the one game that is required by human finitude, while the latter discovered in it the idea of a system of games that lends systematic character to games and gaming character to systems. In the first case, the game of finitude, calculation consumed, as it were, the decision for calculability through its own process. In the second case, the game of systems, formalized calculability devolved back into its presupposition, the formula of justice and economy, and hence of calculability per se. With these alternatives, it would seem that a fundamental condition of modernity—the relation of chance and calculation, contingency and system—has also already been entirely plumbed. But in fact these radical alternatives that manifest themselves with Pascal and Huygens remained outside the history of probability proper. Both neglected to add an interpretation of probability to the structure of chance calculation. They failed to make themselves an example for other things, and thereby remained inapplicable outside of the narrow fields of apologetics and gambling. The calculation of chance did not earnestly kick off its career in modern philosophy and science until an interpretation of probability surfaced. In other words, not until the doctrine of chances was understood as the theory of probability did it become scientifically one of the most spectacular and politically one of the most far-reaching branches of what we call the project of modernity. Its explicit version, the version through which probability became a historical force and developed its own history, first came about through the transcription and reentry of chance calculation into the old language of dialectics and rhetoric, and finally even of the poetics of narration. Long before mathematical probability could put in question the privilege of philosophical thinking (in Quetelet) and the sovereignty of literature and particularly of the novel (in Musil or Pynchon), its constitutive interpretation was indebted to the very vocabulary of philosophical argumentation and narrative poetics.

A history of knowledge like the one with which this book is concerned should take the interpretation of probability seriously and should attempt to analyze it in some detail. The semantic understanding of probability, which shaped its own history as well as its contribution to history in a broader sense, should not be discarded in order to concentrate on a supposedly pure relation between chance and structure. This relation can in fact not be understood independently from how it was interpreted and what its effects in the world were meant to be. A responsible history of knowledge will instead be strongly interested, not only in the final interpretation, but especially in the interpretive process, the negotiations of understanding and computation. In contrast to histories of science as well as of literature, the history of knowledge should not start off its observations and descriptions with a readily presupposed division between literature and science. What should be universally valid for the history of knowledge is, at least in the case of contingency theory and probability, closely tied to the problems of historical analysis. It is true that the mathematical theory of gaming and the tradition of probability are clearly two disparate areas. But only the interpretation of gaming theory as a model of probability can account for the history-shaping potency of probability theory. Only an understanding of the mathematical structure as a model for the calculation of probability shows how the specifically modern shape of reality was brought about, a shape of reality that would find its first expression in the eighteenth century in high-flying expectations of an eventual compatibility between mathematics and logic, aesthetics and statistics, the novel and calculability.

Calculable probability begins to expand from abstract concentration on the one game of life or the systematicity of fair play into a much more comprehensive manner of shaping reality precisely at the moment of interpretation. Not until this turn do the contexts in which Pascal and Huygens had introduced the game of chance become mere examples. The two foundational instances of contextualization—the singular decision of finitude with Pascal and the establishment of justice with Huygens—disappear already in the eighteenth century behind the many instances of judgment in which subjectivity is now seen to model the world through its own formulas. In the statistical probability of the nineteenth century, we finally recognize a complete replacement of norms that had previously been grounded in theology and carried out in law. These norms colonize ever-expanding zones of the real by means of social and scientific implementations of the probabilistic interpretation; and they do so under the auspices of the final decision and the economy of justice.[2]

It is difficult to follow step by step how the interpretation of the game developed into the model of calculable probability. The interpretations proposed by Arnauld, Leibniz, Pufendorf, and Jacob Bernoulli—to name just the first and most important—overlap one another in subject matter and in chronological order. It would therefore be misleading to offer any simple succession of texts and authors. The passages dealt with in this chapter harbor largely fragmentary interpretations of probability, and do not necessarily belong to the proper corpus of probability theory. Subsequent chapters discuss the two full-blown ventures into the field by Jacob Bernoulli and Leibniz, in which they finally attempted to interpret chance in gaming as a measure of probability. This chapter, meanwhile, focuses on the patchy groundwork laid for these projects in connections drawn by thinkers in gaming, logic, and law.

From Huygens to Leibniz: The Anthropology of Judgment

Jacob Bernoulli made Huygens's *Van rekeningh in spelen van geluck* the starting point for probability theory in the eighteenth century by using it as book 1 of his *Ars conjectandi*. It is only in book 4 that the interpretation of probability is unmistakably put on record for the first time in history. Traces of this interpretation, however, are already discernable in certain notes by Bernoulli on Huygens's treatise in book 1.

This foreshadowing is evident, for instance, in the case of expectation. The concepts of hope and expectation (*expectatio*) actually mean different things in Huygens's theory and in everyday usage, Bernoulli notes.[3] Colloquially, they imply the greatest possible abundance of good. In mathematical theory, however, they indicate the mean between what players hope to win and what they fear losing. Expectation and hope in the mathematical sense therefore become calculable as the mean value of what we call hope and fear in a psychological sense. These comments seem necessary to Bernoulli to avoid endangering even the initial statements of the theory of chance from the outset. Hope—or, as Bernoulli now distinguishes it, mathematical hope—must shed its psychological meaning.

Expectation, hope, and fear are elements in any gaming theory for which interpretation in terms of probability suggests itself. Such elements may be called checkpoints along an anthropological transition to probability.[4] Leibniz also discusses these notions with great interest. Particularly in his fragment *De incerti aestimatione*, Leibniz again suggests an anthropological reading of the terms, even if the questions raised are here more emphatic than with Bernoulli and the solution takes the opposite path. The draft, which Leibniz himself dated to September 1678, the time after his return from Paris,[5] clearly derives from a direct encounter with Huygens's ideas. The very first sentence of the fragment brings hope in connection with Huygens's basic term of justice: "A *game* is *fair* if there is the same proportion of hope to fear on either side. In a fair game, the hope is worth as much as it has been bought for because it is fair that a thing should be bought for what is the worth, and the fear is as great as the price of the hope."[6]

Read as an interpretation of Huygens, two things attract our attention here. First, Leibniz seems willing to have people buy ready-made chances on the market without first having to create a cosmos of convertible games of chance that would determine the value of the *kans*. In Leibniz's account, people recognize a reciprocity of fear and hope on grounds of what we might call anthropological knowledge. The fact that both parties experience fear and hope according to the same basis or quotient (*ratio*) is, secondly, also a basic condition for assuming the existence of a market where chances can be bought and traded. In short, Leibniz transforms Huygens's positing of justice and its economy into the givenness of a world in which humans behave according to the proportionality of hope and fear. The model for this world, as another passage in the fragment shows, is the limited trading company. Leibniz sees players as partners (*socii*) of a commercial or insurance society who strike contract deals among themselves that should bring profit to the individual members, but that do not take anything away from their society as a whole ("neque toti Societati quicquam adimant"). The right of each single member remains protected in the sense of *suum cuique* even by his win or loss, because and insofar as the right as such belongs to the society and radiates out from there to its members.[7] The configuration where the company as a whole holds in its hands the right as a whole and transfers it to members of the cooperative is the model of wholes that can be translated further according to one's point of view. We may apply this model to a commercial company, to the state composed along mercantile and sovereign lines, or even to the world of self-regulating streams of wins and losses. In this sense the *tota societas* in which fear and hope stand in a relation of justice is not only a model of society and the world at large, but in a particular way it is also a model of preestablished harmony. The *tota societas* composes a whole and self-contained world for itself. Its order, which raises it to the best of all possible worlds, is the equivalence of the system of values (for all) and the system of affects (of individuals). But it also stands under the condition that other worlds exist or could exist.

Just as for Huygens the arithmetic mean is both numerical formula and semantics of justice in one, the algebraic proportion (*ratio*) in Leibniz corresponds to a semantic concept. Leibniz's first algebraic theorem is called the *ratio* of hope: let S be the hope (*spes*); R the net gain (*res*), and n the number of possible outcomes or cases. Under this assumption we should find it to be true that $S = R / n$. A further condition pertains that the hope be focused on a single advantageous case F ($F = 1$, where F means *eventus favens*) among n possible cases. This point of departure regards hope (or fear) as the expectation of victory, but not the distribution of possible events. Leibniz comes to the general formula for the hope S, under which any number of advantageous cases F may be considered, by way of a telling intermediate step. The argument relevant for this step no longer highlights hope but instead shows the economy of a preestablished society (or cosmic harmony). The formula of the *tota societas* means a relation of relations, a *ratio* (order, reason, proportion) of proportions: S / R / = F / n. Hope has the same ratio to gain as the number of advantageous outcomes has to all possible outcomes.[8] This proportion forms the semantically and conceptually meaningful intermediate step. Through it, Leibniz can cross over from the intuitive hope (for F = 1) to the general symbolism of hope (for any arbitrary F). This latter point is reached when, after several transformations, Leibniz finally extracts from the proportion the formula S = (F / n) × R. Ever since Abraham de Moivre and Willem Jacob 's Gravesande, this latter formula has represented the evidence of probability (P), or, as de Moivre says: the probability of happening (P = p / (p + q) is the formula for event probability where the value of the gain R as a simple constant is no longer noted down).[9] For Leibniz, in contrast, the proportion S / R = F / n, the mathematical and semantic representation of hope, is the more important expression. In the proportion, *notatio* and *numeratio* are closely related: the fractional equation represents a certain value at the same time as it serves as an instruction for performing a calculation. The equation between the psychological left side and the observable cases represented on the formula's right side displays the preestablished order of the *tota societas*, a harmonious order in which hope for gain clearly manifests itself anthropologically and mathematically at the same time.

Leibniz tucks fear and hope even more closely into a psychological understanding than Huygens had done in his tractate—in a move that drew Bernoulli's criticism. For Leibniz, in contrast to Bernoulli's mathematical hope, the numerics do not open up a space beyond proper states of emotion that would then have to be enlightened by the reason of calculation. It is precisely in his pursuit of the affects in the game—of the counterparts to fear and hope—that he discovers to the contrary their algebraic relation already at hand. In the feelings that are generated specifically by gaming, he sees the first indication of a *ratio* of everyday affectivity.[10]

At the very point where hope and fear part from each other, probability makes its entrance in Leibniz's fragment:

Probability (*probabilitas*) is the degree of possibility (*gradus possibilitatis*).
   Hope is the probability of having.
   Fear is the probability of losing.
   The estimated value of a thing is as high as each one's claim to it.[11]

The initial equation between *probabilitas* and *possibilitas* is the most succinct expression for how the theory of gaming in all its specificities relates to metaphysics in Leibniz.[12] The proportion of hope and fear scans, we might say, the transition from the ontological realm of the possible (and its tendency to being or nonbeing) to the measuring and graduated epistemology of probability. In the fragment on gaming theory, the proposition stating the equivalence between *probabilitas* and *possibilitas* interprets what the proportion S / R = F / n demonstrates.

Bernoulli had separated mathematical hope as controlled judgment and a reasonable method of argumentation from common affective behavior, and opposed the one to the other. Conversely, Leibniz understands probability as a rational principle already at work in the affects, which are however modes of behavior within the *tota societas* or the rules of the game. The functioning of probability calculation requires the reasonableness of the affects appropriate to gaming and vice versa. According to Leibniz, probability exists where there is an equivalence of fear and hope; an equivalence of fear and hope exists in the just game of the preestablished totality of a society for its members. The precondition for this argument is the unlimited sustainability of the game. Fear and hope under fair conditions can only be fair when the players know that they can continue the game after each loss and have to after every win. Only under this condition is the joy at winning just as great as the dismay at losing.

The way Leibniz reads and transforms Huygens's theory of chance is clearly the opposite of Pascal's dramatic game that man is forced to play because of life's finitude. We may assume furthermore that Leibniz is also replying in this fragment to Pascal's wager (the *Pensées* appeared eight years before in the Port-Royal edition). The equivalence between the two affects concerning the future, fear and hope, in fact entails a further meaning. As Leibniz spells out in the course of his interpretation, the equality between fear and hope is also that between the burden of having to play the game and the joy of playing. This equivalence, however, is as much a principle of balance as we know it from Huygens as it is a meta-argument on gaming in the manner of Pascal's wager. Whoever joins the society—the society of those who play the game, the trading company for mutual benefit, or the society of a state—acknowledges the totality of that society's law, and agrees to play the game under the condition of that encompassing law. Only those who know that they can continue to play under the same conditions will not regard the acceptance of the condition as an unfair burden. "The need to play" namely becomes troublesome when "the mind is occupied with no prospect of winning." And in return, it is also true that "this is only favorable for those people for whom the occupation of the mind by the game equals the joy which they get from playing."[13] The equality between the burden of rules and the joy of playing the game balances the affectivity in the game, making it a model for calculating with probabilities.[14]

No other author provides an interpretation of this ilk. Surprisingly, the standard interpretation of probability in the eighteenth century would not take off, as we shall see, from Huygens's textbook-like theory of justice, but rather from Pascal's speculative theory of decision. Even in Leibniz, the interpretation based on Pascal's theory predominates in a number of other contexts. Nevertheless, the version suggested in *De incerti aestimatione*—even if not yet carried to its conclusion—is simpler and more elegant than Jacob Bernoulli's theory of argumentation and protostatistics. For it does not need to be wrung from the game and its chances through applications, but can instead be found already tucked away in the preconditions of justice and equality.

Antoine Arnauld: Introducing Probability in Logical Terms

When Leibniz devised the equation between probability and possibility in the 1678 fragment on the theory of games of chance, the interpretation of gaming theory as probability theory was already sixteen years old. Rather than in the context of Huygens, as in Leibniz's fragment, however, probability was first made the theory's key term with regard to Pascal's wager. This was done by the Jansenist theologians Pierre Nicole and Antoine Arnauld in their book *La logique, ou l'art de penser*, whose first edition appeared in 1662.

Two different lines of interpretation suggest themselves in the aftermath of this work and have indeed been pursued by scholars thus far. The first concerns the dramatic relationship between Pascal and the authorities of the Port-Royal convent: how did Arnauld (the author of the chapter in question) understand Pascal, and what did he make of Pascal's fragmentary texts? The second line of interpretation poses different questions: how did *La logique* explain the concept of probability (*probabilité*), and what meaning of probability did it develop in the relevant chapters? The first—philological—interpretation reads the relevant chapters as a dialogue between individual protagonists, between Pascal on the one side and the Port-Royal philosophers in residence on the other.[15] The second interpretation treats those chapters as documents in the history of ideas of probability theory.[16] This book will undertake a third, discourse-analytical, line of investigation. The focus will be on problems at the root of and therefore prior to both of the interpretations sketched out above. This antecedent question is aimed at the act of interpretation that Arnauld performed in using the term *probabilité* in a context where mathematicians had seen no reason or need for such a concept.

The hypothesis to be unfolded has two major steps: (1) Arnauld introduces the term *probabilité* in the same context in which probability had once functioned in the Aristotelian *Organon*, the context of the topoi. To this end, (2) he makes use of the circumstance that a gap had persisted in the discussion of topoi ever since Cicero; and it is exactly here that by bringing the propositional logic of contingency into play, *La logique* locates the theory of games of chance. The interpretation of the theory of games of chance as probability theory, as we shall see, owes its emergence to this discourse-historical juncture. Bringing together the probability from the topoi and the theme of contingency was Arnauld's crucial move, which thus far has either been overlooked or dismissed outright.[17]

(1) Five chapters on method, in which Arnauld discusses the rules reason can employ in estimating the credibility (*foi* or *fides*) of events, close the fourth book of *La logique*. In the prominent chapter 12 of book 4, Pascal's terms *certainty* and *uncertainty* appear. The terminology of probability (*probabilité*) unfolds and prevails from chapter 13 ("Some Rules for the Proper Use of Reason in Determining When to Accept Human Authority") to chapter 16, which paraphrases Pascal's wager argument.

Critics have often argued that the sequence of these chapters forms a supplement that lacks a true conceptual connection with *La logique* as a whole.[18] This is not the case, however. Arnauld's rules for estimating probability are indeed carefully detached from the conventional theme of the topoi as being probable arguments. But it hence becomes all the more clear that they form a counterpart of the traditional topoi designated for storing and relocating probable arguments. They offer a way to judge the probability of those arguments. This meaning of the concluding chapters can be gathered from the architecture of *La logique*. The third book—"Reasoning"—deals first of all thoroughly with the syllogism, and then briefly summarizes the topoi and the doctrine of spurious arguments. The fourth book—"Ordering"—begins with a broad description of the geometric method and concludes with the five chapters on the credibility of contingent events. The symmetrical correspondence between the two books is obvious. In the register of certainty (book 3), the geometric method corresponds to the traditional doctrine of syllogistic argumentation.[19] In the register of uncertainty (book 4), the (new) doctrine of the measurement of credibility corresponds to the (old) art of the topoi to store and relocate probable arguments.[20] The impetus for introducing the calibration of credibility as a parallel in method to the reservoir of probable arguments has its roots all the way back in Cicero's *Topica*. For Cicero, the power of the topos to bring forward irrefutable arguments was what he called its *fides*, that is, its credibility or reliability. And he explicated the nature of such argumentational *fides* by comparing it to the credibility of testimony of witnesses.[21] This in turn is precisely the subject of book 4, chapter 12 of *La logique*, where Arnauld discusses the credibility of witnesses and testimonies. Witnessing is an issue for Arnauld mostly in the context of Holy Writ and things religious, hence *fides* indeed means faith for him, in contrast to science, as far as our senses and our faculty to judge from the senses are concerned.[22] These complex links between witness testimony, *fides*, and probability from Cicero's *Topica* are what supplied Arnauld in the first place with a model for associating the evaluation of credibility of events with probability and its logic.

Another discursive link can be added. According to Cicero, the art of disputation (*ratio disserendi*) has two parts. The first part, the art of inventing probable arguments (*pars inveniendi*), consists of the topoi. Cicero contrasts this part with the art of judging probability (*pars iucandi*).[23]

However we might picture such a doctrine of judging the probable, it is entirely missing from Cicero's *Topica*. This lack remained the rule in the subsequent tradition. The empty slot of the *pars iudicandi* was a highly visible blank space into which Arnauld could inscribe the five concluding chapters of the fourth book of *La logique*.

(2) In order to insert the calculation of chances in this blank space of a doctrine of judgment on the probable, Arnauld has recourse to the Aristotelian logic of contingent events (4.13).[24] But how is it that Arnauld imports precisely the logic of contingency to the place of probability?

"In order to decide the truth about an event and to determine whether or not to believe in it, we must not consider it nakedly and in itself," without its relation to other events. In pure isolation, Arnauld argues, we can determine the truth only in the case of things mathematical and metaphysical. In order to grasp the truth of a singular event, "we must pay attention to all the accompanying circumstances, both internal and external."[25] This clearly brings the topoi into relevance: when we wish to achieve the effect of a true argument in the case of singular events, we cannot proceed by pure syllogistic reasoning, but must complement and fill the event in with additions that are linked by topical relations to the respective matters at hand. In the rhetorical and dialectical tradition, such supplements required for a logic of events would be called topoi, or, more specifically, the circumstances of an event (*circumstantiae*). *Circumstantiae*, the topoi of narration, make up the core of the production of a probability of events in Arnauld. From this inconspicuous moment on, the interpretation of gaming theory as probability theory is intimately related to the history of modern narration.

Certainly, Arnauld lends the topoi of narrative circumstances a new twist, which is connected precisely to the use he makes of them for a doctrine of contingency. Arnauld distinguishes internal circumstances that lie within the event itself from external ones that rely on the statements of others about this event.[26] For Hacking, this distinction justifiably counts as a decisive move in the development of calculable probability.[27] In fact, this division is, again, a citation from the old tradition of the topoi. Ever since Cicero's *Topica*, handbooks have taken their departure from a distinction between internal *loci*, which search out references belonging to the relevant matters, and external *loci*, which treat the reliability of statements concerning such matters.²⁸ This distinction was not an achievement of the verification of sources in modern philology and historiography—as Hacking assumes, following the lead of scholars of history—but determined the art of probability in terms of its operations since antiquity.²⁹ For the ancient topoi, the distinction coincides with the question about the competence or incompetence of topoi and rhetoric: the internal topoi lie in reach of the topical and rhetorical art; the who? how? where? and so on, are the domain of the *narratio verisimilis*. In contrast, the external topoi, Arnauld's external circumstances, are not or at least very little in the hands of the orator or debater. The external locus of a testimony's credibility, for example, marks the point where the art of speaking finds its limits. For Arnauld's project of *probabilité*, however, it is crucial that both kinds of circumstances, along with their distinction, serve the common purpose of estimating the probable in it.

If both the internal circumstances of an event and its witnesses and documentary evidence, as well as their difference from one another, are together taken into consideration, then the space of a fourth-wall stage or a novel emerges before our eyes. It is in this space that events to be evaluated can finally play themselves out for the observer of probabilities. There are no probabilities measurable in the participatory observation of the first order, the situation of rhetoric and the dialectical art. A measure of probability can be constructed and controlled only from the position of the spectator or reader, who is not subject to the events of the probability that he is estimating. This means, however, that the material links to the world (internal circumstances) and the observations of others (external circumstances) may no longer, as in ancient rhetoric and dialectics, fall asunder in a way that cannot be calculated by the same observer. For the spectator or the reader, the second-order observer, hypotheses about observed events and the evaluation of reported events are now only relatively different cases of the same type of data processing.³⁰ This procedure again intensifies the close link between the interpretation of gaming theory as probability, on the one hand, and the history of modern narration, on the other.

The explicit introduction of games of chance as an example for the calculation of any possible chance event in Arnauld takes the form of a narration, of a moralistic anecdote, so to speak, or even almost of a small novel. The first part of the passage in question relates, in a style clearly distanced from the usual wording in a textbook of logic, the following little story: "It was reasoning of this kind that led a princess, who heard that some persons were crushed by a falling ceiling, not to enter a house ever afterwards without having it inspected first. She was so convinced she was right that it seemed to her that everyone who acted otherwise was imprudent."[31]

This narrative nicely composes a scene of observation raised to the second power. In the narrative, a precedent story and eyewitness accounts repeatedly conveyed by anonymous messengers are set together at the level of a present time in which the *princesse* evinces her unshakeable conviction to third parties. It is no accident that the anecdote within the anecdote reminds us of Simonides and the story of the destroyed banquet hall with which mnemonics had exemplified the art of memory.[32] The story that occupies the affects of the *princesse* in the story is burned into her memory, so to speak, as an unshakeable topos. It is this affective or mnemonic structure of the story that matters to Arnauld, and not at all any kind of opposition between hearsay and eyewitness accounts. Even the difference between singular and repeated observations does not seem to matter. The instance of the game, then, responds precisely to this narration—this particular narrative fabric of the anecdote—in Arnauld's *La logique*, which stages and confirms its exemplary character once and for all. According to Arnauld's lesson, we ought not only to judge the good and evil of any event in itself (we should not only let anecdotes solidify into affective mnemonic topoi), but "it is necessary to consider . . . also the probability that it will or will not occur." In addition, we should "view geometrically the proportion all these things have when taken together." All this, moreover, "can be clarified by the following example."[33] The story that follows as the instance that is meant to illuminate rationally the anecdotes and anecdote's anecdotes, is the example of a fair game of chance. Ten people, each of whom stakes one ecu, play to win the whole sum of ten, that is, each one plays for a gain of nine ecus. The example teaches that though the gain of nine ecus may appear tempting next to the stake of only one, players should and will rather calculate and compare the one to nine proportion of winning chances. The example of the game is called "cet example" (this example),[34] because it posits the act of counting against the anecdotal brilliance of the winning sum of nine ecus, and thereby finally against the irrationality of an aristocratic *memoria*.

Completing the example in which a game of chance can for the first time appear explicitly as an instance, *this example*, Arnauld inserts a peripeteia between the literary world of the anecdote and the numeric example of the game. The fascination of "so many people" (*tant de gens*) for lotteries is what intervenes; the many and the average man are making their entrance, in connection to the ever newly repeated staging of chance that is the drawing of the lots. In Arnauld, this world of the average, the multitude, and repetition still remains entirely bland and receives nothing harsher than a warning, moralistic remark. But it already marks the place between the literary dignity of the *sensibilité* of a princess and the enlightened exemplarity of the example of the game. It marks the transition from gaming theory to probability.

The many are admittedly fascinated by the prize, as is the princess by the fear of death. But what in the latter is a literarily imparted affect, is in the former blindness from greed. The lottery is admittedly a game of chance, like that of the band of ten playing for nine ecus. Yet while the latter has to do with a fair game, the perfect balance, the lottery is an unfair game, a game of fraud.[35] The blinded masses, deceived and cheated in the mechanism of the unfair game, prefigure the world that in the nineteenth century will be termed "society." It is this impure, neither poetically stylized nor mathematically pure demi-world—a world between the sensibility of individuals and the preestablished possibility of events in a game—that both divides and binds together the two poles, the anecdote and the example. The interpretation of the game of chance (in the instance of the fair game) as probability (for the princess) functions only when, explicitly or not, it also traverses this field of the many and of the impure. A new relation between the literary and the probable is thereby hinted at. In it, we recognize the possibility of the modern novel.

Leibniz and Pufendorf: Introducing Probability into the Theory of the Game in Juridical Terms

Arnauld's interpretation of gaming theory as probability theory became entrenched in logical discourse mainly through citation, repetition, and amplification. In 1672, ten years after the Port-Royal *La logique, ou l'art de penser*, Samuel Pufendorf's *De jure naturae et gentium* [Of

Natural and Human Law] appeared, whose chapter "Contracts Relating to Chance" discusses contracts based either tacitly or explicitly on games of chance or on lotteries.[36]

Pufendorf attempts to tread a middle path between condemning and simply accepting games of chance. As far as games of this sort as well as lotteries can be regarded as contracts, they are allowed by natural law. The sovereign, however, can and should provide for limitations. Lotteries should be tested according to the measure of probability represented by the fair game. Such regulations could prevent the injustices of deception that Arnauld had cited in his remarks on lotteries. Pufendorf then quotes Arnauld on the fair game for ten ecus among ten people word for word. In the context of the law, the example of the game now becomes the measure for the moral and social reliability of games.[37] Arnauld's logical interpretation of the game of chance as probability thus crosses over into a further interpretation, this time the interpretation of law.

Just as a gap in the tradition of logic had been marked out for Arnauld's logical interpretation, a space in the tradition of law already awaits Pufendorf's legal interpretation. This discursive space was called the theory of "contracts relating to chance." Under this heading, Pufendorf spreads out a panoramic view of the whole range of policies in a nation-state with regard to contingency—from war to duels and from gambling to insurance. The legal core of this political complex is the conditional contract, the mutually qualified promise, "where chance plays a role, since whether or not an event or a thing exists does not depend on the players' efforts."[38]

Pufendorf merely made the legal reinterpretation of Arnauld's probability interpretation habitual, however; he did not invent it. Leibniz had hammered out this legal construct in 1665 in the *Disputatio juridica [prior] de conditionibus* and the *Disputatio juridica posterior de conditionibus*.[39] In the opening sentence of the second of these, we read: "Certainty [the certainty that an agreed-upon condition comes into effect so that compliance with the contract becomes binding] can be evaluated according to hope, with hope understood as a measure of general acceptance."[40] By further describing the evaluation and measurement of the probable as *logica probabilium* or *veritable topique,* Leibniz also shows himself to be a careful reader of Arnauld. The interpretation of gaming theory as probability theory in book 4 of *La logique, ou l'art de penser* bears the same relation to the old topoi in book 3 as the geometric method in book 4 does to the old syllogistics in book 3. Leibniz spells out what in Arnauld is hidden in this disposition of materials. For him, the new probability theory exemplified in the game of chance now becomes the *veritable topique*—the true, better topic.

With this move, Leibniz obviously carries further, not Huygens's interpretation, but rather the older Pascalian line of exemplifying the game of chance, the line continued before him by Arnauld. In the papers in which Leibniz draws directly on Huygens, he offers a typically mathematical argument. Here, in the juridical context, taking up Arnauld's line of thought, he considers a mathematization of probability only cautiously and from a distance. This consideration finds expression in 1669 in the reworked version of *Specimina juris* even more clearly than in the *Disputationes juridicae de conditionibus*.[41] In contrast to the game theoretical fragment from the 1670s that, following Huygens, led to a legal foundation for probability, Leibniz here takes the sphere of law as his point of departure in order simply to probe the idea of calculating the probable.

The first step in a version of the treatise on conditional contracts that Leibniz composed for a subsequent new edition concerns the possible forms of the condition. Conditions on which contract obligation is based can be bound simply to the existence of a person or a particular declaration of will. This is so, for instance, in cases of inheritance. Conditions can also be linked—think of an advance on the arrival of a ship—to the mere fact of the occurrence of an event or the deadline of its occurrence. This distinction, which only exists in law, can be evaluated and expressed in numbers. The scaling of legal entitlements is hence so far based entirely within the frame of legal processes and can only be represented by numbers in hindsight. Secondly, Leibniz writes in his *Specimina juris*, the conditional legal entitlement representing the greater value has priority in business agreements. If the condition for the fulfillment of a contract says that my ship should return safely from Asia, then, in the case of the occurrence of this condition, the claim of the party to whom 200 guilders were promised is greater than the claim of one to whom hardly 100 guilders are owed. The numerical representation for the relations of legal entitlements to each other in this case is provided simply by their monetary values. So far the numerical valuation is normative and can only be applied in hindsight, or it is economic and the numerical expression of the legal entitlement is only borrowed from the monetary ratio. After addressing these cases of indirect numerical representation of contractual conditions, Leibniz comes thirdly to the fundamental principle of the juridical theory of measurement. In this section of his argument, he announces the new logic of probability, the true topos. And here the relation of semantics to numerical expression is brought into a new balance.

The *principale fundamentum* reads as follows: "The greater the probability that a condition exists [*probabilitas existentiae*], the greater the conditional legal entitlement."[42] From this rule, Leibniz draws a series of conclusions, which lead eventually from the semantic relation, as indicated in the construction "the greater . . . the greater," to mathematical proportion. At the center of this transposition from semantics to calculation, hope functions. Hope brings the legal norm, the economical measure, and the psychological gradation into one and the same process of calculation. There is only hope, however, when and if the condition to which it refers (under which, for example, a contract is fulfilled) has a certain possibility of becoming reality (*possibilitas existentiae*), a possibility that can be determined according to its probability (*probabilitas existentiae*). In other words, hope is the transferability or proportionality of ontological possibility in epistemic and psychological probability. This is the core principle for calculating probability in Leibniz. In the narrower sense of juridical measurement of probability, this principle has the following implication: only those conditions count as the fulfillment of a contract that occur or do not occur in the frame of explicit and analyzable conditions. The constitutive function of hope for probability theory, however, points also to the sanctuary of Leibniz's metaphysics. Leibnizian metaphysics requires, together with the (not necessarily optimistic) hope inscribed within it, that being has a sufficient reason—a rational ground and a proportion—of the possibility for its becoming real or not. The conditional contracts of the young jurist are the later metaphysician's drilling ground.

Leibniz unfolds the juridical probabilistic translation of possibility into probability in the *Specimina juris* with affectionate casuistry. If someone who is harmed by the condition can prevent its occurrence, then this probability is of more relevance; if someone to whom the condition is advantageous can assist its coming about, then the probability is of less importance. Leibniz then reformulates this semantic-normative version of probability in such a way as to bring legal norm and mathematical measure into balance. If the realization of the condition is left to the discretion of someone who has to demonstrate its fulfillment, then, Leibniz says, it results in minimal probability (noted with the value 0); if it is left to pure chance, it lies in the middle (with the value ½); if it is left up to the one who determined the condition in the first place, then it has the maximum probability (or the value 1). This coincidence of semantic typology and numerical evaluation is all the more interesting inasmuch as Leibniz generally suggests assessing legal claims on a scale of 0–1 / 2–1 according to intuitive estimation. The three types of conditional contract, corresponding to the three numerical grades, in Leibniz's seminal work on legal theory hence constitute a kind of reference system for juridical judgment in general.

In the three types and values of conditional contracts, law and calculation are brought into a fundamental balance. The possible cases of conditional contract law fulfill as if by themselves the predetermined numerics of the three degrees of legal claims.

This observation shows that for Leibniz, a fundamental analogy holds sway between law and number. This, in any case, is demonstrated by the summarizing formula that speaks to greatest effect in the language of mathematics, and that nevertheless also confesses the analogy of the metaphor most clearly: "The greater the breadth of a condition is, the greater is the variety of forms it can exist in, that is, the more it counts. The greater the length of the condition on the other hand, and the more that is necessary for its fulfillment (for we must traverse the entire length of a path, but need cover no more of its breadth than we wish), the less it counts."[43]

The metaphor of the path whose breadth one can use to one's own advantage, but whose length must be covered in order to reach the goal, stands ingeniously between law and number. It translates—as a metaphor of method—the chances and conditions of legal process into the calculability of measurements.[44]

We can understand the later fragment on the doctrine of chances after Huygens as the precise counterpart to the continuation of the logic of conditioned law after Arnauld and Pascal. For Leibniz, conditioned legal claims and games of chance establish zones in which normativity and calculability support and require each other. The two zones do not simply dissolve into one another, however. Hence there is no conditional logic (a logic of entitlements and expectations in an if-then structure) that is simply and literally identical to probability (gaming theory understood as probability theory). This fact becomes evident particularly in the role calculation plays in each case. In juridical conditional logic, calculability is always metaphorical in the end. Leibniz shows how a gradation with which one can calculate is prepared for in the logic of hypothetical propositions. But the numbers that he writes and the calculations he carries out are only quasi-numbers and quasi-calculations. The case is different in gaming theory probability following up on Huygens. Here Leibniz understands the equation of two proportions ($S / R = F / n$) as the equation between the psychology of judgment and the observable events that are to be judged. The proportion can also be interpreted, however, as the formal representation for an underlying semantic expression or metaphor. According to this reading, hope that is directed toward gain is the psychological expression or mental metaphor of favorable cases in their relation to all possible cases. In other words, probability takes its point of departure from a mathematical formula, the proportional equation. The proportion, however, must first of all be understood as the representation of a metaphor.

Open Interpretations

Probability never returns to get behind these first acts of interpreting the theory of games as a theory of probability. Arnauld, Leibniz, and Pufendorf paved a broad path for others to follow: they established the theory of the game in the language of logic and of law. Certainly, the respective acts of interpretation are self-evident only in hindsight. If we observe more closely, we easily recognize the cracks. Strengths and weaknesses are particularly interesting if we compare Arnauld and Leibniz. Arnauld's exemplification of the game of chance for probability in the framework of traditional logic offers a comparatively rough link between gaming and judging. The link does not extend into the fine structure of the two sides of example and exemplified. With Leibniz, the case is different. But precisely because Leibniz gets drawn into the details of both the probability of Huygens and conditional contracts, for him the analogy of calculation in conditional law, on the one hand, and the proportion of affects in gaming theory, on the other, do not merge together in a simple center of the probable. The if-then expectations of contract law, which he uses for virtual calculation, and the proportion of gaming chances he sees built into hope and fear as probability calculation, do not meld together into a single theory. This does not mean, however, that they contradict each other. They are like two pieces of a theory that in its entirety has been lost or is yet to be discovered—of a theory of the possible and the probable, a theory of the legibility of the mathematical formula in the psyche and the psychological process as proportions we can express and operate with in numbers.[45] In both of the partial theories, calculability of chances and events can only exist on the basis of coherencies in assumed systems and in spaces of prior standardizations. In the case of conditional logic, the normative constitution of probability and of the probable events overshadows the mathematical representation. In the theory of games of chance, in contrast, the aspect of rule-following and the laying down of laws arises only through the process of gaming. In playing the game, we experience its hidden norms through the equivalence of hope and fear. By the same token, it is clear that fear and hope are only calculable when we are dealing with them *sub specie* the regularity of the game. Despite the convergence of game-theoretical and juridical probability, Leibniz does not present any completed theory that fully integrates the two sides. Between the quasi-calculation of conditional contracts and the quasi-psychological interpretation of chance calculation, a gap remains that is not bridged anywhere.

To sum up, the interpretation of gaming theory as probability theory should be described as a discursive process. We need not fall back on an intuition of Arnauld, nor do we need to call upon a zeitgeist bringing together the most disparate elements of mathematics, logic, and law. Most important, we do not have to try to understand how the interpretation in view of gaming theory suddenly recognized a new meaning for probability in order then to lend it to the game of chance. Logicians and jurists made use of the possibilities in the discourse of their respective disciplines with unforeseeable and complicated interconnections in order to impose a meaning on the game of chance beyond its own. Analyzing the act of this odd and momentous bestowal of meaning should begin to offer a solution for the many attempts we have seen to understand anew and better the meaning unfolded in each effort. It was the meaning of probability that led mathematicians into so many paradoxes, until they erased it at the beginning of the twentieth century in the axiomatization of probability; and it is the same meaning of probability that, from the eighteenth century on, has allowed the modern novel to become so much a synonym for the experience of reality that despite all their experiments and revolutions, the avant-garde of the twentieth century could hardly shake it off.

5

Probability Applied

ANCIENT TOPOI AND THE THEORY OF GAMES
OF CHANCE—JACOB BERNOULLI

Bernoulli's *Ars conjectandi* and the Place of History in Probability Theory

It was only the interpretation of gaming theory as probability around 1700 that finally opened the sluice gates to allow a cascade of applications for the mathematical structure of chances to begin flowing. This trend got its start by applying probability to the natural history of man: to his life expectancy (the insurance business); his reproduction (population policies); and his physical and ethical provisioning (moral statistics). The same tendency continued later on with the further employment of probability in the observation of nature (theory of errors); and finally in the natural processes themselves (physics and chemistry).[1] By means of its many applications, probability theory made history until it began to leave truth itself in the shadows—the very truth (*Wahrheit*) that probability or *Wahr-schein-lichkeit* was supposed to be the semblance (*Schein*) of. But before jumping ahead of ourselves, we first must understand how application has been constitutive for the theory of probability and the historicity inscribed in it right from the beginning, ever since Jacob Bernoulli. Condorcet begins his history of probability theory, after a brief summary of Pascal and Méré, with these words: "Geometers have long contented themselves with a theory of games of chance, in itself a trivial and frivolous subject. Only they themselves were able to see even so long ago that these methods would one day be applicable to more important areas of research and that, in dedicating themselves to such an amusing occupation, they were laying the foundations of a science as equally far-reaching as it is useful."[2]

In the phase of its frivolity, or so Condorcet implies, the example of the game was not yet understood as an example for anything else. Interpreting and applying probability, however, means forgetting the very example that is interpreted and applied. The two strategies to deal with probability, interpretation and application, thereby define in one move the theory's scientific rigor, applicability, and historicity. This is quite striking in the case of the interpretation of gaming theory as probability theory, since the example and its prescientific and prehistorical frivolity, as far as the game of chance is concerned, amount to pure, meaningless structure. Probability, however, is precisely what creates scientific knowledge and history and, thereby, meaning—meaning as the semblance (*Schein*) of truth (*Wahrheit*) and as probable (*wahrscheinliche*) history.[3] In short, the interpretation of gaming theory as probability is the rededication of structure to meaning.

Application and interpretation of gaming theory are the themes and the stakes of Jacob Bernoulli's *Ars conjectandi*, which was published after the author's death by Nicolaus Bernoulli in 1713. For this reason, the history of probability first takes off for Condorcet with Bernoulli's book: "Jacob Bernoulli in his posthumous work entitled *The Art of Conjecture* was the first to demonstrate the connections of this branch of mathematics in nearly all areas of philosophy."[4]

This view, which would later be passed down by way of Adolphe Quetelet and Auguste Cournot to the twentieth century,[5] had already been worked out preliminarily in the preface that Nicolaus Bernoulli wrote for his late uncle's *Ars conjectandi*. In his introduction, he mentioned Pierre Rémond de Montmort and Abraham de Moivre, two authors who had published theories of games of chance after Jacob Bernoulli, thereby endangering his primacy. Unsurprisingly, Nikolaus criticized them sharply and exactly because they had "forgotten the most important aspect, namely, the application of probability theory on questions of economics and politics."[6]

Applying mathematical formulae and the project of an applied mathematics in general are for Condorcet the motor of historical progress in knowledge, a progress that proceeds by means of the alternating processes of despecification and specification. Thus the example, the frivolous context in which knowledge was hidden or trapped is released in order to become applicable in new fields, and the new fields thereby turn into examples of its theoretical-historical potency. The specificity of the first context—the exemplarity of the game of chance itself—should disappear as mere frivolity; and the process of application moreover also brings under its control the specificity of the new contexts that it defines and occupies. Juridical mathematics, which Nicolaus Bernoulli presented as one of the first cases of application for his uncle's theory, counts for Condorcet as the pattern of such a release.[7] For him, jurists are the quintessential caste of scholars and functionaries who have chained the common good of thought—judgment—with their pedantry and social conceit. The law in its Roman form alongside its "obscure commentators" can and must be delivered from itself by means of applied probability theory.[8] Jacob Bernoulli's *Ars conjectandi,* in its architecture of gaming theory, combinatorics, and the application in ethics, law, society, and economy, carried out such an interpretation of the theory of chance as probability. If there is any history of probability that does not simply follow and repeat the tale told by Condorcet of the bestowal of meaning to frivolous structure,[9] then such a history must be brought to light by analyzing the historicizing gesture of this foundational text.

The figure of belatedness, as it inheres in the very process of interpretation, is already apparent in the genesis and editing of Bernoulli's *opus postumum.*[10] Bernoulli had died in 1705, and Leibniz, Montmort, and other interested parties who knew of the long process of its development kept pressing for publication. His student Jacob Hermann placed the text at the disposal of Fontenelle for the Académie royale des sciences' obituary and used it himself for an appreciation of Bernoulli in the *Acta Eruditorum*. The scientific public thus knew of the *Ars conjectandi,* but it remained "just barely short of completion."[11] Years after Jacob's death, Johann Bernoulli refused to allow its publication, and especially its revision, because of stubborn disagreements with his elder brother.[12] When Nicolaus Bernoulli, their nephew, finally brought it out in 1713, he had already published his own *Dissertatio inauguralis mathematico-juridica, de usu artis conjectandi in jure* (1709), which he introduced as an exemplary application of his uncle's theory.[13] Belatedness thus becomes a hallmark for the publication and reception history of *The Art of Conjecture*.

This is especially true with regards to the fourth book of Bernoulli's work. In book 1 of *Ars conjectandi*, Bernoulli annotates Huygens's theory of games of chance. In book 2, he deals with combinatorics, and in book 3 with particular problems arising in games of chance. It is then only the fourth book, "On the Use and Application of the Previous Theory to Civil, Moral, and Economic Affairs," which Hermann held to be "nearly" completed, and Nicolaus claimed just barely arrived at the decisive point, that introduces the *usum & applicationem* of the theory and with it Arnauld's interpretation of probability in the world of moral decisions.[14] What Jacob Bernoulli here calls the application of the previous theory would first become for Nicolaus and for the tradition instructed by Condorcet the theory itself: the theory of measuring probability. That theory, at the time of its posthumous publication, however, had already been applied in exemplary fashion in Nicolaus Bernoulli's juridical-mathematical dissertation.

The game of completion and incompletion, the game of posthumous publication, had already given the stage directions in Bernoulli's own treatment of his *Ars conjectandi*. On the one hand, the evidence of letters and documents from the literary remains suggests that the work had already been extensively completed in 1690 or at latest 1692. On the other hand, Bernoulli wrote to Leibniz on October 3, 1703, that though his book was largely finished, the most important part was still missing. Bernoulli is clearly hinting at the application of the theory to civic life, social relations, and the economy. It is not without irony that Bernoulli alludes to the application of the theory as the only thing lacking from the completion of the fourth book, given the fact that, even in his own view, only the application would found the theory in the first place.

When he wrote these lines to Leibniz, Bernoulli well knew that Leibniz had for quite a while been promoting something he called the *logic of the probable*.[15] Leibniz would continue to talk up this project— more daring and fundamental than Bernoulli's—to the end of his life. Bernoulli's book, conversely, proved itself in hindsight to be complete, even though he was no longer able to conclude it.

Loci topici: Arnauld's Construction of Logic in Bernoulli's *Ars conjectandi*

In the two concluding chapters of the fourth and last book of *Ars conjectandi*, Jacob Bernoulli announces the solution of a problem that he had worked on for twenty years and that surpassed everything else in meaning and usefulness.[16] This problem is framed as a challenge to provide a mathematical model of what Bernoulli calls "a posteriori" probability. Speaking of a posteriori probability means to determine the frequency of events whose distribution is not already known and that cannot be represented by either combinatorics or any arithmetic formula. Such, however, are the cases of the application of probability in civic life and ethics that Bernoulli had promised to present. They lack what Ian Hacking calls a "fundamental probability set": we cannot recognize in these games the specific condition that would render them fair and just. The task arising from this situation for Bernoulli consists in determining the ratio between the number of observed events and the fluctuating range of the distribution we may postulate for these events, and determining it in such a way that, for observations repeated any number of times, the difference of the limits of the postulated distribution falls below any given value.[17] To deal with this problem, which is no longer arithmetical or combinatorial but a problem of limit analysis, Bernoulli first considers the structure of the binomial coefficients whose suitability for distributions of chance had already been recognized by Pascal. He then formulates a new limit theory on the basis of the theory of series.[18] The fourth book dedicated to the application of probability begins, however, with considerations that have been termed a calculus of argumentation. The connection between this theme in the first chapter of book 4 and the development of a theorem for determining the limit value at its end has remained in the dark. Nevertheless, this question marks the precise spot where we touch upon the central difficulty connected with Bernoulli's introduction of an application of the theory of chance: how does the theory of probability in argumentation relate to a theory that could determine the limit value of statistical observations?

Bernoulli discusses the application of the theory of chance to argumentation in chapters 3 and 4 of the fourth book. In the second chapter, he divides the *argumenta* into intrinsic or artificial, on the one hand (that is, elicited by considering the cause, the effect, the person involved, the related events, and the indications); and external or nonartificial, on the other (that is, statements that must be evaluated by the *auctoritas* and reliability of the witness). Bernoulli calls the former arguments *loci topici* in the narrower sense. Even in the very details of the example and commentaries (a murder involving the discovery of a dagger coupled with a certain person's blushing and turning pale), this amounts to a citation of the Ciceronian tradition of the topics, or dialectics,[19] with which Bernoulli, as a former student of theology, would have been familiar.

Even the division itself into "intrinsic" *argumenta* (the *loci topici*) and external *argumenta* (which Bernoulli does not designate as topoi) originates in Cicero's *Topica*. Clearly, Bernoulli inherited this citation of the *Topica* from Arnauld and Nicole (he refers elsewhere to the Port-Royal *La logique, ou l'art de penser*). He evidently looked up the passages in Cicero silently cited in Arnauld and added more Ciceronian references of his own. A small but important shift took place in the process, however. In order to interpret gaming theory as probability, Arnauld employs the topical distinction between intrinsic, properly rhetorical topoi and nonrhetorical, external ones. Bernoulli, who enlists Arnauld's probabilistic vocabulary as a matter of course right from the beginning of the first book's commentary on Huygens, and then throughout the fourth book, employs the probability theory developed from the *Topica* once again on the *argumenta*—the intrinsic and external topoi. At stake for him is the challenge of actually measuring the probability of these *argumenta* or topoi.

Even scholars who have taken note of the Arnauld citation at this juncture of the *Ars conjectandi* have invariably overlooked a citation of the *Topica* in it.[20] If this second quotation is taken into account, however, the much-debated question of what should be understood by *argumenta* is easily answered: *argumenta* are clearly reasons brought into play in procedures of proof according to the *Topica*.[21] *Argumenta*, in other words, are reasons that can be derived from *sedes argumentorum*, Cicero's *loci*. The follow-up question, as to whether such *argumenta* are fundamentally a matter of *in re* or *in verbis*, is just as misleading, in light of the Cicero quotation, as are attempts to reconstruct Bernoulli's supposedly altogether original theory of argumentation in modern terms.[22]

The citation from Arnauld persists even further. In the third chapter we read a little story that seems to define a new type of argumentation that founds, for its part, a tradition in exemplifying probability from Nicolaus Bernoulli up through Quetelet. The example highlights the respective weight of different probability reasons—*argumenta*—when we evaluate the status of a particular fact or the report on that fact. According to the example, I am waiting on a letter from my brother, who has begun a journey by ship. The letter, which may have been promised but at the very least is expected, fails to materialize for long beyond the expected time. The question, then, that will decide on the further analysis of the *argumenta* and their probability, can be framed thus: what are the possible *argumenta* for the letter's absence? Probability is here defined by how the present brother, in the world of his knowledge and expectation, estimates the situation of the absent brother. Is the brother, whose letter I am now looking forward to in vain, merely lazy? Has he forgotten his promise? Is he ill or even dead? Is he perhaps overburdened with business? The hypotheses that result from this example, ensconced within its web of telecommunications, form a logic of observation. In the construction of this example, Bernoulli takes up and advances Arnauld's achievement by leveling out the opposition between intrinsic and external topoi (the intrinsic and external circumstances of an event). Only for an observer outside the world (of the absent brother) are the circumstances of an event and the reports about that event two masses of data in the same logical space, facts that can be balanced out and checked against one other. And it is only because events and reports share a common nature as data that any kind of balance and checking against each other becomes possible. This is what Bernoulli expresses in his little example story that again confirms Arnauld's revision of the ancient topics: for the observer at home awaiting a report, the probability of the reported event cannot be separated from the report of the event. The world of probability is a world of interpreting telecommunications.

To explore the structure of the world characterized by the knowledge of the waiting addressee, Bernoulli uses a matrix of two parameters: one defined by necessity and contingency and the other by existence or indication. For instance, laziness—as far as the stay-at-home brother can know—is either existent or not existent in the character of the absent brother, and in each case it exists in itself with necessity. If the character trait is present, then it can either be the factor preventing his writing a letter, or not (if one of the other possibilities happens to be the case). In other words, its indicative value, or the value of its being an indication for letter writing, is contingent. Corresponding considerations apply to the further *argumenta*: the death of the absent brother is contingent in terms of existence (he can be dead or not) and necessary according to its indicative value (if he is dead, then he cannot write; if he is alive, then he could write a letter), and so on. The matrix of existence/indication and necessity/contingency carves up the *argumenta* into graded yes/no combinations with decidable yes/no alternatives for necessary cases and nondecidable yes/no alternatives for contingent ones. The seemingly idiosyncratic analysis of the *argumenta* (or circumstances in Arnauld's sense) is instrumental in turning those "reasons" into factors for the probability of events.

In other words, in book 4 of *Ars conjectandi,* Arnauld links the topics to contingency with an operational use. Analyzed according to the parameters (existing/indicating and necessary/contingent), the example of the missing epistolary report chops up the topical *argumenta* into the fundamental example of the coin toss, the strict binarism of heads-or-tails (yes or no). Bernoulli's story of the missing letter thus depicts and performs what was contained only in the formal construction of Arnauld's text. There, the game played by the ten people for ten ecus became the analytically illuminating and psychologically healing example offered to the traumatized *princesse* and her sentimental story. Bernoulli again brings the narrative topics together with a contingency analysis. He even takes the exemplarity of the game of chance for the probability of the topics quite literally: topical *argumenta,* to paraphrase his assumption, should be immediately subject to a game of chance.[23] Precisely this is the point upon which the unity of the fourth book of the *Ars conjectandi* hinges. If the analysis of the *argumenta* according to the binarism of contingency is taken literally, then the project of a calculation of probability, as developed at the beginning of book 4, does indeed belong strictly together with what can be called the proto-statistics of the limit theorem at its end. Evaluated arguments and counted events take place in one and the same world of applied probability.

The transition from the second to the third chapter, from the citation of Cicero's *Topica* to the necessity/contingency matrix that brings the game of binarism into play, is its epistemic breaking point. This move on Bernoulli's part opens up the history of probability theory. It is, however, also motivated by the cited topics with which Bernoulli breaks here. Without the citation of the ancient topics, which Bernoulli effectively cancels out in the moment he seems to rely on them, the game that becomes apparent in the example of the epistolary expectation would not have become the fundamental example of probability.

Because of this important nexus of ancient topics with the analysis of contingency, a brief outline of the basic formula of *probabilitas* that early modernism developed out of the ancient tradition becomes desirable and even necessary here. It is this basic formula of *probabilitas* that Bernoulli finally takes up in the fourth book of the *Ars conjectandi*. The sketch will take us as far back as the sixteenth century with the humanist scholar and logician, Rudolf Agricola, because his work laid the basis for early modern treatments of the topics and first inserted the logical figure of contingency into the discourse of probability.

Excursion in Topics: Epistemic and Juridical Elements in the *argumentum probabilitatis*

In his book *Topica universalis*, Wilhelm Schmidt-Biggemann argues that an entire type of scientific organization and a foundational connection between knowledge and literature changed around 1700 with the critique and rejection of early modern topics.[24] He does not tell a single-stranded story in which the topics suddenly came to an end, to be replaced by hermeneutics, modern historical thinking, and epistemology. Instead, there was already a tension inherent within the topics. By the end of the seventeenth century, it was not so much the case that the topics died out as it was that their intrinsic tension became manifest in the competition between the "ancient" topics of argumentation, and the "true" topics of contingent events. The competition between traditional topics and a mathematical theory of probability was far from being merely an external occurrence; it did not simply contrast a new, stricter methodology with the rhetorical-topical knowledge of old European *historiae*.[25] The complexity involved here is captured in the title of the German historian Arno Seifert's book *Cognitio historica. Die Geschichte als Namengeberin der frühneuzeitlichen Empirie* (*Cognitio historica:* How History Gave Empirical Knowledge its Name and Title).[26] "History" stands here for the early modern notion of *historiae,* the narrative knowledge of natural histories of all sorts in the Aristotelian tradition; and "empirical—or factual—knowledge" is the code name for the modern scientific project whose origin and nature in recent years have been traced in different ways by Lorraine Daston, Peter Galison, Mary Poovey, and others.[27] Seifert's book, however, reminds us that the place of the fact and the singular occurrence as an object of scientific knowledge was marked out, even if framed differently, by the Aristotelian *historiae*. The structure and presentation of *historiae,* however, were provided by the topics.

For the context of the history of science as well as the relation between the history of science and literature, it is equally important to verify the points of discursive intersection between both the topics and the analysis of contingent events. Arnauld's citation of the topics is precisely such a crossing point; another—not quite so visible but with deeper and more far-reaching consequences—intersection is the citation of this citation in Jacob Bernoulli, when he applies probability to arbitrary events of the social world, thereby enforcing the statistical extension of probability.[28] In the context of their respective interest in the topics, both Arnauld and Bernoulli aim for a concept of the probable in which the epistemic question of what approaches truth and the juridical question of what has been reliably testified intersect with each other. They thereby come across a noteworthy ambiguity in the traditional understanding of the topics.

We can roughly distinguish three models of classical topics, the third of which may be understood as a fusion of the first two: (1) the dialectical model, in which the justification of statements is made subject to the rules of disputation and probability emerges in relation to ontological principals; (2) the juridical model, in which trustworthiness is represented as a more or less regulated procedure between parties, so that probability must be described as a component of the trial and its institution; and, finally, (3) the psychological and epistemic model that entangles probability in an intermediary position or even in a certain indecisiveness between ontological and procedural references. This is the model that shaped the early modern era in the history of science as well as of literature. And this third and intermediary model is also the reference target for the citation of the topics in probability theory.

(1) For the sake of brevity, the first model can be represented by Aristotle's *Topica*. In Aristotle, the dialectical operation develops between the universality of philosophy and the specificity of the individual sciences. Dialectical discussion with the help of the topoi is useful in philosophy, says Aristotle, because "we shall more readily discern the true as well as the false in any subject." The principles or premises of individual sciences, meanwhile, can only be treated by means of universal topoi. "For if we reason from the starting-points appropriate to the science in question, it is impossible to make any statements about these (since these starting-points are the first of them all)."[29] Aristotle's dialectical topics clearly make up, on the one hand, a self-contained field—discussion based on principles leading to other principles—but on the other hand, it is also an intermediary field. The all-embracing nature of the topos expels it both from the logical coherency of philosophy and from the material particularity of individual sciences. For one thing, the universality of the topos reacts to an impossibility in the individual sciences. This is the impossibility of transcending the principles, the *prōta*. For another thing, however, it turns certain intrinsic difficulties in philosophy into an advantage for philosophical presentation. This is accomplished by shortening the discernment of truth and falsehood in cases when such discernment would take much too long by way of the philosophical principles. The virtue that leads back behind the principles of the individual sciences and helps us out of the difficulties in representing truth in philosophy is called sufficient reliability (*pistis*) and is generated by probabilities (*endoxa*): "Those things are *true and primary* which get their trustworthiness through themselves rather than through other things."[30]

The dialectical topics of Aristotle has its place between logical demonstration (*apodeixis*) and contentious deduction (*eristikos syllogismos*) and differs from both.[31] The *prōta*, or statements on first things that one comes across in the course of the apodictic proof, are true. The acceptable statements, which one introduces in an eristic debate, seem believable, but are not true. Going back to the first things in the elucidation of truth simultaneously sets the frame and the rule of the truth discourse: the search for the first things implies a scene and a regularity in which either true or false comes out. The victor of the eristic competition is placed within a scene in which only semblances dominate and no path leads into the openness of truth or falsehood. The semblance of eristic probability (the *phainestai* of the *endoxa*) is always and necessarily mere semblance, something that seems what it is not. The probability of dialectic, the *endoxa* of dialectical discussion, in contradistinction to apodictic speaking as well as to eristic competition, takes place in what can be described paradoxically as an open frame. Dialectic speech operates in a game whose rules are not fixed in advance but generate themselves while playing the game. Truth and semblance are at issue in this game; and with it, dialectic poses the question of the probability of the probable.

Dialectic is a game played with truth statements, which, however, includes additional features other than those of truth statements. The internal structure of the propositional statement and the performative frame of the game do not correspond to each other from the outset. A tension exists between the apodictic aspect of the topos as a first thing to which to appeal and the eristic, all-embracing, incursion-allowing aspect of topoi that do not refer to an already presupposed domain or system. For both apodictics and eristic, the structural analysis can easily be supplemented by imagining the respective scenes of their performance, without any new points of view resulting from this—the scene of discussion in both cases is equally predetermined by the demand for complete truth (apodictics) as it is by the precondition of ruled-out truth (eristic). Only the intermediary area of the *endoxa* makes the tension palpable between pure statements and the staged arrangement of making them, between the propositional construction and the performative character of statements.

It is nevertheless true that we find hardly any information about the kind of dialectical scene or the institutional nature of the debate in Aristotle's discussion of the dialectical topics.[32] Determining the nature of the topos[33] remains entirely restricted to the linguistic and logical relation between the proposition (*problema*, problem) and the performative construction (*protasis*, proposition).[34] We might call this the minimal institutional staging of the debate: the *problema* is the question posed to the adversary; while the *protasis* is the first performative move of the one who has to defend or contradict the problem.[35]

(2) Cicero translates the Aristotelian topics—which he cites like an esoteric and dark background in his own *Topica*—into the manifest explicitness of the rhetorical and juridical scene, as if it were a matter of mere self-evidence. Cicero defines the topics from the point of view of the law and the kind of psychology that comes with performing the roles of accuser, defendant, or judge under the law. His guiding concept, which corresponds structurally to Aristotle's *endoxa,* is termed *fidem facere*: to provide something with the force of belief and acceptance; to validate something. *Loci*—according to the definition at the beginning of Cicero's *Topica*—are needed for arguments, which should "lend belief [*fidem facere*] to a doubtful issue."[36] Cicero characterizes the implied doubt hardly anywhere in theoretical or epistemic terms. Neither is it determined by its relation to truth, or to any constitutive grounding for this or that science. Cicero is concerned with the doubt of the judges and the audience at court—a remote echo of Greek sophistry as we know it from Plato's interpretation. This becomes particularly clear when Cicero takes up the concept of *fidem facere* in the context of *testimonia*. Providing belief and trust relies here on the factual evidence of documentary proofs, attestations, and witness statements.[37]

Validation and reliability are fundamentally linked for Cicero with the juridical concept of testimony and hence of *auctoritas*. The same context of *fidem facere* in the end also characterizes its use and meaning for the topoi. We can grasp this linkage in particular when it comes to the rhetoric of narration, the representation of the facts of a case or the course of an event: "Proof [*fides*] follows on narration [*narratio*], and because proof is primarily brought about by persuasion, it has been said in the books which deal with the whole theory of public speaking which Places [*loci*] have the strongest force for the purpose of persuasion."[38]

This remarkable passage shows why topical probability in the Ciceronian tradition revolves around the *narratio*. In the *narratio verisimilis,* the probable narration, Cicero brings together *persuasio,* the affective influence, and *fides,* the reliability of the statement; and they relate to each other as the pragmatic and the referential aspects of the same thing. In working over the plot of his story by means of the circumstances of an event, the speaker as narrator turns into a *witness,* as it were, in the second degree, someone who demonstrates the reliability of his report through his own *gravitas*. The *persuasio* of the narrator/witness is his *fides* made psychologically effective in the frame of the trial. The *fides* of his narration/statement relies in turn on *persuasio*. In the *probable narration* and through its means, *persuasio* and *fides* presuppose and stage each other.[39]

Two sorts of topical probabilities can thus be said to exist in the ancient tradition. Each of them is oriented around its own different framing, yet at the same time each contains the other according to its respective logic of constitution. On the one hand, there is the probability of Aristotle, the *endoxa*. This takes its departure from knowledge and its representability in philosophy and science, implying a performative space of disputation but never describing its implementation in fuller detail. Decision and conclusion, performance and propositional logicality merge indissolubly in the use of this topos. Cicero, on the other hand, openly exhibits the scene of the courtroom trial. The meaning of the probable—of the *fidem facere*—in his *Topica* is tailored after the pattern of *probable narration*. Narrative verisimilitude in turn connects the poetics of storytelling with witness testimony. An institutional and, in particular, a juridical relationship between ignorance or doubt—on the part of the judge—and procedures of confirmation and assurance molds the construction and the pragmatics of probability. In doing so, this relationship shapes what might be called the inner-institutional psychology of probability.

(3) Boethius passed down the topics and its theory of probability to the Middle Ages and early modernity. He translated Aristotle's *endoxa* with *probabilia*[40] and inserted Cicero's conception of trustworthiness, *fidem facere,* into this probability concept.[41] This connection of the Aristotelian disputational probability with juridical probability results—for the first time in the history of logic and epistemology—in a psychological setting for the primarily logical issue of probability.[42]

The novelty of Boethius's conception of probability initially emerges in his determination of the *argumentum*. First of all, the argument is *ratio*, a logical procedure. With the development of this concept, Boethius brings in a healthy share of Aristotelian thinking, though the *argumentum* nevertheless remains insufficiently determined. An argument, Boethius adds, in point of fact, is the kind of a *ratio* that answers a doubt preceding it. Argumentation for Boethius, to be precise, means "validating" or "proving" something in the face of doubt. He uses *probare* in this context simultaneously with Cicero's expression *fidem facere*. The doubt and the validating attempt at proof reveal the Ciceronian and juridical descent. On the other hand, Boethius sharpens the derivation of the topoi to syllogistics, that is, to the logic of conclusion, even beyond Aristotle. While Aristotle's topoi had their specific place in difficulties of foundation and conclusion between philosophy and science, Boethius evinces no such consideration regarding the construction of philosophical or scientific discourse. Instead, doubt and attempts at proof make their entrance as derived but also already removed from the juridical context. With Boethius, doubt and validation become a pair of gestures that shape the argumentation in a psychological manner. The juridical conception of testimony penetrates, as it were, through the Aristotelian field of syllogistic argumentation. The result is an ad hominem conception of the argumentation—an argument related to or even leveled at a person.

With this turn, the topics, the forging and use of probable reasons, have become the quintessence of scientific logic.[43] Like every rational argumentation, probable arguments can be represented in a syllogism. But the effective connection between rational argumentation and the space of doubt and trustworthiness, according to Boethius, can only be found in a much more specific linkage. A detailed reconstruction of the syllogism as a process that should produce a conclusive proposition of subject and predicate leads to this connection. Expressed in syllogistic structure, we are concerned with finding the middle term that can be connected with the subject in the predicate position and the predicate in the subject position.[44] Where do we find this middle term, however, and how can subject and predicate be combined such that formal as well as material probability is satisfied?[45] According to Boethius, the topos is the place where one finds a middle term suited to produce probability in the sense of trustworthiness (*fides*).[46] Boethius defines the topos in this specific sense as *sententia maxima*; this *sententia* is universal and self-evident (*universalis et per se cognita*) and brings a final end to resorting to ever further supporting arguments (*neque indigens probatione*).[47] Boethius provides an example: the proposition to be proven states that the envious man is not wise. The syllogistic chain runs thus: an envious man is one who is grieved by others' good fortune. But a wise man is not grieved by others' good fortune. Therefore an envious man is not wise.[48] The formal conclusion still leaves open—or so Boethius assumes—how to obtain true trustworthiness, *fides*, by it. It therefore only represents the argumentative side of the topics. The formally valid conclusion, however, can also be interrogated about which maxim or which universal proposition it relies on. The maxim and its universality represent as it were the juristic side of the topics. Both sides together, as Boethius claims, first constitute the question about the *fides* of the conclusion. The *sententia maxima*—the wise man banishes all envy—provides the grounds or the principle of the conclusion in view of the middle term brought into the play of argumentation.

The psychological and epistemological orientation of topical probability can be seen as a combination of ontological and juridical probability. In modernity it has expanded ever more in the direction of measurement and methodological confirmation of what later would be called the empirical sciences. The Rhenish humanist Rudolf Agricola holds a key position in this story. Many of the important reformers of rhetoric and science in the sixteenth century, such as Erasmus, Melanchthon, and Pierre de La Ramée (Petrus Ramus), take their departure in one way or another from his rhetorical-topical work *De inventione dialectica* (On Dialectical Invention) (1479). The method of argumentation—in this point Agricola does not differ from Boethius—is that of the probable speech.[49] For Agricola, the concepts of what is subject to disputation and of the self-evident statement that brings disputation to a halt enter, however, into a delicate situation of mutual dependence and support: "nothing subject to doubt can be proven from itself [*per se probari*]—which I think is too obvious to require any explanation: no less obvious than the fact that if darkness is bound to give access to our eyes and sight, it requires a light from somewhere."[50]

The passage, which amounts merely to a stage direction in *De inventione dialectica*, is emblematic for the communicative paradox of probability in the sense of trustworthiness. We can observe a regress of levels of argumentation: the difference between what causes doubt and what explains itself from itself is, in its own turn, something that explains itself from itself, in contradistinction to what remains in doubt. In other words, it can only be explained by resorting to the selfsame difference for support in a spiraling infinite regress. This situation, however, follows precisely from retracting the disputational and ontological foundation, on the one hand, and the juridical frame of topical probability, on the other. "Thus there only remains the possibility that everything that must be supported argumentatively [*confirmandum est*] draws its trustworthiness [*fidem*] from something else."[51] In summary, the resolution of the difference between *confirmatio* and *fides*, between logical and juridical probability, appears as the systematic tendency of the humanist scholar's work. Such a resolution would finally make it possible to transfer the probabilities of either logical disputation or institutional psychology, the probabilities of an either implicit or explicit scene of performance, into the one probability that refers monologically to the "historical," that is, the empirical things and events of the world.

Wilhelm Schmidt-Biggemann assumes this transition to have been complete in early modernity. For him, the term *topica universalis*—universal topics—means precisely the organization of *historiae* that deal with singularities in nature and history. The theatrical and performative implications of probability disappear behind the *historiae* and their representational accomplishment.[52] Agricola indeed completes the turn of probability to a process of representation in the second chapter of his first book: "Thus the truth we have shown will not be able to hide from anyone who reads what we have said about invention closely: everything that can be said for or against everything coheres and is bound together with it, so to speak, by a certain community."[53]

The coherence (*cohaerere*) that the use of the topoi produces in narration and the relation (*cognatio*) of the topical circumstances with the thing or event themselves are treated by Agricola as if they were one and the same. More precisely, the two forms of congruence—the logical coherency between topically ascertainable statements and the ontological relation between the circumstances and the facts of a matter—presuppose a third congruence. This third and encompassing form of congruence concerns the very relation between certain areas of topical circumstances and the facts of the matter. It operates beyond the divide of narration and ontology. With this final consequence, Agricola and the tradition that follows him lead us even further out of the realm of ancient topics and into the notion of a "narrated world," a notion that Bernoulli picks up in his citation. Agricola in any case does not yet solve or even address the problem that opens up here for every theory of knowledge. He leaves the matter by declaring:

Aristotle may well contend that all true things are in harmony with each other, and that there can be no truths that contradict each other—logical consistency [*consentire ipsa*] is one thing; the support of trustworthiness [*fidem astruere*] is another. Therefore in order that something can be employed as a confirmation of something else, it must have some connection of a certain kind to the matter in the proof of which it is employed, and must be related in some way to it.[54]

(146) A study of the topical coherence of things among each other (*coniuncta*, by the way, make up their own topos in Agricola)[55] thus amounts to an analysis of how topoi relate to things and events. "According to number," Agricola summarizes, "things are . . . immeasurable, and equally immeasurable are their idiosyncrasies and differences. . . . Nevertheless, a certain outer form clings . . . to them all, and all of them have a tendency to a similarity in nature: they all have their determined substance; all result from some cause; all have some effect."[56] Topoi open up the relevant surroundings for the facts of each case, and they do so half logically, half juridically.[57]

The preceding brief outline should bring into relief that segment of the history of the topics to which Arnauld and through him Jacob Bernoulli refer in their citation of the topics. Arnauld and Bernoulli, as we have seen, take up Cicero's division of *inventio* and *iudicium*. By doing so, they make the relation between the narrative topoi of the circumstances of an event and the external topoi of its attestation their point of entry for the new contingency logic of probability. They introduce the theory of the game into the topics exactly at that place where the fusion of the Aristotelian with the Ciceronian, the epistemic with the juridical-rhetorical probability had also already taken place. At this point, the transformation of the topics into the early modern order of an empirical ("historical") knowledge became possible.

Contingency and *topos a contingentibus*

The early modern development of the topics into an analysis and reconstruction of the "narrated world," as it becomes manifest in Agricola, provides the background for Arnauld's and, especially, Jacob Bernoulli's citation of the topics. In the fourth book of *Ars conjectandi*, Bernoulli takes up the theme with the division and the parallel he assumes between the calculus of argumentation and the development of statistical models. He thus accomplishes the calculization for the two topical domains of dramatic argumentative probability, on the one hand, and referential probability, on the other, the two facets of probability that had been fused together since Boethius. He then implements in detail the transition from the topics' citation to the introduction of the concept of contingent events, as modeled by Arnauld. A telling piece in this context is the reference to a *locus* of contingency that occurs within Bernoulli's general reinterpretation of *loci* in their entirety as logic of contingency: the case Bernoulli adduces in the second chapter for explaining his take on the topoi refers to the death of a certain Titius, the discovery of a bloody dagger, and the blushing and blanching of Maevius.[58] This example refers us once more back to Cicero. In his *Topica*, Cicero had dealt with a criminal offense only once, taking up the instance of a murder. The *topos* at issue here is called the 'concomitants' (*adiuncta*)— the matters connected to the event. Alongside such preceding events as the arrangements for the symposium at which the murder takes place, Cicero also mentions things following the event, such as the possible culprit's blushing and turning pale and the bloody sword discovered near the crime scene. Cicero ascribes a meaning specific to criminology, as opposed to logic, to this topos of the *adiuncta*.[59] For the same reason, nonetheless, he also calls it the rhetorical topos par excellence, since this topos is applied when the lawyer deals with a case according to the *quaestio facti*, that is, the question of whether a certain event has happened or not.[60] This question of the adequate legal viewpoint, however, is called the *status conjecturalis* in the language of juridical topics. The topical tradition that resurfaces in this passage of Bernoulli's fourth book has clearly coined the title of his entire enterprise, the *Ars conjectandi*.

Cicero's topos of the *adiuncta* has a rich and long tradition of commentary. Important for us here is the fact that Rudolf Agricola, in a very exhaustive and emphatic section of the *Dialectic*, renames it the *topos de contingentibus*. Agricola justifies his terminological choice by opposing the new *contingentia* with a topos called *eventa*, events. Viewed according to the *locus* of *eventa*, Agricola explains, events are assigned a cause or even an intention.[61] In contradistinction to *eventa*, the *contingentia* concern relations in space and time without any causal claims. Agricola calls these—typically narrative—connections *contingentia* because "they can continue to exist with a given thing or without it."[62] Or, with a clearer reference to the temporal and spatial spheres of narration, *contingentia* for Agricola may "take place [*contingere*] in the sphere of a certain thing to the extent that this thing can exist even if what has been designated as contingent does not take place, and vice versa: that, even if the thing does not exist, the thing thus designated can take place."[63] Accordingly, contingency is at issue in the intermediary space between something that exists and something that takes place. In this sense, Agricola also includes the probable (*eikota*) and the sign or indication (*signa*) in the topos of contingency: a broad chest thus indicates bravery (physiognomic sign); a person without means who wishes however to live in luxury steals (criminalistic indication); a woman who walks through the streets alone at night is on her way to a lover (moralistic judgment).[64] Thus, even the transition from the citation of ancient topics to the modern logic of contingent events turns out, again, to be a quoted topos: a citation of the *topos de contingentibus*. The terminologies of conjecture, of the probable, of signs, and of contingency are all modeled according to the topics. As great a mathematical achievement as calculating the outcome of games might have been around 1660, it did not gain the efficacy of probability buoyed by the optimism of boundless application until it cited the old probability of the topics.[65]

Ars conjectandi 4.1–3: The Calculus of Argumentation, or, When Does the World Correspond to What the Case Is?

How does Bernoulli develop the application of the calculation of contingency in the trace of early modern topics? How, in other words, does he formulate such an application of probability theory for singular facts and things in the world, and, in particular, for the situation of reports from the world reported to an outside observer?

Regarding the letter that does not arrive, whose telecommunicative situation allows the difference between event and report to be observed and their respective values to be calculated with each other, Bernoulli has this to say:

> The above makes it clear that the force of proof peculiar to some argument depends on the multitude of cases in which it can exist or not exist, provide evidence or not. . . . Therefore, the degree of certainty, or the probability engendered by this argument, can be deduced by considering these cases in accordance with the doctrine given in Part 1 [of this book] in exactly the same way as the fate of gamblers in games of chance is usually investigated.[66]

With the reductive conversion of the arguments into yes/no alternatives from the perspective of the absent observer, Bernoulli accomplishes the essential application of gaming theory. Compared to this application, all the primary exemplifications of games of chance in the texts of Pascal and Huygens will fade away. Arnauld had only reconciled the spheres of the sensitive princess and of the game of chance with the appeal to overcome fear rationally, but he had not depicted the literal and even literary language in numerical representation. Bernoulli takes the demand or reconciliation literally. He approaches the singularities of Aristotelian *historiae*, which are organized in a topical manner, with the contingency of events that take place or do not take place. Bernoulli calls this the task of *moderatio*: "I suppose that all cases are equally possible and can take place with the same facility. Otherwise they ought to be moderated by assuming, instead of each easier occurring case, as many others as that case is easier to happen. For example, instead of a thrice easier case I will count three such cases that can occur as easily as the others."[67]

The *moderatio* is the fundamental methodical gesture of probability theory in the phase of its application.[68] The singular facts and events whose probability we want to calculate have to be mustered for conditions that can exist or not exist with equal ease, or for events whose occurrence or nonoccurrence is equally probable. Since the binarism of an equally strong yes or no is not naturally evident in all singularities, the coin-toss situation must be introduced by method. Bernoulli combs through the singular facts and events for those whose presence or occurrence befits the binarism of an equally strong affirmation or denial. Out of these binary facts, he then tailors the elements for a probabilistic measurement of the occurrence or event in question. This operation is called *moderatio*: a process of modulation or tempering; of restraint or fitting. *Moderatio* is the introduction of a measure where there is none. The sense of discipline and normativity implied in *moderatio* can be said to be the last trace of institutional logic in probability. In fact, *moderatio* is not restricted to those unruly cases where it is difficult to uncover elements of facts or events measurable as contingencies. The *moderatio* rather occurs in every case where probability is applied to historical singularities. According to the Aristotelian concept of *historiae*, whatever exists or occurs in a singular way (*to hekaston*), is essentially invisible under the aspect of contingency. Singularity requires "moderation" in order to be measured as contingency.

On the basis of this *moderatio*, Bernoulli initiates two analyses and with them two modes of measuring the *argumentum* in a game of probability. These two analyses allow us to study to what extent the *Ars conjectandi* carries out an interpretation of the topics in terms of the game of chance.

The first analysis has already been mentioned. In this context, Bernoulli discusses the opposition of necessary to contingent on two levels: first in view of the existence of certain facts of the matter ($argumentum_e$); second in terms of the indicative force ($argumentum_i$) for the facts of the matter up for assessment. If, for example, I am waiting for a letter from my brother, I may adduce the following *argumenta* for the absence of the letter:

(a1) The brother is dead; he has passed away in some far-off place during the period of our separation. Since this argument is a supposition of factuality in the moment in which I assume it, it is either existent or not, that is, it exists contingently. But it is necessary regarding its indicative force: dead men do not send letters. (a2) The brother is lazy. Considered as a character trait independent of time, in the moment I assume it in my conclusion, this argument is necessarily existent. But it indicates the absence of the letter only contingently: laziness can but need not lead to someone's failure to write. (a3) The brother is busy. This *argumentum* exists contingently in the moment when I employ it, because I cannot know my brother's present workload. It is also contingent in its indicative force, because the brother's amount of work can but need not prevent him from writing. The fourth possible case, that of double necessity, is omitted because it is trivial. Necessity, or full certainty, has the value 1. The only interesting cases are those in which the *vis probationis* of the *argumentum* is a fraction due to the contingency of existence, indication, or both parameters.

To determine the probability of the *argumentum* Bernoulli uses the rule for interdependent probabilities expressed by the product of both partial probabilities: P (*argumentum$_e$*) × P (*argumentum$_i$*).

Bernoulli constructs the tableau of possible cases carefully: the theorem of chances is simply inserted in *a*1 and *a*2, because in these cases respectively one part of the *argumentum* has the value 1 (for full certainty). We do not see the probability of the *argumentum* expressed as a product until *a*3, the case of double contingency. For (*a*1), the quotient (a / b) or (a / b) × 1 results[69] as the probability of contingent existence (and necessary indicative effect) of an argument (with b = the sum of the cases in which the argument is existent, and a = the number of cases in which the *argumentum* indicates for or against the matter). For (*a*2), the quotient (α / β) or (α / β) × 1 results, which represents the probability of (necessary existence and) contingent indicative effect (where β = the sum of all cases in which the indicative argument is correct or not, and α = the number of cases in which the argument as an indication is positive). In case *a*3 for the degree of certainty for contingently existent and contingently indicative arguments, according to the rule of the combination of interdependent probabilities, Bernoulli gives the product of the fractions (a × α) / (b × β).[70]

This result represents an adequate algebraic notation for the formation of the topics that had been possible ever since Boethius. It had already been suggestively conceivable with Agricola, and became, as it were, unavoidable after the closing chapters of Arnauld and Nicole's *La logique, ou l'art de penser*. Expressed in traditional terms, Bernoulli's *argumenta* of existence are located on the plane of making reports and testimony: Is the brother dead or not? Is the brother a lazy person or not? The *argumenta* of indication, on the other hand, are located on the plane of narrative circumstances, that is, of relations that the narration claims to be spatial and temporal factors of an occurrence in the world—the topoi in the narrower sense. In the decision for the multiplication, Bernoulli formulates the dependence of both *argumenta* on each other: P (*argumentum$_e$*, *argumentum$_i$*) = P (*argumentum$_{e,i}$*). The difference is thus honored between the artificial *argumenta*, the topoi proper (particularly those of the *narratio*), and the nonartificial *argumenta*, the *auctoritas* and other factors of the *testimonium*. At the same time, in this scenario, both sides are treated as factors of the same argumentative space, and they can be calculated in their relation to one another. In modern terms,[71] Bernoulli's first analytical step merges intra- and inter-propositional relations. He transforms them into complex inter-propositional relations that as such can be calculated. In indicative arguments, the relation between two facts is at issue. Both facts can be expressed in propositional form. Their relation can therefore be constructed as inter-propositional, that is, as a relation between statements. The arguments of existence, in contradistinction, concern relations within a statement; in other words, they are intra-propositional. The relations between a subject and a predicate present an instance of this. In the example of the missing letter, however, the arguments are also simultaneously placed in a common space with the inter-propositional relations. Every indicative argument receives an existence indicator in the complex formula P (*argumentum$_{1,2}$*). This move ensures their common legibility as propositions that stand in relation to other propositions.

Thus, in any case, the analysis begins; but Bernoulli takes a second step. The first part of the analysis remains valid in most modern interpretations of Bernoulli's fourth book, and scholars accept it as unproblematic, though perhaps somewhat eccentric in its manner of expression. The second part, in contrast, has received quite exhaustive critical commentary. The difficulty for the modern reader here can be dubbed the return of the relation between the mathematical structure of the calculation of chances and its interpretation through the probability of the topics. The tension within modern topical probability makes itself noticeable again in this relation: the tension, namely, between the probability of the logical proposition and the *fides*, which the performative space of the courtroom is meant to bring about.

Bernoulli introduces the second part of his analysis as a mere appendix to the first. He speaks about how the degree of certainty of arguments can appear, on the one hand, as a so-called pure case: this happens, if, for example, an *argumentum* with regard to a given matter indicates either this matter of fact or else nothing at all. Or the probability of the argument appears, on the other hand, as a so-called mixed case: if, for instance, an argument with regard to a given matter of fact can indicate either this or that condition of its existence. In the pure case, the *argumentum* exhausts its indicative power in the designation of the matter. If it does not indicate it, then it remains unused and unconsumed. The purity of the pure case is just like its exclusive relation to the one fact that the *argumentum* designates. The mixed case attaches the indicative power of the *argumentum* to two things; if it does not display the one thing that is searched for, then it shows another. In the example Bernoulli's provides, Titius's murder must be explained. Gracchus is one suspect. Gracchus turns red when questioned. His blushing—the old example of the *topos a contingentibus*—is a pure argument. For if we can trace the blushing back to his consciousness of the deed, then it is an indicator of his guilt. If turning red has any other cause, then it is no argument at all in terms of the question of who committed the murder. The probability of the pure argument is hence measured according to the ratio of cases of blushing from consciousness of guilt to the physiological causes for blushing when interrogated. Bernoulli meanwhile gives another example for mixed argument probability: it is known that Titius's murderer wore a black robe. Four men were seen wearing black tunics at the site of the crime; one of them was Gracchus. The black robe shows either Gracchus's guilt or his innocence, which is then a complement to the guilt of one of the three other men wearing black cloaks.

The differentiation between pure and mixed cases can be represented in algebraic terms as follows: in the pure case for the degree of certainty of the given event we may write x/y (where x = the positive cases; z the negative cases; and $y = x + z$). The same relation holds true for the mixed case. In the mixed case, however, the reverse of this formulation is also true, in contrast to the pure one. In mixed cases, as Bernoulli explains, the degree of certainty for the opposite comes to z/y. What is true for the mixed case never happens in the pure: in the mixed case, a rule applies that will later be termed the addition law of probability, meaning that the sum of the probability for the hypothesis's correctness and for its incorrectness is always equal to one: $x/y + z/y = 1$, because $x + z = y$. The difference between the pure and the mixed cases of probability thus proves to be the exact point where argumentation theory enters the realm of algebra. The pure case—which precisely herein proves its purity—is located as it were only halfway in the algebraic order. Indeed, Jacob Bernoulli does not shrink from advancing emphatically to the consequence that, in specific cases of their combination, pure probabilities can allow the value of both the positive and the negative probability to become larger or smaller than ½. According to Bernoulli's definition, this consequence means that the pro and the con suppositions in the same case can both be mathematically probable—even if to different degrees.[72] In other words, Bernoulli's idea according to which the addition law of probabilities applies in some cases and not in others, allows formulas to arise in certain combinations that can be seen later as nonsensical, that is, as formulas that push the mathematical representation itself *ad absurdum*.[73]

What then is at stake with this distinction for Bernoulli? To begin with, the examples of turning red under interrogation and of the suspicious black cloak make it just as clear as the mathematical treatment that Bernoulli is deploying the theme of pure and mixed probabilities in the context of his first step, the context of indicative and existence arguments.[74] We may even ask whether the two distinctions overlap with each other. Reformulating certain examples for the distinction between pure and mixed cases with the help of the distinction between arguments using existence or indication is indeed thinkable. For example, we may start out from a type of event such as *blushing from consciousness of guilt* instead of from the type *blushing*. The pure case of blushing in the interrogation corresponds—we might argue—to one necessary indicative effect (guilt-induced blushing points to the deed) in contingent existence (the contingency of the presence of guilt-induced blushing among all possible other cases of blushing). The mixed case of the indication of the cloak would then correspond to a contingent indicative effect (the circumstantial evidence can point toward the deed or not) in necessary existence (the presence of the circumstantial evidence is confirmed or not).

But even if we agreed that the two types of distinctions—argument of existence versus argument of indication and pure versus mixed probability—articulate the same material complexes, they still do so in different ways. Ian Hacking, for one, cites an overlap and a confusion of the type distinctions.[75] Be that as it may, it is certainly true that the distinction between the pure and the mixed cases repeats the distinction between existence and indicative efficacy in a different way.

The second part of Bernoulli's analysis, which interprets the chance theorem one more time, introduces a new dimension of framing probability. The opposition of pure versus mixed cases brings an additional and stronger determination into play as compared with the first opposition of existence versus indication. In the mixed case, we are told, Gracchus's black cloak indicates his culpability in a ratio of 1 to 4 when it is known that the culprit was clad in black, and that besides Gracchus, three and only three of the people present at the crime scene wore black tunics. In this second version of the *argumenta*, their indicative power or *vis probationis* is bound more prominently than in the first analysis to a hypothesis of what took place. Taken as events for themselves, such things as the donning of black cloaks and the number of people who wear them have very little to do with who is a murderer and who is not. The construction of a story—a hypothesis with the formal skeleton of a chain of proof—must be asserted first. Then the question comes up about whether and to what "degree of certainty" the *argumentum* can take on the position assigned to it by this construction. The fact that Gracchus is wearing a black cloak is now integrated into the narrative construct, whether it indicates Gracchus as the murderer or not. The negative case for Gracchus is simultaneously a positive clue regarding one of the three other cloak wearers. The same is not true for turning red. Only blushing from a consciousness of guilt would be valid in the construction of the murder case. If Gracchus turns red because he is sick or annoyed, then it has to do with incidents in another world, in a world outside of the final relation to the criminal story. In short, the second part of the analysis contains first the specific criminological and juridical plane of analysis. The pure case is determined by the fact that the negative cases have no defined place in the world of the case history. Neither the sickness nor annoyance of Gracchus come up in it, nor do they testify to cases in which he does not blush from a consciousness of guilt, as for instance his innocence. If Bernoulli nevertheless allows and calculates this *argumentum*, then this decision reveals the topical pragmatism of the juridical process of proof: we know and reckon with the fact that not all *argumenta* can be integrated perfectly in the world of the case.

Or vice versa, we know that one must fall back on *argumenta* that one cannot integrate fully. And this situation is now precisely the pure case. It is the case that shows in purity how the world and its reconstruction by the hypothesis of the case history relate to each other. The malfunction occurs in the pure case when we start out from the first world, the world of immediate *argumenta* determinable by their indication, and hold the world of the case history separate. In the pure case, the possibility remains that the *argumentum* plays out exclusively in the first world.

This last consequence, however, means nothing less than that the "existence" of the *argumentum*, its nonartificial topical power, cannot be fully dealt with in the same calculation as the artificial topical power of indication or circumstance. The observer is influenced and fascinated by the occurrence of the *argumentum* without being able to integrate it seamlessly into his *narratio verisimilis*. The corresponding case can also be construed, where one ascertains a circumstance—Gracchus is wearing a black cloak, and we know that the murderer was wearing a black cloak— without being in a position to procure sufficient data about the "existence" of the *argumentum*—for example, about how many relevant people were wearing black cloaks. Obviously, only criminal psychology and investigation can bring us any further at this point. We must strive to translate the world into the world that is the case.

Such considerations certainly did not occur to Jacob Bernoulli. Quite the contrary, with the distinction between the pure and mixed cases, he makes himself at home in a provisional world in between. In a first order of things in this world, we may be able to translate the world of our common experience—the social and physical world—into a closed cosmos of calculable data, but we still cannot further integrate this ensemble of data into the world of the case, of its construction and its hypotheses. In this situation, Bernoulli again accepts the irrevocability of the difference between arguments' existence and indication, between *testimonium* and *circumstantiae*, the quality of testimonies given and the testified stories. The tension between the contingency calculation and its interpretation through topical probability returns behind his back—invisible but structurally irrefutable. Bernoulli's conception of probability does not lead the interpretation of the mathematical theory of chance as probability theory quite to its end. It remains stuck in the formalization of argument calculation in between a complete calculability of the world that is the case and a pragmatic calculation that partially violates the mathematical laws and that holds fast to overlappings between the world of the old topical probability and a rigorously formalized world of calculable contingencies. We may see in this tension the problem that prevented Bernoulli from completing the *Ars conjectandi*. As far as Bernoulli's motives are concerned, we cannot progress beyond pure speculation. We can, however, conclude that although his move to interpret the calculation of contingency as the measure of probability indeed made probability theory possible, calculation and probability still did not simply coincide in his account of things.

6

Continued Proclamations

THE LAW OF *LOGICA PROBABILIUM*—LEIBNIZ

Taking Distance from Bernoulli's Calculus of Argumentation: Probability and Game

Leibniz only left behind incomplete beginnings and fragments about probability theory. His disagreement with Huygens's algebraic theorem of chance remained fragmentary; the quantifying theory of if-then combinations in conditional contracts was nothing more than a hypothesis.[1] Nevertheless, the question of probability belongs among the central and recurring themes in Leibniz's work. The list of fields in which probability plays a key role for Leibniz is long and varied; we shall only deal with the most important here. Like Fermat, Leibniz assigned probability to *ars combinatoria*, the foundational discipline in mathematical logics. The comprehensive projects of medical and economical statistics, with their relation to Leibniz's academy programs, comprised a part of applied probability,[2] while theology and ontology revolved around the question of cosmological contingency.[3] The chronological leaps in Leibniz's involvement with probability theory throughout the constitutional process of its early history are also striking: the work on conditional law started quite early, in the mid-1660s, immediately following Arnauld's inscription of the game of chance into logic; the closing notes stem from the 1670s, when Leibniz was able to gain insight into the whole Parisian discussion of the 1660s. On the other hand, Leibniz still belonged to the group of scholars who were involved with—both delaying and hurrying along—the posthumous publication of the *Ars conjectandi*.

Probability assumes a precise function in the various moments in which Leibniz's multifarious thinking verges on crystallizing into a system. The philosopher's debates with the greats of English science and philosophy are of particular importance in this process. In a 1697 letter, Leibniz makes an illuminating remark in this context to his Scottish correspondent Thomas Burnett de Kemney, who was charged not only with imparting the general scientific and literary developments in England to him, but above all with the task of arranging contacts with Locke and Newton:

Philosophy has two parts: theory and practice. Theoretical philosophy is based on true analysis, of which mathematicians have provided preliminary samples. It can also be applied to metaphysics and natural theology, however. . . . Practical philosophy in contrast is based on the true Topics or dialectics [*la veritable Topique ou Dialectique*], that is, on the art of measuring the strength of arguments [*probations*]. This art cannot yet be found among logicians. Only legal scholars have provided samples that should not be dismissed outright and that can serve as starting points to establish a science of proofs [*preuves*] with the help of which historical facts can be verified and the meaning of texts can be determined.[4]

These phrases anticipate the better-known passage on probability that we find in the *New Essays on Human Understanding*. By proclaiming the measurement of proofs as the *true* or *veritable* topics, they make the structural underpinning for the discursive fabric of Jacob Bernoulli's *Ars conjectandi* explicit. The old Aristotelian-Ciceronian topics (dialectics), as Leibniz intimates, were no logic at all. Their replacement, or rather the first realization of a true and veritable logic, can be read as Leibniz's desired scenario for Bernoulli's as yet unpublished *Ars conjectandi*..

Legal rules of evidence, however, play a different role for Leibniz than they do with Bernoulli and his followers. For Leibniz, evaluating proofs implies more of an institutional pragmatics than the actual calculation of arguments in an *ars conjectandi*.[5] In fact, Leibniz never speaks of applying gaming theory in civil, moral, and economic matters as Bernoulli does in his *Ars conjectandi*, where the rhetorical-legal proof in the *topos de contingentibus* takes pride of place. Instead, Leibniz introduces the juridical doctrine of proof, whose French name *probation* serves to announce the term "probability" in his letter to Burnett, as an intermediary between the old, false topics and their true, veritable form. According to Leibniz, the jurists' rules of evidence do not wait around to be discovered as a case for applied gaming theory and, thereby, of probability. Instead, they are already the first level of a kind of calculation in themselves. Probability is characterized by an activity that, though it is only metaphorically calculative, receives its preconscious measurement through the framework of the legal institution and the psychology inscribed in it. In Leibniz's world, calculation, even if in a rudimentary and psychological manner, has already begun in the probative process before it can and must be expressed in numbers. In his correspondence with the author of the art of conjecture, Leibniz touches upon this matter only six years later. In the postscript to his April 1703 letter to Bernoulli, he writes: "As I hear, you have extensively worked out the theory of measuring probability [*doctrina de aestimandis probabilitatibus*] (the importance of which I consider to be extremely great). It would be desirable if someone could treat the various ways of playing mathematically (whereby beautiful samples for this theory can be found)."[6]

Almost as if in an unrelated aside, the second sentence tags on a request that describes precisely the foundational plan of an *ars conjectandi*: taking up the game in order to find an example for probability's mathematical treatment defines Bernoulli's entire enterprise. Keeping in mind that we cannot know how much Leibniz actually knew at this point about Bernoulli's theory, two readings are possible. Either Leibniz has made a good guess about Bernoulli's plan for the *Ars conjectandi*, and is securing for himself the distinction of having already come up with the connection between probability and gaming theory without Bernoulli; or, on the contrary, he is showing that though he sees such a work as a "desirable" employment "not unworthy" of a great mathematician, he considers it only accessory to a doctrine of measuring probability, a doctrine whose true principle lies elsewhere.

The postscript is located in a noteworthy place in the letter. Leibniz had originally drafted a different postscript, in which he justifies himself to Jacob Bernoulli for a remark that had obviously not been circumspect enough regarding the latter's scholarly feud with his brother Johann. Leibniz subsequently moved this remark into the body of the actual letter, but not without first crossing out a long connected passage.[7] In this deleted section Leibniz relates how he himself had come to Paris as a "self-taught but thoroughly well-versed geometrician." The strapping young hero learns the ropes with hints from Christiaan Huygens, though all the while he works independently on new problems. Finally, as the tale concludes, he develops the general theory of curves, whose core principles were confirmed by Huygens. In short, Leibniz's famous outline of his *vita* as a mathematician originally would have occupied the place of the passage that then in fact was inquiring into Bernoulli's measurement of probability.[8] This tale of scholars continuously harps on questions of authorial priority. In it, Leibniz portrays himself essentially as an autodidact: entirely on his own, he rediscovers propositions developed earlier by Pascal or Descartes. Then he presents the design of his general theory of curves to Christiaan Huygens, who, having formulated them earlier himself, is certainly very surprised. Leibniz even admits that Huygens had initially inspired him to his investigations. The portrayal of his relation to Huygens is particularly significant. It seems to insist upon Huygens's priority as a biographical fact, but nearly erases it entirely as a scientific event. We must bear in mind that the intended addressee of the deleted passage was Jacob Bernoulli, whose own avowed relation to Leibniz regarding the calculus was very similar to the Parisian friendship with Huygens that Leibniz sketches out in the postscript. Leibniz had doubtlessly written this outline as a moral example for the brothers Bernoulli, whose arguments about priority were notorious. The nod to others, however, by its deletion becomes the hidden matrix of a renewed involvement on Leibniz's own part, this time regarding Jacob Bernoulli and the question of who invented the measurement of probability.

Regardless of our interpretative choices, however, in the 1703 postscript of his letter to Bernoulli, Leibniz definitely wants to keep open the question of the definition of probability theory. Whoever invented the theorem in the first place, there is something in the discursive construction of probability theory that does not allow its being understood and formalized once and for all. The material reason can be found in the passage from the letter to Thomas Burnett. Leibniz implicitly outlines the general program of an *ars conjectandi* in this letter when he speaks of measuring proof as *veritable topics*, and when he assumes at least the possibility of their being modeled in a mathematical theory of the game. But such modeling does not yet provide the final solution for Leibniz. He seems much more concerned with posing the problem and keeping it open. Leibniz advocates a probability theory that, while not contradicting, still antecedes Bernoulli's logic of applications. In such a theory, the legal rules of evidence contain the interpretation of gaming theory as probability in a rather provisional and, at the same time, more fundamental way. This approach to a theory of probability leads Leibniz far beyond mere probative guidelines or even the *veritable topics*. If in fact the probability of legal grounds of evidence, on the one hand, and the calculable structure of contingent events in games of chance, on the other, could be shown to be one and the same, then Leibniz's great project of theodicy to reconcile moral freedom with providential order would be complete before it ever began. It would not only be unnecessary, but also impossible. The project of a *logic of contingency*, a term Leibniz uses alternatively with *veritable topics*, contains the entirety of the philosophical system and hence the procedure of metaphysical thinking within it. But for this reason, Leibniz cannot simply treat it as a local project of probability logic. For Bernoulli, the *art of conjecture* is an Enlightenment enterprise meant to win a certain measure even from ignorance. The enduring difficulty of the split between the interpretation of probability and the theory of chances inadvertently becomes manifest here in the absurd consequences of the mathematical calculation. Leibniz returns to this problem again and again, and always with the same air of being able to announce its solubility and thereby the completion of his theodicean project of demonstrating that God is just.

A first version of this announcement, as expressed in the letter to Burnett cited above, negotiates the juridical and mathematical details of a probability, on the one hand, and the elusive wholeness of a metaphysical solution, on the other. In an attempt to close the gap, this version points to the political desirability of introducing a measurement of probability for the sake of mental pacification.[9]

Establissemens and the Announcement of Contingency Logic

In a letter to his Scottish correspondent Thomas Burnett soliciting help, Leibniz expresses his goals even more frankly than he usually does in more systematic treatments of the matter. His project concerning the measurement of proof and evidence aims to align it with the role of the old topical arguments that had actually been simply rhetoric and persuasion. Leibniz even understands his principal philosophical mission to be such alignment. The completion of this mission is announced, though it always remains incomplete. It proves the quintessence of what is yet to be done: "The greatest flaw of our logic lies in this. . . . Thirty years ago I made a public statement to this effect. . . . If God allows me sufficient life and health, [I] desire to make this my principal occupation."[10]

The development of a logic of probability makes up a "principal task" in Michel Serres's sense of the term. According to Serres, there are several different, but equally justified access points to Leibniz's philosophy, and these points of entry reveal various centers or sundry views of a virtual center in Leibniz's unwritten system.[11] The measurement of proof and evidence, of trustworthiness and probabilities, makes up one such center. It arises in Leibniz's writings particularly between 1690 and 1710, that is, in the time when Bernoulli's tractate is nearly complete but awaiting its posthumous publication. It presents itself in the mode of extreme deficiency: an occupation reaching far back and a task yet to be fulfilled.

This drama, which Leibniz emphasizes more than elsewhere in his letter to Burnett, results partly from the fact that in it he sets the scene for the logical problem of contingency in two very distinct contexts: the general, historical backdrop of the contemporary European situation on the one hand; and a specific, personal intellectual ambition on the other. The context of safety and pacification reminds us strongly of the urgent theme that occupied writers in the wake of the devastating civil and religious wars of the seventeenth century.[12] The letter to Burnett echoes this concern quite explicitly, inasmuch as it deals with "the most important and serious matters of life that have to do with legal order, the peace and commonweal of the state, the health of the people and even religion."[13] On the other hand, Leibniz's first and ongoing attempt to come into contact with Locke provides the pragmatic context for his letter. A memorandum by Leibniz on the *Essay on Human Understanding* forms the cause and means of this attempt. As Leibniz remarks in the letter, however, Burnett, who had been charged with the memorandum's delivery to the famous English philosopher, could report no more gratifying news than that Locke had set the text aside with a polite comment. But the theory of measuring proof and evidence as a logic of practical philosophy and of "historical" (empirical) knowledge is precisely the theory that Leibniz would use to counter Locke's probability of evidence in his later *New Essays on Human Understanding*. If the explicit context Leibniz engineers is a logic that should help end disputes with legal and political consequences, then the pragmatic context is a dispute about logic and epistemology, with which Leibniz accuses Locke of refusing to engage. In the open context, Leibniz aims for a theory of resolving disputes; in the hidden one, he maneuvers for the great philosophical argument that he would like to initiate vicariously through his critique of Locke. In both of the letter's contexts, the moment of measuring probability plays the central role as a task simultaneously normative and utopian—and it is equally normative and utopian as a continuing life goal that Leibniz had publicly set for himself thirty years before.

This task consists in the establishment or institution (*establissement*) of a comprehensive logic or doctrine of argumentation that should span both theory and practice, the entire field of philosophy: "I call a proposition established if at least certain points have been conclusively determined [*determiner et achever*] and certain theses are pushed beyond any reproach, so as to gain ground and a foundation upon which to build."[14]

Leibniz's term *establissemens* refers to an aspect of successful foundation that comes in addition to what Huygens had already called a foundational basis. In Huygens's theory of games of chance, the foundational aspect is the notion of equity. The *establissemens* in Leibniz, however, not only mean the grounds laid down or the reasons given for something, but also imply a decision about whether something can function as territory and foundation. An acceptable basis thus involves not only the "status" of reason or grounds, but also the will and the activity to make something "stand" as reason or grounds. The operative *stare* (*stabile, status*) is still quite literally visible in establir, the old spelling of the modern French word établir.[15]

The *establissements* aim first off at a theoretical philosophy that is mirrored in and ensured by the establishment or institution of the revealed Word and its interpretation.[16] From a methodological point of view, however, they derive from mathematics: "Mathematicians in particular have a method that differentiates the certain from the uncertain [*certum ab incerto*], what has been discovered from what remains to be discovered [*inventum ab inveniendo*]."[17]

This two-pronged formula contains or is the *establissement*. It presupposes as unproblematic the very thing that Leibniz raises as a problem under the heading *establissement*: How can the difference between certainty and uncertainty be ascertained once and for all? How can we say with certainty and not just likelihood where the boundary lies between what has been found once and for all and what is yet to be discovered?

Leibniz's reference to the validity of *establissemens* for mathematical method arises from Pascal's point about the interplay between logical validity and persuasive power in forcing an argument.[18] Leibniz's *establissement* can be understood as assuming a final unity consisting of the logical determination and the act of concluding (*determiner, achever*), however problematic or even paradoxical such a unity might be. He uses a number of phrases and metaphors in order to illustrate how an ideal—logical—validity can be achieved through specific operations, which necessarily already act upon the acceptance of certain rules: a more conceptual way of concluding an argument once and for all (placing it *hors de dispute*); and the rather metaphorical rules of the military occupation of an area (*gagner terrain*) and a builder's foundation laying (*fondemens*). As these expressions show, the operation of *establir* exhibits a normative aspect because it acts upon the assumption of certain rules and values to be enforced. Its other side deploys a utopian moment by envisaging a final unity of such rules or values and merely logical evidence. Already at the beginning of the letter the notions of territorial gain and foundation find their way into a kind of normative utopia. It is just such an idyllic establishment of norms that Leibniz interprets in biographical terms toward the end of the letter to Burnett when he speaks of his announcement thirty years earlier and the task of developing a logic of probability by means of *establissemens* that still lies ahead.

While Leibniz thus narrows the task of his *establissemens* to Pascal's paradox of a simultaneously logical and persuasive evidence, he opens up a view from there to the relation between the theory of science and the political order, a fundamental issue in the last decades of the seventeenth century.[19] This second point is developed in Leibniz's letter in terms of demonstrative truth and probability. Under the heading of *establissemens*, Leibniz names the demonstrative proof and the probable proof as two kinds of the same thing: "I divide the propositions I would like to see established with certainty [*establissemens*] into two types. Some can provide demonstrative proof absolutely with metaphysical necessity and in an incontrovertible manner; others can provide moral proof, that is, in a way and manner that confers so-called moral certainty."[20]

This division reflects the inherited Aristotelian difference between *scientia*, the philosophical knowledge of universals, and *historiae*, the knowledge of singularities. It comes out of the primary distinction of topical dialectics from philosophy and its understanding of general laws in nature and being. Leibniz's *establissements* are hence directed toward the point in early modernity at which the *historical* (empirical) knowledge of the singularities intersects with the philosophical knowledge of universal truths; and they concern such a point in its fundamental bifurcation and utopian unity of logical and pragmatic concerns. In its epistemological construction and practical influence on university organization since the Middle Ages, Aristotelian science had for centuries excluded such a point of intersection where *vérités de raison* and *vérités de fait* are united, as Leibniz says in his *New Essays on Human Understanding*.[21] A new kind of unity of knowledge and its organization was on the European agenda around 1700,[22] uniting anti-Aristotelians such as Locke in England and Thomasius in Germany in its most general expression, even if the solutions they suggest are still quite different.[23] Leibniz elaborates on the unity of this division particularly with regard to theology, the area that had in any case been the departure point for the logical-performative problem of *establir* in the first place. On the side of philosophically necessary knowledge in theology, Leibniz names the principles of metaphysics and of natural theology; and on the side of probable knowledge, he cites history, empirical things or factual occurrences (*les faits*), as well as the interpretation of texts.[24] In the general expression of the division of knowledge and its modern project of unification, metaphysics and natural theology thus appear as areas of theoretical philosophy. Practical philosophy is subsumed under *probable topics*, the juridical rules of evidence being for Leibniz the first and only extant sketch of the requisite "art of estimating degrees of evidence" (*art d'estimer les degrés des probations*).

The *establissemens* have their roots in the specific, detailed problems of the juridical rules of evidence in regards both to their systematic point—the paradoxical relation between performative determination and constative basis—and to their intention in political science—the reorganization of the old division between *scientia* and *historiae*. In noting in his 1697 letter to Burnett that he had publicly committed himself to working out the logic of probability thirty years before,[25] Leibniz can only have had in mind his juridical dissertations between 1665 and 1669. The fact that, in the same year as his letter to Burnett, Leibniz began to rework another text from his juridical juvenilia from 1667, the *Nova methodus discendae docendaeque jurisprudentiae*, also speaks in favor of this claim. The logic of the probable plays an equally central role in this outline of a reformed science of law.[26]

In his 1697 letter, then, Leibniz was referring mainly to the abovementioned juridical juvenilia on conditional law. As we have seen above in chapter four, Leibniz considered the hope of players in a game of chance to function as a sort of subconscious measure, a psychological beginning of calculation before any numbers come into play. At one point he explicitly speaks of certainty that "can be evaluated according to hope as if to a universally binding measure."[27] Such a formulation, however, still does nothing more than make explicit the implications of natural law. A truly proclamatory air can, however, be sensed in the *Specimina juris* of 1669, where Leibniz reworks the conditional law again. In this revision he decisively undertakes to overhaul the definitions and rules of conditional probation taken from the law with philosophical terminology and according to philosophical method.[28] This time, the theory of conditions at one point makes express use of numerical representation. Leibniz notates "1" for pure right in necessary conditions, "½" for the conditional right of contingent conditions, and "0" for the null right of impossible conditions. And at the precise moment when Leibniz stops writing out the numbers in words, but switches instead to Arabic numerals, he immediately draws our attention to the analogous character between legal validity and mathematical structure: "Just as the fraction between 0 and 1 lies in the middle, so too conditional justice lies between pure law and law that is null and void. And just as the fractions differ—a fraction below ½ approaches 0; a fraction above ½ approaches 1—conditional justice attracts different estimations and approaches in varying degrees either pure or null justice."[29]

Whether it is the case that the law takes up residence in human souls as a kind of proto-mathematics or simply that the legal practice of evaluating proofs is determined in numerical ratios, the result is the same: Leibniz's performative-constative *establissemens*, the institution of a new order of knowledge and pacification of the state, has to take place somewhere between law and mathematics. But it can never be beheld directly as itself. The establishment of that order and pacification remains a task whose fulfillment is yet to come.

If we imagine Jakob Bernoulli reading these passages,[30] we will recognize that his *Ars conjectandi* works well in the framework designed by Leibniz in these announcements. Leibniz avoids conclusions that lead to Bernoulli's problems, however. He does not have in mind a mathematical theory of games whose exemplarity for argumentative moves and decision making would have to be proven. Instead, he takes as his point of departure a cluster of argumentation and decisions whose intrinsic character of gradation he investigates and represents according to numeric ratio analogies. The explicitly mathematical theory is a project whose fulfillment comes into view only at the end of a longer process. Leibniz begins administering this project with a loud proclamation in the *Specimina juris*. This proclamation goes further than the *Logica juridica* and announces a "logic of degrees of probability" that "as far as I know, has as yet not been treated accurately by any logician."[31]

It lies within the characteristic style of the proclamations that Leibniz determines the juridical-mathematical problem with more and more exactitude, on the one hand, but on the other, keeps sketching a larger horizon in which to place this precisely described problem. Above all, the first version of the announcement in the *Specimina* of 1669 makes it unquestionably clear that Leibniz, like Bernoulli later on, would like to propose a logic of the probable as a general theory of argumentation, that is, a *veritable topics*. The cases brought forward from the theory of conditions, however, present argumentative and decision probabilities exclusively in given, legally defined relations. Even if Leibniz eventually undertakes a measurement of the "condition itself" and its probability, such probability is always determined in regards to the legal agreement in which it functions as a condition. The world of measured probabilities is neither an "objective" one of events nor a "subjective" one of argumentation, but rather the world of contractual assumptions. Their assumed validity may then be measured according to legal norms and degrees of hope. What Leibniz's *logic of degrees of probability* measures, therefore, are neither mere events in the world nor strengths of argumentation in life—as Bernoulli did in his theory of application—but rather evaluations within assumed frames of procedure. The logic of probability finally depends on a general frame of its validity, a frame that in its generality has yet to be set. Jacob Bernoulli admittedly also poses the problem of the assumed standardization in his theory of probable *argumenta*. This problem even makes a dramatic appearance as the necessity of a politics and an implementation of measurement: Bernoulli proposes a publicly introduced and guaranteed measure of *moral certainty*, that is, of the kind that counts as sufficient to take a reasonable decision in social life and in court. The appropriate value he wants to bracket seems to fall between 99/100 and 999/1,000.[32] The standard calibration of *moral certainty* in Bernoulli obviously raises the question of its epistemological accountability. Leibniz's question, in contrast, concerns the very introduction of the measure itself, the institutional logic of its existence. For this reason Leibniz remains, in respect to a conclusive mathematization, more indecisive and conservative than Bernoulli.

Leibniz's announcement of the logic of probability nevertheless goes beyond the legal frame, or conversely it extends the validity of its frame to include other nonlegal areas. The decision between the two solutions is not easy. Does the meaning of *degrees of probability* as it comes to light in legal thinking remain the determining factor for a more extended task, or does it lose its all-determining importance in a generalization that carries it further and pushes it back to the status of an example?

Logica probabilium: Law and Measurement

This question occupied Leibniz intensively toward the end of the 1690s. He even worked at this time on retrospectively strengthening the proclamation's continuity. In 1667, he had dedicated his juridical doctrine of methods (the *Nova methodus*) to the Mainz Elector Johann Philipp von Schönborn, who was planning to reform Roman law in the spirit of natural law. There is no indication of juridical logic at all here, at a point between the first 1665 version of conditional rights and the programmatic announcement of the logic of probability in the 1669 revision of the *Specimina*. Not until a revision, begun between 1697 and 1700 but never brought to completion, was the announcement retroactively inserted.[33]

There is good reason why Leibniz might have understood the full importance of *logica verisimilium*—the "work that still remains to be accomplished"—only after the fact of its first formulation; and this reason leads us back to the observation that the "logic of probability" acquired its true impact only when revealed in its quality as a new edifice on the ruins of the old topics. In 1667, Leibniz mentions the two parts of Ciceronian topics, *invention* and *judgment*, just as Arnauld had before him and Bernoulli would after him. In the usual fashion of early modern scholars, he only calls the first part, invention, the topics; the second part, judgment, appears under the name "analytics." In the newer text, however, the distinction between invention and judgment acquires a universal meaning that subsumes both parts. They appear together under the title of a general *heuristics*. Leibniz finally refers to these generalized topics with the peculiar name *Logocritica* in a revision that is exactly contemporaneous with his first attempt to contact Locke. *Logocritica* consist first of the geometric doctrine of method, which is also called the logic of necessary true and false. With this, Leibniz obviously recalls the logic of Arnauld, book 4, chapters 1–11.[34] Second comes a *logica verisimilium*, which is yet to be carried out; Leibniz also calls it a "logo-critique of contingency" (*logocritica contingentium*). The name implicitly refers to chapters 12–16 of the fourth book of Arnauld and Nicole's *La logique, ou l'art de penser*, although Leibniz does not even mention them and seems to erase their work on contingent events in introducing his own *logocritica contingentium*.[35]

With his reshuffling of terms, Leibniz makes explicit what Arnauld and Nicole had only implied in the architecture of *La logique*. Leibniz gives the invention/judgment distinction, which had been current in the topics since Cicero, precedence over the traditionally higher-level distinction between demonstrative truth and topical probability. The topoi, in the traditional sense of Aristotelian and Ciceronian "places," are now referred to as the probability of invention; regarding the probability of judgment, Leibniz calls for a new kind of probability logic. The emphasis has been shifted here from invention to judgment. All the traditional terms and instruments are allocated without exception to the art of invention, such as the syllogisms and *loci*. New concepts are assigned to the art of judgment: here we find in particular the contingency logic that has yet to be carried out.

We can sense the pains and the care Leibniz took in these years before the turn of the century to determine a proper place for his newly proclaimed logic as he staged its proclamation as the great continuity, the initial desire, and final goal of his work. The statements on probability in his exchanges with Bernoulli and Burnett in these years prepare for the passages on probability and its logic in Leibniz's *New Essays on Human Understanding*, which would still count as his lasting contribution to the field later in the eighteenth century. The probative measurement, the *veritable topics*, which Leibniz develops in this nexus of writing, summarizes and brings together these competing ways of thinking probability around 1700: Leibniz clearly relates to Bernoulli's translation of the old topics into a mathematical theory (however detailed his information about it might have been). At the same time, Leibniz clearly joins Locke (and in another way Thomasius) in transferring the traditional canon of subjects divided in *scientiae* and *historiae* into a new epistemological order of disciplines that fundamentally rework the old difference between demonstratively necessary and historically probable cognition.[36]

The revision in the *Nova methodus*, the letter to Burnett, and the discussion with Locke together describe a dramatic act full of tension and suspense. In these passages, Leibniz extends probability measurement, as it existed within contract law, with its analogous structure of semantics and number, into two different but related layers that definitively rewrite the topics: (1) a probability theory for argumentation, and (2) a probabilistic philosophy.

(1) The theory of proof and evidence that comes out of Leibniz's early work on jurisprudence finds its most thorough expression in the *New Essays*. In the *Nova methodus*, Leibniz had still argued that the "logic of probabilities has not been worked out by anyone more thoroughly than by jurists, who deal with complete and incomplete proofs, presumptions, and evidence."[37] This statement appears amid a discussion of the grading of evidential proof in the early modern Inquisition, a practice with which Leibniz as a jurist would have been very familiar. He adds a polemical

note, however, in a manner common to natural law's critique of Roman law, which occupied Leibniz in his youth. The normative structure of the juridical theory of evidence in law, Leibniz says, lacks certain criteria for measurement. Grading proofs with numbers is obviously taken quite literally here, and hence exposed to critique. Taken as a model for argumentation logic, the full, half, and quarter proofs of the jurists seem to be only a groundless pseudo-mathematics. The case is different in Leibniz's answer to Locke, the *New Essays*. Here Leibniz presents in detail how the scales of evidential proofs in a trial work: if a fact of the case is declared to be common knowledge, then any further demand for evidence is relinquished. In the case of a full proof, the court can make its judgment based on what the contents of the evidence indicate (as far as civil suits are concerned). Evidence that is declared to have a greater proof value than one half can be complemented by the party who appeals to them by means of a supporting oath. Proofs that are to count as less than half can be refuted by whoever argues against them through a purification oath, and so on. If in this context Leibniz does not point out the insufficiency of the basic criteria of gradation, the failure is due to a particular, overriding concern of his at this point. For Leibniz, it is crucial to state that the degrees are allowed in precisely defined procedural circumstances. "The entire form of the procedure" is a logic embodied in legal categories.[38] The procedural framework here is in no way the factor that mars the purity of the foundation of the logical enterprise. It rather brings about a logic that at this point is not even strictly considered "measuring." The relation to the juridical doctrine of proofs shows that Leibniz around 1700, like Bernoulli, wants to develop probability as a logic of arguments. But in this attempt, he is troubled by the relation between norm and measurement. The interplay of these two very concepts in the logic of conditional contracts had provided the precise impetus to inspire the *logica juridica*.

(2) The attempt to relate the measurement of probabilities to epistemology is an equally visible move in both Leibniz's letter to Burnett and his *New Essays*. In his *Essay Concerning Human Understanding*, Locke had radically rewritten the traditional system, which divided academic disciplines into areas of philosophical universals and historical singularities, into an epistemic-psychological system of certainty and probability.[39] Probability for Locke is a function of judgment, which differs from

and faces up to the state of certainty, the ensemble of valid knowledge. Certainty in a given area means the perception of connection or agreement between several ideas, a perception that comes about in intuitive or demonstrative evidence. Probability acts as a substitute for certainty in cases where there is only the appearance of connection.[40] But what is the status of appearance when we are dealing with the phenomenology of the perception of internal ideas? What does it mean to speak about the appearance of a connection of ideas when their coherence "is not constant and immutable, or at least is not perceived to be so" (a phrase that translates the old problem of singularities into the language of epistemology)?[41] Locke introduces a noteworthy temporalization of the phenomenological quality of appearance of connection: an apparent connection differs from a true connection by its instability and the fluidity of impressions, not by any material quality that would differentiate them as merely appearing images from the things imagined. It is suggestive to think of temporal representation in the theater: probability in judgment has the same relation to the certainty of knowledge as a theatrical play, the entrances and exits of actors, has to representations independent of time. We can imagine the scenario as a mental stage for the entrance of ideas. For the spectator who observes such inconsistent and mutable sorts of internal appearance, Locke assigns degrees of assent and a measure that ascertains them.[42]

Leibniz inserts his notion of probative measurement at the corresponding place in his *New Essays*. In fact, traces of the same implication of establishing and reading the measure can be found in both Locke and Leibniz. The principles of probability—as Locke formulates them indecisively halfway between standardization and registration—consist in "the measure whereby its several degrees are, or ought to be *regulated*."[43] Locke, too, has a juridical model for this implication, which, however, is more limited and diffuse in legal theory than Leibniz's example of conditional rights: witness testimony. Locke proposes to measure degrees of probability according to rules for judging testimony and witnesses. These rules are procedural guidelines and have no legal contractual implications; Locke moreover keeps more to the procedural rules of historians than of jurists.[44] From a typological point of view, we are confronted here with a generalized form of *fides*, the trustworthiness of witnesses. Cicero had used the term for his juridical-psychological reformulation of the Aristotelian *endoxa,* and subsequently Boethius and later Agricola introduced the model into their thinking about topical probability. Locke's epistemological judge of probability thus observes a stage and a play where witness statements demonstrate their conformity with certain knowledge and must reveal themselves and their connections with each other. In doing so, it finally becomes irrelevant—as 's Gravesande concludes—whether the origin of the observations lies in one's own senses, the testimonies of others, or analogies to one's own or others' observations.[45] The probable judgment appraises one's own and others' testimonies (or observations) and particularly the witnesses (the agents of these observations), their play and their appearance. It does so according to the measures of the external world of judges or observers including the stage and the play of testimonies. While Bernoulli's argumentation theory, with its basis on games of chance, reverses the topical semantics of *endoxa* into the probable measure of countability, Locke's epistemology replaces topical semantics with the uncountable measure of a quasi-aesthetic observation (judgment) of observations (testimonies). Locke's solution, however, poses the question of the consequences of such reversal even more pointedly. If the original observation and the observation of that observation both count equally for assigning degrees of probability, in whose name, then, is the whole process accomplished? Since observer and observer of observer—witness and judge—both seem to be implied, the question arises of whether or not we have to assume a final and all-embracing observer of all observations. In other words, the question becomes: whose probability is it?

The position Leibniz claims to have taken from the beginning in the belated announcement of the *Nova methoda* leaves the decision between the two translations of the topics unresolved. It remains undecided whether the juridical model authoritatively informs a specific theory of argumentation, or whether such a model and its juridical implications are only instances of a general epistemological or even metaphysical problem over which they do not have any final authority. Thus, after touching upon the juridical doctrine of proof and evidence and making a brief reference to medical semiology, Leibniz cites gaming theory and its development by Fermat, Pascal, and Huygens in the chapter on probability in his *New Essays* (an allusion to Bernoulli's still unpublished *Ars conjectandi* is missing). Here again Leibniz connects the mathematical theory of games of chance with law. This time he explicates traditional laws of inheritance as an anticipation and illustration of the division problem. Again, such interconnected instances seem to point in the direction of a final closure and unity of probability, which then would embrace mathematics and law, scientific measure and social norm. But the hints once more fall short of conclusive actuality. The mathematical and juridical examples remain instances that refer to a center without this center ever being revealed in itself. If we take this to be Leibniz's precise message on probability in the *New Essays* and throughout his writings, we can say that Leibniz reflects the situation implicit in Jacob Bernoulli. Bernoulli, as we have seen, claims to formulate a theory that finally unites juridically relevant argumentation and computation based on games of chance into one and the same structure. But then, at the end of his mathematization of probable arguments, a gap between argumentation and mathematics appears to return in the guise of a strange consequence of the mathematical formulation. Such an alternation between the desire for closure and the return of the original distinction seems to be reflected and exhibited in the manner in which Leibniz announces the future establishment of his theory of probability. Leibniz in fact admits that his doctrine of probability can only be presented by a group of interrelated but homogeneous instances, while implying that a universal theory still could and would have to be finalized. This theory would then provide an end to argumentation and conflict and bring peace to the state.

The Other Side of *logica verisimilium*: Contingency

Between 1695 and 1700, Leibniz continuously holds out the prospect of a final unity of the logic of conditional clauses and the calculation of chances. Adjustment to a legal framework gives support to respective individual instances of *logica verisimilium*, but it also fixes them firmly in their peculiarities. The fact that Leibniz makes the grounds of probability in law so emphatically, contrary to the trend in de Moivre, as well as in Jacob and Nicolaus Bernoulli,[46] hinders its coherent formalization. Precisely this legal grounding, however, characterizes Leibniz's probability logic. Its ultimate tasks of pacification and the resolution of all disputes about legitimacy require a formula that still corresponds to the discursive text of law, but that cannot be calculated in pure mathematics.

With his pragmatic and political instances of probability logic, Leibniz gives the exterior view of his concept of contingency, which, on the contrary, forms the onto-theological keystone of his unwritten system. Leibniz aims at a concept of contingency that would make the creation of the world and the maintenance of its order philosophically thinkable. The unity of the distinction between law and measure would make it possible in the *best of all possible worlds* to combine the contingency of choice among various possible worlds with the necessity of the realization of the definitive one that is the best.[47] With a view to the onto-theology of the unity of contingency and necessity, we can see that the task of probability logic presents itself as solvable, even though its calculations can never be carried out to the end. We may even say that the abrupt switch between the pragmatism of *logica verisimilium* and the esoteric speculation on contingency is an essential part of the concept of probability in Leibniz. Probability is precisely that which manifests itself in imminent situations in political and social life, while being at the same time a task of conceptual thinking that withdraws into the depths of the system. Probability is the very immanence of the metaphysical principle within the world.

This basic design of Leibniz's theory of probability becomes evident in his seemingly stubborn but in fact fundamental resistance to the concluding section of the *Ars conjectandi*. Bernoulli just barely manages to arrive at this ending of his work, which remains as it were under the stage direction of its posthumous condition. He forms the conclusion of book 4 by bracketing the distribution of events between certain limit values, a distribution whose range and precision depends on the number of observations. This operation makes possible the central tenet of Bernoulli's treatise, the methodical interpretation of gaming theory as probability theory.

The two thinkers' brief debate over this question is short but illuminating. In response to Leibniz's request, Bernoulli elaborates on his method for determining the limits within which the supposition of a frequency distribution would be certain, despite its being a posteriori in regards to the number of experiments or gathered tests.[48] The method Bernoulli sketches out is based on the theory of sequences and series.

Leibniz, however, objects in principle. For him, there can be no method at all to measure the probabilities by means of experimental testing, as he rephrases Bernoulli's claim. Infinitely many factors of an event cannot be determined by a finite number of experiments. This is the reaction of someone who proposed and organized as many statistical surveys as Leibniz had done, but who could not see sufficient grounds for forming a posteriori laws from them. With this reasoning, he responds to the mathematician Bernoulli, who in his turn says that he lacks data for the application of his thesis. Leibniz argues that men believe nature's limits to be determinable once and for all. In truth, however, the distributions of events can change in time. Perhaps, for instance, new diseases will appear in the future. Statistical conclusions on the grounds of mortality tables could then not be continued in the same way as previously. A finite amount of data does not yet determine how one should theoretically draw conclusions from the individual observations. We assume for instance that the points that have been ascertained for the path of a comet can be connected by drawing the simplest, most regular figure. But in fact the comet may conceivably have taken an infinite number of other paths between any two points. "Nature does indeed have her habits, which come about through the return of causes—but only *hōs epi to poly* [in most cases]" (Aristotle characterizes poetic probability as lying between the singularities of history and the universals of philosophy with this very phrase in the *Poetics*).[49]

Bernoulli gives the objection short shrift. As he says, with a certain degree of methodological justification, Leibniz's remarks have nothing to do with the a posteriori or empirical character of data and theoretical model.[50] Bernoulli's argument, in contrast, is valid from a methodological point of view alone. It is not the case that Leibniz failed to understand the exemplary character of the calculations. He is, rather, asking where the modeling character of the examples comes from; in other words, he wishes to investigate the implications of Bernoulli's methodical stance. Neglecting time as a factor and assuming the simplest hypothesis both presuppose an order of things that is governed by rules and calculates with rules, that employs norms and functions with norms. Such an order of things first of all cannot be found in physical nature and natural science, but rather in morality and the practice of jurists and administrative officials. "The difficulty seems to me to consist in the fact that the contingent [*contingentia*], namely, that which depends on infinitely many circumstances, cannot be determined by a finite number of experiments," Leibniz observes.[51]

Contingency presupposes infinity—an infinity of circumstances on which the contingent event depends, or an infinity of observations needed to bring the upper and lower limits of observed values into line. Bernoulli had proposed the series-theoretical method of approaching (as close as one likes to) the upper and lower limit as a solution for precisely this question. Here, Leibniz is not presenting an objection against the method, but rather against the methodization of the problem. The onto-theological peculiarity of contingency resides exactly in this paradox, the paradox of stating and writing *infinity* as a number of circumstances or observations while not being able to name infinitely many circumstances or to calculate the probability equation with an infinite number of observations. God's competency in contingency is preserved in this impossibility of naming and calculating; at the same time, the identity of the distinction between contingency and necessity is preserved in the symbolic manageability of infinity. A limit theorem, in its methodological design, would result in the destruction of this balance.

Bernoulli's *moderatio*, the methodical deduction of singularities to contingent events, is a kind of pragmatism and politics in the precise sense in which Leibniz does not tolerate such practical considerations. *Moderatio* is a methodical procedure in Bernoulli's thinking. It means to "temper" or "regulate" empirical situations and things in the world in order to make them objects to which the mathematical theory may be applied. Leibniz's philosophy does not allow this sort of accommodating of the world for the sake of theory. He holds instead that the pragmatic-political factors should be considered within the foundational concepts of theory itself.[52]

In his understanding of contingency, Bernoulli combines two kinds of methodization. One of them amounts to probable narration according to the discursive method of the topics. With its help an event can be shown to have entailed a following event or not, and can retrospectively be understood as its harbinger and sign or not. The other type of methodization lies in the mathematical calculation of probability. In this case, quantities of events are structured and simulated in situations modeled for the sake of scientific observation. For Leibniz, in contrast, contingency is a concept that, for various bits and pieces of a projected theory, holds out the promise of their unity without ever being able to demonstrate this theory in its entirety. The fragments of a gaming theory according to Huygens, which take the ratio of hope and gain to represent the proportionality of affects implied in the contract of the game, have nothing in common with the passages on conditional contracts. But they both tie their operation to issues of legal validity, and they both refer to contingencies that are created and administered in the framework of such validities. We begin to recognize the outlines of an integral theory of contingency that has its argumentative-calculable unity in a sphere where justice makes calculability possible and calculability postulates justice.[53]

The measurement of probability for Leibniz always implies a judgment about contingency. In his revision of the *Nova methodus* from the late 1690s, he calls the juridical rules of evidence a "critique of contingency" (*logocritica contingentium*).[54] At the beginning of his fragment on games of chance, he alludes to the concept with the phrase "judgment about an event in the future." In the fair game, fear and hope are equal, where "every distinction that can be drawn between the players has to be assigned to the outcome alone [or just to the event happening—*in solo eventu*],"[55] that is, to chance. Leibniz's axiom speaks the careful language of jurists: in fair games, all possible grounds for winning or losing must be seen as chance, the bare event. Speaking of chance is thus a matter of ascription, a wording valid according to a certain agreement. "Where the appearances are the same, the same judgment can be formed about them, that is, the way of thinking about the future outcome, is the same; and the thoughts about the future outcome are hope or fear."[56]

Judgments about probability relate to the rules of the game and the correlated patterns of behavior in the players, not to physical investigations of dice or lottery drums. The game takes place in a sphere of appearance, in the jurisdiction of certain rules and assumptions. Contingency judgments are performed in a normative-aesthetic sphere in which jurists' laws and geometers' numbers communicate. The normativity of justice bears the appearance that makes the calculating behavior of gaming possible in the first place. Probability exists—from a methodological point of view: only—in the appearance of this equity. The game of contingencies is the exemplary space underlying the various partial methods of establishing areas of probability, but it is also partially removed from them.

Leibniz designs contingency logic in specific distinction from the logic of necessity, demonstrative logic. He does so with a view to the unity of logic that embraces both. In knowledge and in the logic that organizes knowledge, necessity and contingency form the characteristic difference between qualitative-semantic truth and quantitative-measuring ratio (between *veritas* and *proportio*). Contingency (as grasped by the *historiae* of nature and history) relates to necessity (as exhibited in metaphysics and geometry) as dullness does to transparency. The determination that manifests itself unhindered and with the character of compulsion is necessary. The necessity of geometric propositions is thus established by and through itself. It is defined by this self-establishment, which reflects the final unity of ontology and epistemology in this case. Contingent determination in the worlds of physical things and moral events, on the other hand, offers an obscured image of this unity. The unity in this case remains a mere postulate of reflection. In moral actions we do not grasp "the entire play of our spirit and its thoughts [*tout le jeu de nostre esprit et de ses pensées*], which in most cases are imperceptible and confused." But this play has determined or will determine persons in their decisions. Similarly, in physical motions we do not succeed in "sorting out all the mechanisms which nature puts to work [*fait jouer*] in bodies."[57] In this passage, Leibniz formulates his sentences with extreme care. It is clear how he combines moral contingency with physical contingency by using the metaphor of the play in both contexts. A concept of contingency becomes graspable in these metaphorics, even if we are not able spell out exactly what makes for the identity of contingency in both cases.

The mode of causation changes from necessity to contingency according to their respective ontological and epistemological statuses. Necessary conclusions in the realm of geometry and metaphysics are carried out with compulsion, according to the model of causation in physical movements. Compulsion is necessity's characteristic form of action in terms of causation. Contingencies in nature and of moral life, in contrast, bring tendencies about. They make us incline toward certain consequences or actions.[58] When compared to the transparency of causation in the case of necessity, contingent causation shows a certain obfuscation in ontological and in epistemological respects. Above all the relation between ontology and epistemology is obscured in the case of contingent events: our way of thinking contingent events does not easily reveal what contingency means in being. In our dealings with contingent causation, for example, we reckon with parts of the causal chain that remain beneath the threshold of perception; or we use the binarism of yes/no choices in structuring contingent events.[59]

In the case of contingency, knowledge sees itself face to face with a phenomenal world that it can and must model in terms of a game. Determining events in the realm of necessity, which can be called active or passive according to one's point of view, unifies logical determination by predicates and causal determination by causes. In the play of contingency, the active or passive determination becomes self-determining regarding reasons, and as such it hovers between the two. This act of self-determination should be thought of as a judgment in moral decisions and as acts of God in the case of events in nature. At issue here is the question of how reasons can be accessed and used in making judgments.

What does this mean for learning how to represent and how to express our knowledge of the play of contingency? Necessary truths can be imagined as finite processes complete in themselves. For Leibniz, the finitude of the process of determination in necessity lies under the aegis of speaking or of literal discourse. Necessary truth—one may think again of geometrical figures and arithmetic formulas—is predicable by a demonstrative proof. And one can "predicate" the same truth in ratios of natural numbers. In short, necessities can be expressed in a *proportio effabilis*. Between the discursive text of the proof and the mathematical representation, a parallel meaning gains ascendancy that is based in the "predicability" of both. The contingent truth, on the other hand, comes about where we deal with the problem of the infinite and where the methodical process of determination cannot come to an end. Here, as in cases of necessity, we find truth and apriority; and this assumption clearly distinguishes Leibniz from Jacob Bernoulli's empirical conception of contingency through statistical frequency. But we also switch from the realm of speaking and discourse that ruled in necessary truths to the realm of seeing and of mere numbers. The inexhaustible problems posed by the infinity of data demand constant repetitions of the analytical process, which lead to infinite series.

Infinite series as such, however, are not accessible to apperception; a multitude of acts that cannot be completed also cannot be comprehended in a unity of thought, that is, in units of ascertainment or representation. They cannot be objects of demonstrative explanation or discursive predicability. God can grasp them at once. But it is His view alone that comprehends them as in a picture. Contingent truth is exclusively an object of what Leibniz calls a *scientia visionis*. The mathematician can write the infinite series and compose a formula thanks to the differential method for series. But then he is no longer writing down natural numbers; he is writing numbers that are "deaf numbers," or signs for the deaf and dumb, *numeri surdi*: irrational numbers. What is necessary presents itself, due to the transparency of thought, in a discursive-numerical discourse, a predicability of proof and a speaking of numbers. The contingent event causes the discourse that was based in the parallel orders of discourse and numbers to break apart and collapse into an image. This image is formed by the letters for a higher reason and by those numbers with which mathematicians calculate without understanding them any more. For Leibniz, however, what can be written remains connected to the reference of an object. Writing is never pure operation without any understanding possible.[60] But neither is the ability of the irrational numbers to be written a discourse meant for humans. Only God's view can read such writing like a book in his *scientia visionis*. Irrational numbers, *numeri surdi*, offer what one could call an abstract image, or an image that, paradoxically, we humans cannot imagine as an image of something. But even if we cannot imagine how this image could refer to anything in the world, in looking at the writing, we nevertheless perceive that it is an image, and hence that it must refer to something. We see that the image can be read as an image by another onlooker, God. In God's view, the writing of the *numeri surdi* is legible like a letter of the alphabet, while to us it offers only instructions for calculation.[61]

7

Defoe's *Robinson Crusoe*, or, The Improbability of Survival

Probability, Games of Chance, and the Figure of Reality

To an extent surpassed by no other narrative text, Daniel Defoe's *Robinson Crusoe*—whether we call it a romance, merely a narrative, or already a novel—straddles the poetics of allegory and the aesthetics called realism. In the nineteenth century and through the 1950s, most readers displayed a fascination with the wealth of detail in Crusoe's account and the pragmatic bravado with which he copes with his island existence. For readers from Karl Marx to Ian Watt,[1] this makes for realism plain and simple. What they mean by this classification is a kind of improved or duplicated probability. For them, realism elaborates on the poetical rule in effect since Aristotle for authors to observe the regular sequence of events in their works. This rule of basic verisimilitude is then thought to be enhanced by a new and more serious attention to detail, and it is interpreted as the appearance of things in the literary text. Realistic probability thus aimed for is not supposed merely to bring the play of chance events closer to the true being of things, but also to refer to the fact that the true being of things is represented in a game of chance events as modeled by poetics. The kind of probability they comment on amounts to the appearance of truth, and references the reality in which we live. In literary criticism of the later twentieth century, in contrast, Defoe's clear claims that *Robinson Crusoe* is an allegorical text, as voiced in his foreword to the novel's second installment, regain a measure of validity. These later critics generally share the related view that Defoe narrates Crusoe's life in the style of puritanical biography, that is, according to the model of conversion.[2] The probabilities of every adventure of such a life would accordingly have their meaning and their basis in the exclusive case and adventure of conversion. For them, the religious figure of this singular event annexes all other probabilities. Both of these interpretations can be grasped as exemplary for the alternative that allows the legibility of the modern novel to exist at all: they show the alternative between a relation to reality (reference) and the novel's internal, figurative function. It can be shown, however, that the realism of reference, at least in *Robinson Crusoe*, is itself in fact a figural operation. The reality to which the text of the novel refers is an allegory of what is probable.

This allegorical realism is seen most pointedly in the novel's figures of probability or in the mathematics of contingency interpreted as poetical verisimilitude. Defoe tends to employ idioms from betting or gaming in his later work. Narrators and characters remark, for instance, that a chance stands at ten to one, or that one could bet a thousand to one for the occurrence of this or that event. Sterne parodies this preference in the "chapter of chances" in *Tristram Shandy*. In it, he stages Uncle Toby's utter inability to calculate probabilities, which does not keep him from evaluating events with the probability value of a hundred to one.[3] In Defoe's 1719 novel, however, the explicit formulas of chance calculation mark out three distinct stations in *The Life and Strange Surprising Adventures of Robinson Crusoe* after the protagonist's shipwreck on the island at the mouth of the Orinoco.[4]

Three dangers, three rescues: Crusoe sums up the chances of the damaged ship having been unreachable for him as 100,000 : 1. Yet he manages to swim over and bring tools, provisions, and seeds back to the island with him. The sinking ship becomes a Noah's ark for him. Crusoe does not make this calculation until after he succeeds in securing the initial conditions for his survival on the island. His consideration, ornamented with the probabilistic formula, marks the first phase of his island life. In later passages, our retrospective autobiographer reckons the chances of his death as a result of falling into the hands of the cannibalistic savages on the nearby mainland as 1,000 : 1. This calculation functions as a comment on the end of the first and the beginning of the second phase of his island life. Having established himself as self-sufficient, Crusoe now considers how he can leave the island and reach the land that he can make out on the horizon. All contact with other humans—with the savages who come from this neighboring island, with Friday, and with the crews of Spanish and English ships—at this point still lies in his future. Finally, in a later episode of the book, farmers in Languedoc assess the probability that badly chosen defense measures would leave Crusoe and his companions vulnerable to an attack of wolves during their crossing of the Pyrenees as 50 : 1. This incident takes place after Crusoe returns to Europe and begins to look into restoring his fortune.[5]

The three numeric formulas provide a falling curve of fatality chances, lending the numbers—which in each case may well be arbitrary—exactitude in their relation to one another. Their precision is their proportion. They can thus provide a scansion of the events of the narration, a narration that, as literary critics have noted time and again, is so much in need of form and design;[6] and they do so by marking out decisive turning points in the novel's allegorical meaning. The chances, given as they are in arithmetic figures, reveal the likelihood of Crusoe's death on a falling curve and hence in reverse numerical signs from God for Crusoe's salvation. The smallest chance of survival belongs to the deliverance from the shipwreck, or more precisely: the deliverance from the chance, lying far below ½, that Crusoe's survival of the shipwreck is not merely temporary. He does not, in contrast to his nineteenth-century successors in Jules Verne's *Mysterious Island*, reengineer culture in nature in the spirit of technical ability. Crusoe is rather a survivor who is delivered from hostile nature, the nature of the Fall of Man, by means of highly improbably provided aid. He is able to remove just enough remnants of culture from the stranded ship to survive in raw nature with their protection, maintenance, and further development. The first rescue, distinguished by the tiniest chance quotient, does not consist of temporary rescue from the waves, but rather of enduring deliverance in nature. The deliverance in nature, however, requires a certain amount of culture from the sinking ship, rather like a care package, as Crusoe's guideline.[7] This initial salvation illustrates the minutely small chance of the survival of man in nature after the Fall.

The second rescue, figured by the middle chance quotient, consists in a deliverance from social danger. Instead of cultivating a secure survival in nature, Crusoe keeps a lookout for how he might leave the presumably uninhabited island and return to the inhabited mainland, the continent of society. In doing so, he deliberately overlooks the dangers resulting from the fact that all paths back are facilitated, and hence obstructed, by middlemen on the edge or outside of society. Savages or mutinous ship crews make up this margin of human society. The people whose companionship the lonely island inhabitant has longed for turn out to be rebels and cannibals: the Hobbesian paradox of the law of nature.[8] In short, the fringes of human society represented by rebellion and cannibalism result in small if not minute chances of survival.

The third deliverance comes about in a situation of tactical avoidance of danger. Its chances, though much greater than the previous ones, still lie well under ½. Crusoe, now back in Europe, the continent of society, leads a small group of travelers to cross the Pyrenees at the beginning of winter, who are attacked by wolves during the crossing. Here we are dealing with real tactics of fending off an attack of hostile nature from within a functional society.[9] The inhabitants of the mountain village have an experiential knowledge about the behavior of the wolves. By using their probability formula, they make Crusoe aware of the extent of the danger, which he otherwise would have overlooked. Crusoe makes a tactical mistake here out of simple ignorance about the nature of wolves' aggressive behavior (which could just as easily be human behavior: *homo homini lupus*). The third and smallest improbability of survival is thus that of coping with dangers in a situation where wild nature or the natural state of society re-emerges within a society and its security.

The Life and Strange Surprising Adventures of Robinson Crusoe narrates the odds of holding onto life in the face of constant danger in the adventure that is life itself.[10] These adventures are defined by the Christian nature of the Fall, the formation of society, and the strategies of coping with situations when nature reappears in society. The distribution of these chances results in something like an anticlimax. In this calculation, life first presents itself to be read by Crusoe, who becomes the narrator of his life through this lesson. Through the small rhetorical triptych of gambling formulas, the story of the narrator's adventures turns into an allegory of salvation history and of God's particular interventions in it. John Arbuthnot, a friend and collaborator of Jonathan Swift's, quite similarly drew the conclusion, in the same years, of the intervention of God, from the improbable distribution of gender, as shown by birthing tables. Under the assumption of monogamy, God ensures—as the text behind the numbers reads—human survival by engineering the slight predominance of male over female offspring in a world where men are subject to greater fatality chances than women.[11] Probability figures for human survival thus prove to be of primary importance in understanding God's work in nature. This observation is particularly significant for the controversy about how to read *Robinson Crusoe*. The two fundamental alternatives of the novel's interpretation can be represented by Ian Watt and his concept of a realism of details (or "circumstantial realism") with his emphasis on rational action, on the one hand, and the allegorizing of conversion programs inherited from the Patristic tradition, on the other.[12] In the figures of probability the opposition between the two interpretive stances is carried to its extreme. In numerical form, however rhetorical it may be, a mode of representation is revealed that undermines the opposition of the two interpretations and perhaps of what can be interpreted at all.[13]

Allusions to the parable of the prodigal son run through the entire first part of the novel. Crusoe, the narrator, relates his own life as the story of the lost son not merely, however, from a transcendental distance. Instead, he also narrates the story of the allegorical reading of his life. During the very first storm of his first sea voyage, the oath that immediately occurs to Crusoe would have him take on the role of the prodigal son from the biblical parable should he escape the danger.[14] In the novel, Crusoe is a prodigal son (*filius prodigus*—Luke 15 in the Vulgate) because he leaves his home for distant places and squanders the inheritance (*substantia*) that should provide for his living. This first and most important case, in which the allegory is offered ripe for the plucking, already demonstrates an important point about allegory in this novel. The allegory does not so much lend the text and its readers a moral interpretation as it is the attempt of a character in the text to understand his own actions and to give himself guidelines. The allegory clearly just does not fit. Robinson Crusoe is precisely not the squandering inheritor, but rather the third son, who in any case was prevented from learning any trade by his none too wealthy father.[15] Groping for an allegorical interpretation, Crusoe foists a foreign situation upon his own very different circumstances in order to act in his own story as if he were in someone else's. The "circumstantial realism" (Ian Watt) and the allegory do not only fall apart; their split is the mode of narrating a story in which Crusoe's narration of the story always already appears. Hence he naturally never finds his way back to his father's house in the fiction. In writing the book, however, as the narrator of his life story, he finds his way back over the far seas and from the place where he does not belong, the island, to the allegorical imputation of the heavenly Father's house. The narration, in other words, passes through the hierarchical layers of allegory and the rapidly increasing values of survival chances to come back to the deus ex machina of *providentia specialis* familiar to Protestant thinking.

Calculation of contingency and allegorical exegesis coincide in survival chances that lie far below ½. They do so due to the same causes that make these formulas themselves have their meaning in a point of extremity in which chance events and allegorical imposition no longer comprise an opposition.

Precisely for this reason, the rhetorical calculations of chances in *Robinson Crusoe* can be said to function as allegories with respect to Robinson Crusoe's tale of his life.[16] The proofs for divine intervention from the improbability of an event at first do not offer any hold for probabilistic calculation in a particular situation. Yet they do circumscribe what a particular situation will have meant for Crusoe. The tripartite succession of probabilistic formulas articulates this nature of the formulae's functioning even more exactly.

They are calculations in retrospect or from an external point of view, but they work as such in different ways. Crusoe assesses the chances of the wrecked ship having been driven aground at some unapproachable place or having sunk before he could reach it at 100,000 : 1. The first-person narrator presents this calculation as his first reflection that closes the phase of instinctual and uncommented survival behavior. It is a reflection of the hero in a story narrated by himself, but it is no maxim for acting. No amount of reflecting on chances will lead to any new impulse for action if one can neither influence the occurrence of the event nor choose a different course of events. The narrator then figures the chance ratio for not surviving an encounter with the cannibals under his given situation as

1,000 : 1. This calculation could conceivably have steered his behavior. In this case, Crusoe is indeed faced with a choice: search out society or remain on the island. But this calculation occurs only in the reflection of the narrating Crusoe. In the moment of the hero's decision, it would have been impossible. The unique nature of the choice—to search for society or not—had prevented him from choosing deliberately. The experienced inhabitants of Languedoc, finally, estimate the improbability for Crusoe's party of escaping the wolves with their poorly chosen tactics to be 50 : 1. Crusoe is only confronted with this frightening balance after the fact. He had not only underestimated the danger, but had not even noticed the direness of his chances in the first place. Crusoe, who is the one who has to make the choice, is also immediately distanced from the calculation of the improbability of his survival chances. But the reason for this is quite simply pragmatic; it results from the acting parties' insufficient experience. The figure of reflection on survival chances that corresponds to the formula of a geometrical slope approaching 1, or the quotient 1 : 1, is therefore thrice removed from Crusoe. The first kind of distance between Crusoe and his calculative reflection is quasi-ontological and resists individual influence. In the absence of any alternative or any freedom to model favorable or unfavorable events, the only remaining possibility is to know one's odds without any possible consequences for current actions. In this situation, one's chance calculation is fundamentally retrospective. The second gap, the choice of whether to search for society, is based in anthropology or pragmatics. Acting in this context is revocable only with great difficulty because the existence of calculating reason depends precisely on the object of its calculation. Knowledge and action disrupt each other in the decisive moment. The third remove from calculation is of a pragmatic and simply contingent nature, because Crusoe, the person in question, suffers from a lack of knowledge, while others can know and act accordingly.

The formulas of gambling chances indicate the status of circumstantial realism and allegory in *Robinson Crusoe* even more clearly than the biblical allusions, but they do so in their own unique way. The assumption that calculating chances and interpreting one's life according to Christian conversion could go hand in hand may seem surprising at first sight. For starters, we know very well of the intimate interrelationship that existed between early modern bookkeeping and practices of recording one's life in diaries, a requirement for and basis of conversion biography. In this context, however, the structural correspondence between the two functions matters even more than the practical relations. In the decisive moment, the moment of decision, the calculation of different kinds and strengths of reasons is never accessible. But neither is decision making the calculus's proper task. Crusoe manages to act in all cases, with God's help, more or less successfully. As we have seen, calculation is an activity that, by its nature, is always performed in retrospect, and the act of calculation thus resembles the process of allegorization. Only observers at a distance from an event—whether others looking on from outside or the actor reflecting in hindsight—can know anything by calculating chances or by allegorical interpretation. In this sense, all calculation of chances can be said to be an ongoing process of practicing conversion. By means of the neutral foil of chances, action divides itself from itself. Seen from the point of view of a game of chance, the success of actions proves precisely not to be grounded in any intelligence or rational deliberation that seemed to guide such action at first. But it is only possible to know this fact and base the plan of our lives on such knowledge if we remove ourselves from our own rational deliberation. First, we must keep distance from our first impulse of thinking, and second, we must renounce the idea of actions governed by individual intention in general. This double exclusion of rational deliberation is required in any given case in order to draw the immaterial limit of an event for chance calculation and to circumscribe the corresponding distribution of success and failure. Crusoe begins to make his life on the island narratable with the numbers of the first calculation of chances. In those of the second calculation, he accounts for his story to the reader. In those of the third calculation, he becomes the story of others. The first part of *Robinson Crusoe* progresses no further than these three steps.

Two Versions of the Game of Survival in the First Part of *Robinson Crusoe*

The three steps discussed so far define the extent to which the *aleae geometria* governs what the title calls the hero's *life and adventures* in *Robinson Crusoe*. The extended title phrase might be read in a hendiadys as "the adventure of life," or "life as the misfortune of minimal survival chances." This scansion of *Robinson Crusoe* through the probability formulas is entirely congruent with the exemplary structure of the game in the computation of chances. The first part of the novel—the part that has more and more been taken for the whole in the history of its reception—in particular concerns the text of the game of chance in its foundational layers. Only the continuation then turns to the derived spheres of applied probability and its interpretations.

The alternative between Pascal's and Huygen's underpinning of gaming calculation can help us to picture two possible readings of the novel. The first reading, corresponding to Pascal, takes the process of calculating survival chances all the way. The proportions of chances are available to Robinson Crusoe for making a decision, but as in Pascal's wager, there is only one single game to play and one single decision to make. Robinson Crusoe plays for the final return home to his father; he bets on God's help and intervention for the one defining moment of homecoming. What he decides for is trust in paternal interventions. According to this reading, the odds of survival are at stake in the endgame about the good life, with survival meaning here allegorically to outlive mere temporal existence. Achieving such spiritual survival stands for the life beyond life that represents the stakes and winnings in Pascal's game as well. In this reading, therefore, games and decisions exist, but they do not constitute pragmatic situations in life. The place that gives us the point of view for this reading would be the first arithmetical figure, the maximal chances of dying. In this ontological type of situation, then, Crusoe can indeed calculate his chances. But from this calculation, he can only derive the one conclusion of how small the odds are that he should still be alive in order to carry out the calculation.

A very different reading is also possible, however. Even if none of Robinson Crusoe's three calculations can facilitate an instant decision in pragmatic situations, they still provide models for other cases in which Crusoe chooses and decides between alternatives. Though he lacks sufficient knowledge to enlist the locals' calculation in planning his defense tactics against the wolves, nothing speaks against him or others enlisting such calculations in other such cases. The fact that the figure of arithmetical proportion can never play a role in the context of a current decision would lead to an entirely different difficulty in this reading. A game in which the chances are exactly as the calculation indicates is never the game that is taking place right now with us participating in it. To be more precise, the situations of split-second decisions are not games at all. It is necessary first to recognize some kind of game in those situations; and then we may be able to recalculate that game into the odds of a truly calculable game, that is, into Huygens's standard fair game. For this reason, it is always only someone other than the player, or only a player in a different position, who sees the chance distribution in a given situation. In this reading, the place of the text's second arithmetical figure is the crucial factor. It provides the type of situation in which the one who finds himself in a situation of current action is therefore precisely the one who cannot see its gaming character. In order to calculate, one must know that one is playing a fair game in its precise meaning. But the extent to which we are currently playing a game of chance at all eludes us as we act in the moment, and this is especially true to the extent that we are playing a fair game. Crusoe retrospectively recognizes the structure of chance distribution as a conversion autobiography only in order to realize that he could and should have recognized it sooner.

According to Pascal, the finite being's decision plays in the one and only, all important game; Huygens's corresponding primary attribution of meaning indicates a large contract system to be the theater, the stage of the game, a system built on the assumption of a norm of fair distribution. A great disparity in style and dedication of thought lies between these two choices of emphasis: the finite human existence and the juridical-technical formula of foundational equity. Because humans only have one finite life, an entire lifetime is condensed into the time of a single game of chance in which one decision is at stake. Life has the distinct form of a game, like a cardboard cutout, because decisions must be made in it, only in it, and necessarily in it. Huygens, on the other hand, constructs the artificial time and artificial space of the game for making decisions to act, a game in which some events either occur or do not occur. He slices islands of framed event distributions out of the multifaceted, chaotic lives of humans; and he ascribes to them the game of Aristotelian justice, the same chances for and against the occurrence of the event as their illuminating taxonomy. Two forms of the reality of the game in life emerge from these two concepts: the axiomatics of finitude and the axiomatics of social technology. Both concepts, as their common body of gaming rules proves, and as Heidegger made clear even against his will in *Being and Time*, are two sides of the same coin. The two possible readings of *Robinson Crusoe* stage this divided context of both exemplifications of the game of chance before the arrival of applied probability in the narrower sense. In comparing the two possible readings of *Robinson Crusoe*, we recognize what the axiomatic scene of Pascal's game of existence and Huygens's axiomatic stage of the foundational game of justice have in common. Both are basically games of survival and, while interpreting survival in different ways, both provide the structure of allegorical narrations of such life games. In other words, the two readings inform both the narrator Robinson Crusoe and his narrated life.

Crusoe's minimal chances of survival define the novelistic scenario in which the two primary interpretations of gaming calculation—Pascal's singular game of human finitude and Huygens's social system of games as based on the formula of justice—intersect in a particularly intimate way. It is also and for the same reason a citation of discourse history. Survival chances in the wake of a shipwreck also cite both cases in which the theory of games of chance first hit upon statistical material: marine insurance and annuities. Huygens and his friend Johannes Hudde worked on this problem in the 1690s, and Jacob Bernoulli had asked Leibniz for Edmond Halley's calculations regarding the Breslau death tables. Among the suggestions in Daniel Defoe's *Essay upon Projects*, which is concerned with taxes, pension funds, promoting the sciences, educating women, and the construction of roads and insane asylums, is an accident and widows' insurance for sailors, the "Friendly-Society." The book appeared in 1697 as the first publication distinguished with Defoe's name. The loss of eyes, hands, arms and legs, broken bones, ransom money for people captured by the Turks, sicknesses and inabilities to perform one's profession, and finally death in battle or by drink are all included in Defoe's suggestions for potential insurance services.[17] In the language of the *Essay upon Projects*, which holds political arithmetic and literal irony in balance, Defoe leaves open the question of what had remained unclear between Huygens and Hudde and controversial between Bernoulli and Leibniz. Between the arithmetic of contingency, the calculation of the dice, and the numerical tables of political arithmeticians, there still remains a gap in the theory prior to Bernoulli's fourth book. Bernoulli's own limit values, however, then pose the question as to how a priori calculation, a posteriori determination of event distribution, and statistics relate to one another.

The final accord of Defoe's momentous novel *The Fortunate Mistress . . . Lady Roxana* (1724) results from probabilistic figures in the mathematics of trade, insurance, and interest, which can be constructed on the formula of civil law, *regula societatis*. After Roxana has been parted from the table and bed and been removed from the will of her last husband, who has finally bought her the longed-for noble title, only a fraction remains of the fortune she herself brought into the marriage, acquired as the "fortunate mistress" of the king, among others. She invests this remainder partly in the cargo of a trading ship and the rest in the yield of a commerce raider, which proceed to sink each other off the coast of Africa. Lady Roxana's last gamble, and thereby her whole life, turns out to be an unfair game.

The Third Formula of Probability and the Continuation of *Robinson Crusoe*

Robinson Crusoe can be regarded as a candidate for the title of protonovel or even the first modern novel. Anyone who speaks about novels in their modern sense or about novels in contrast to romances is typically making a presupposition of the following kind: in the modern novel, in contrast to all previous poetological genres, the problem of representation must already be solved in the choice of its contents, the history, in order to be able to appear in the appropriate linguistic realization, the narrative.[18] Or, conversely, whatever claims or promises to be a stylistic and discursive technique of narrating novels necessarily already falls back on a preceding stylistics and discursivity predefined by what it aims to narrate.[19] This formulation can be terminologically enriched by doubling both sides: the text's content, then, is understood to comprise the material and the plot representation at the same time, whereas its expression is seen to be in tune with the discursive representation as well as the stylistic manifestation. In any case, an ultimate identity of construction method exists between the allegorically marked novel and the novel that relies on narrative probability. Alone from the tripartite division of the first volume, which can be viewed preliminarily as an independent entity, a structural connection emerges between the allegorical and the proto-realistic readings of the novel. The allegorical reading starts from the quotients of the overwhelming chances of dying and inserts them as signs of paternal care into the frame of the parable of the prodigal son. The reading of narrative probability places its bets on the small quotients of survival chances, which, despite all improbability, eventually bestow an internal consistency to narrative events from the point of view of the re-calculating narrator.[20] In the crypto-statistical motif of quotients of death or survival, the openly presented direction of events by allegory intersects with the ratio of calculable chances ascribed to events in the course of narration. But so far both readings of the text—the allegory of a conversion and the realism of calculable details—still remain two opposed seesaw views of the same structure.

With the third figure of betting on one's chances, the pragmatic situation of improbable survival, the first part of *Robinson Crusoe* comes to a close already having secured the conditions for its being continued. Such continuability is what characterizes *Robinson Crusoe* as a novel in the broader sense of the word, a sense that has become crucial for the history of literature.

Knowledge and ignorance are closest to each other in the third betting formula of the tactical situation. In this instance, Crusoe is given the danger quotient by others, but in contrast to the first two cases, the impossibility of calculation in the moment of action is not connected to the a priori distribution of chances, which only a narrator can see in retrospect. In the third case, the gap between the ignorance of the actors and the knowledge of the calculators is smaller by far than in the earlier ones, just as the odds of 50 : 1 likewise turn out to be much smaller. This proximity inaugurates the contact between the game that can be calculated and the actions of the characters. If we wish to employ the term "probability" only in the narrower sense of its being modeled by the game of chance and its interpretations from Arnauld to Leibniz and Bernoulli, then this is the move that initiates the game of probability. For this reason, the novel in its modern sense also begins only at this point in *Robinson Crusoe*.

Considered in the hindsight of observation and reflection to be a distribution of chances, the option becomes a probability judgment in the instant of action for the acting person. In this narrower sense, "probability" means seeing the contours of a "chance setup," a state of affairs for which events are defined that can occur or not occur, in the current situation; and at the same time, it means to know that and in what way it is a setup, a contrived situation. Ever since Arnauld connected the anecdote of the traumatized princess with the game for ten ecus, this configuration has found a place in knowledge of literature. Most important, Jacob Bernoulli's application of calculation to the moral sciences and juridical decisions implements just such a configuration. But Crusoe's adventures in the second volume are also experiences in such a reflective form. With regard to them, we might speak of second-order adventures. The beginning of the second volume emphasizes this reflective aspect of the novel quite clearly: Crusoe dreams night after night about his island and, speaking in his sleep, he expresses a desire to return.[21] First he is sent on a path of direct repetition. On the occasion of a business trip, he returns for a brief visit to the island—*his island*, on which Crusoe has never even bestowed a name. The adventures of this journey are now those of others. In them, however, Crusoe recognizes his own past adventures (the narration switches occasionally into a narration about third persons while the first-person narrator Crusoe takes on the superiority of the authorial narrator). After the return visit to his island, Crusoe sets out one last time, but now sailing east instead of west. This trip becomes a journey of reflective adventures from the start: first, Crusoe is entangled in the exploits of others (the ship's crew goes on looting sprees in Madagascar, while Crusoe supposedly tries to prevent the carnage by arguing that their lives do not belong to themselves but rather to the shipping company).[22] Then he becomes an observer who nonetheless cannot avoid participating (after the crew maroon him once again in Bengal,[23] he begins the long trek back to Europe with a caravan that becomes enmeshed in adventures especially in the "no-man's-land" of the Tartars). In the continuation, we thus see Robinson Crusoe in the situation of calculating the chances of the game of others while simultaneously deciding and acting on his own. He hence becomes able to follow the reflective view of probability and calculation in and for his own actions. This consideration shows that the second volume, though neglected in most interpretations, is no mere encore of the first. By repeating the first volume, the second one also reflects on the repetition.

Despite how episodic events in the second volume may appear on their own, when read together with the first volume, the continuation inscribes itself precisely in the repetition and reflection of the original adventures. The continuation reflects on a narration that is characterized by allegory as well as circumstantial realism, and in doing so it becomes the prototype of what we call the realistic novel, which is in fact the allegory of *narratio verisimilis*.

Once again, we read the rhetorical figure of the bet three times. Here, however, it is located in much less prominent places and it no longer refers to immediate questions of life and death. They nevertheless make up miniature models for the problem of realistic narration; and they thereby provide an epistemologically substantial commentary on Jacob Bernoulli's *Ars conjectandi*.

On *his* island, Crusoe advises a young man against the boy's supposed plan to marry a girl without means who, even worse, is a bit older than he is. The youth may very well one day return to Europe, where "it would be a thousand to one but he would repent his choice."[24] The incident is less of a sideshow than one might think at first. Crusoe himself has brought along the young man and the older girl, the youth's mother's lady's maid after rescuing them from a wrecked ship during his own crossing to the island. With them—and in particular with the young man—Crusoe repeats the deliverance that he himself experiences in the first volume. With this advice, bedecked as it is with the figure of probability, Crusoe reminds his own youthful mirror-image, despite being thrown back into a state of nature, not to forget what his decision means in terms of European civilization. This episode is reminiscent of the situation of survival in nature by means of culture's debris.

The figure of calculation appears a second time in Madagascar. Having gone ashore, the ship's crew discusses attacking a supposedly rich inland settlement. Once they approach the villages, however, instead of a wealthy town, they find only a few huts. They are hence faced with a decision: should we attack them in order to extort from them the location of the rich city; or should we go around them and continue our search without any information about our actual goal? The second choice carries the risk of the search becoming laborious and perhaps fruitless. The disadvantage of the first is the danger that the interrogated natives may warn the inhabitants of the wealthy town. One would thus have to kill all of them, "and it was ten to one but some of them might escape."[25] The first-person narrator Crusoe reports the strategic debate, although he does not take part in it. Like a narrator, he presents it in an interior monologue; the responsibility for editing the text remains open. His interest in the Madagascar episode is juridical: he criticizes the crew in terms of the natural law that had once forbidden him on his own island to kill cannibals.

The third occurrence takes place in the land of the Tartars, where the Russians can enforce the law only in the vicinity of their garrisons. Crusoe wants to intervene here in a case of idolatry like an Old Testament prophet, but a traveler from the caravan who knows the region warns him that the Tartars will complain to the Russian governor. If the latter is unable to satisfy them, "'t is ten to one but they revolt," leading to a new Tartar war, the traveler concludes.[26] The Tartars in this situation are clearly structurally equivalent to the wolves that Crusoe manages to defeat in the first volume despite his ignorance of lupine behavior. The probability of the Tartar revolt coincidentally has the reciprocal value of what the improbability of surviving the wolves had been.

Instead of the three improbabilities of survival in the first volume, the second volume is marked by three instances of calculating the probability of dangers. The strategic turn from survival to danger deduces in each case a chance setup from the given factors of a situation, and on its basis the planners or advisors construct conditional arguments. *If* the young man returns to Europe—then the game of chance Crusoe speaks of will take place *over there*. *If* the ship's crew attacks the huts preemptively—then the game of chance intervenes on *this very spot*. *If* the governor cannot pacify the Tartars—then the game of chance exists under *those circumstances* of insecurity. The analysis of the situation proceeds through consequences of conditions and thus converts the deciding person partially and on a trial basis into the observer of his own decision and its framing situation. This operation, by the way, is particularly highlighted in Defoe. The first appearance of numerical figuration in the continuation is characterized by the important feature that Crusoe has control over the condition himself: Crusoe can make it possible for the young man to return to Europe or not, which would be the condition for the probability of his regret about a wrong choice of wives. The first figure of the continuation is hence also closest to those of the first volume. Though it is a calculation before the fact, the pre-calculating Crusoe, as master of the factual occurrence of the condition for the game's existence, remains a character external to the game.

The figures of calculation in the second installment are used to calculate in advance, while those in the first part always occur after the fact. Characters calculate in advance in the many situations that come about through law and relations in war and peace, that is, in the sphere that the eighteenth century referred to as that of moral matters or of civil and political life. It thereby becomes clear at a stroke how the probability of the topics shifts when the theory of chances is applied in its numerical representation. In the realm of topical *argumenta*, all partners in the debate have the power of topical persuasion equally at their disposal (the Aristotelian model of the *endoxa*), and in the judicial contest of speech, the access of the speaker to the potential of persuasion and the openness of the audience to such persuasion are conceived of as fundamentally symmetrical (Cicero's model of *fides*). The calculation of probability at first does not change anything with respect to probability's precision or reliability; it does not bring about any more probable probability. Before it is possible to calculate, it instead changes the space of argumentation much more through the unnoticed introduction of the example of the game. The example of the game draws fine yet precise borders between the argumentation and the matter at hand. In this more elaborate sense, only the figures of chance in the second installment are the first true rhetorical parallels of calculated probability. The difference at issue can be very clearly expressed with the help of the distinction between observations of first and second order, a distinction familiar to systems theory.[27] As illuminating and vivid as this differentiation between first- and second-order observations is, however, it renders invisible precisely the process and the transition at stake here in historical and systematic respects. The difference between a first- and second-order observation always presupposes its own mechanism as a given. It cannot itself comprehend the question of the transformation of a first-order space into a second-order space. It cannot show as a rhetorical process how the first volume of *Robinson Crusoe* continues into the second volume as its repetition. This can, however, be seen particularly in the figures of gambling odds. If we always already know where the game and its precise rules begin and the seriousness of mere life ends whenever we launch the example of the game and the wager, then in fact, as the sociologist Niklas Luhmann thinks, everything will run smoothly because it has always already run. For this to be true, however, games and wagers would have to work before anyone had provided their rules and decreed their validity.

The mixed cases in the second volume of *Robinson Crusoe*, which combine observing and acting, and which aim at what can be called dangers, allow for pre-calculation in the sense of mathematical probability. In the chance ratios of the first volume with their fixation on survival, this possibility does not exist. Acting and observing there are always divided. If in fact calculative foresight is understood as necessary for personages in the modern novel, then only the second volume of *Robinson Crusoe* can be said to count as Defoe's first *modern* novel. More precisely, only the second volume's continuation and reflecting repetition of the first, which is characterized by allegorical figuration as well as the logic of chances, can be called a novel in the modern sense. Jacob Bernoulli proceeds in a similar way when he interprets odds calculation as probability. He tries to understand the situation of gaming as certain scenarios in life, and vice versa. Accordingly, in the kind of novel realized in the second volume of *Robinson Crusoe*, the first-person narrator and the narrated "I" also always, but never entirely, take the perspective of someone who could already experience his strategic position as that of a game. In the passages of the continuation marked by probability figures, the everydayness of probability judgments becomes apparent. They incessantly repeat either the impossible decision of the finite person about the infinite or the groundless ordering of lawlessness through justice. In this sense, all the adventures of the second volume are adventures in a situation of pragmatically reckoning with chance. The possibility of such adventures emerges at the end of the first installment with Crusoe fending off the wolves in the forests of Languedoc. The situation is that of danger. Danger then can be understood as a situation in which the values of probability that are valid for narratives are accessible within the process of narration, that is, at least translatable between the narrator and the narrated figures. Modern novels are narratives of dangers that are no longer actual, but only repeat and reflect on the contingency of survival.

Allegory of Contingency

The adventures of the second volume, starting out from the third figure of the wager at the end of the preceding volume, and carrying on into the figures of probability, are adventures that mimic those of the first volume as an original scene, or that play out on a stage that uses the first volume as a backdrop. The long chain of meditations about divine providence continues on from the first appearance of the figure with the smallest chance quotient for survival. They eventually come to a crisis, which Paul Hunter understands as a conversion, in a feverish delirium Crusoe suffers. He cures the fever with the *medicinae mentis et corporis* he has taken from the Noah's Ark of the sunken ship: with tobacco, which Crusoe recalls the Brazilians having used as medicine; and with a Bible, whose European use he recalls for the first time.[28] Crusoe ends the biblical meditations with a practice of pre-probabilistic holy contingency, the random choice of scriptural passages: "I daily read the Word of God . . . . One morning, being very sad, I opened the Bible upon these words, 'I will never, never leave thee, nor forsake thee.' Immediately it occurred that these words were to me."[29]

Crusoe adds an abbreviated form of the apologetic wager: "if I had all the world, and should lose the favour and blessing of God, there would be no comparison in the loss." This consideration allows him to undertake the following intertwining of possibility and probability: "From this moment I began to conclude in my mind that it was possible for me to be more happy in this forsaken condition, than it was probable I should ever have been in any other particular state in the world."[30]

A comparison of possible states of the world according to considerations of probability displays the possibility values of the blessed life. This or some other state of the world is possible, and a blessed life is more probable than improbable in this state of the world. The contingency of states of the world and their respective shares of happiness and unhappiness provide the structural basis for the allegories of the improbability of survival. Neither Crusoe as character or as narrator, neither the author nor Defoe, had to have read Leibniz in order to give voice to the argumentative framework of the doctrine of the best of all possible worlds. Hans Blumenberg sees the possibility of the modern novel in Leibniz's Christian metaphysics of contingency. No matter to which of these four persons

Defoe's Robinson Crusoe, *or, The Improbability of Survival*   191

(Crusoe the narrator or the character in the novel; Defoe or the author) we ascribe this sentence, it shows that the introduction of the example that understands the game as a model of life occurs in the text of the novel itself. Everyone—narrator, character, writer, and author—participates in it.[31] In this sentence, the interpretation of the game of chance as probability becomes the foil for that particular novelistic reality in which we, according to Blumenberg, live to this day.

PART II

VERISIMILITUDE SPELLED OUT

The Appearance Of Truth

8

Numbers and Tables in Narration

JURISTS AND CLERGYMEN AND THEIR
BUREAUCRATIC HOBBIES

Probability as Verisimilitude: From the Semblance
to the Appearance of Truth in the Novel and the
Statistical Table

In novels of the eighteenth century, the final outcome of a series of poetological redefinitions that had been carefully prepared for since the Renaissance finally crystallized into a theory of aesthetic form. Probability no longer denoted merely the limited kind of knowledge that brings poetry closer than history to the philosophical cognition of laws and causes. Probability—or verisimilitude—we might say, was recognized no longer only as the semblance of truth but also as its form of appearance.

With this hypothesis, this study has reached a turning point with regard to interest in and understanding of probability. Probability now comes under scrutiny for its aesthetic quality, or rather its transformation into an aesthetic quality. In order to explain this turn further, a few remarks on terminology will prove helpful. The term "probability" is used throughout this study as a general designation for both of the two Latin words *probabilis* and *verisimilis*. Though they connote different meanings linguistically—*probabilis*, the search for confirmation and evidence; and *verisimilis*, the reference to truth and its manifestations—there has never been a clear-cut conceptual differentiation between the two terms, in either rhetoric and poetics or philosophy. This state of affairs is aptly reflected in the German situation, where *Wahrscheinlichkeit,* the literal rendering of *veri-similis,* is indeed used for mathematical probability as well as for rhetorical and poetical probability or verisimilitude.

In part II of our investigation, the *verisimilis* function of probability will become the dominant aspect, that is, the reference in probability to truth and its forms of manifestation. Put more precisely, the differentiation at issue here plays between two different sorts of such manifestation. The traditional rhetorical and poetical concept of the *-similis* in *verisimilis* can be understood as a kind of likelihood that implies the semblance of truth. Eighteenth-century aesthetic thinking increasingly begins to conceptualize the *-similis* instead as appearance, an appearance that may be either false or true but that is seen as the way things manifest themselves. Again, these two understandings of *-similis* coincide felicitously in the German word *Schein* (as in *Wahr-schein-lichkeit),* where *Schein* can mean both semblance and appearance. As appearance, *Schein* could be understood as the way in which observers perceive how law and causality manifest themselves in the play of phenomena. In this vein, Alexander Gottlieb Baumgarten, in the first volume of his *Aesthetica* (1750), could proclaim aesthetic truth to be probability, the semblance and appearance of truth.[1] Baumgarten, author of the most famous compendium of metaphysics in eighteenth-century German philosophy before Kant and connoisseur of ancient poetry, never explicitly refers to either probabilities or novels in his writings, but he nevertheless composes a framework for their configuration. The figures from ancient poetics in his *Aesthetica* become schemata of intuition (*Anschauung*), whereas probability is conceived of as graded, calculable truth. The modern novel, however, is the very genre where—both literally and figuratively—gradable and calculable probability could schematize the perception of a reality that offers itself through narration.[2]

Two ensuing discourses emerged from the logical and poetological tradition of probability in the eighteenth century. One of them is mathematical probability, the semantics of the probable modeled on the structure of the game; the other is the semblance of truth converted to its very appearance, or the phenomenologization of poetics in aesthetics. In the novel of the eighteenth century, these two descendants of the probable revealed their intrinsic connection. The probabilistic formula of this connection had already become evident in the Port-Royal *La logique, ou l'art de penser.* There, the trauma of Arnauld's princess finds its rational resolution and healing in the ten players' probabilistic calculation of chances. Trauma and the probabilistic model of gaming remain separate in Arnauld's work, but a portrayal of their unison is foreshadowed in the many whose passions follow the statistical probability of win and loss in the collective game of chance, the lottery.

The kind of phenomenologization that comes to light in the transpoetological, aesthetic constitution of the modern novel complements the process of interpreting the game as probability. In probabilistic epistemology, probability is the construction of meaning according to mathematical models; in the politics of the semblance and appearance of truth, probability functions as the perception of meaning. As early as Aristotle's *Poetics*, to be sure, the notion of *eikos* as the domain proper to literary work connects the aspect of poetic representation (*mimesis*) with the way the many live and act, and hence with counting. But the *Poetics* does not pose the question of what portraying the actions of men has to do with determining the probability of such actions through counting. It is only with the uncoordinated and mutually independent elaborations of Aristotelian probability in the eighteenth century that the deeper conceptual relationship between—aesthetic—representation and—probabilistic—counting becomes thinkable. Probability, according to Baumgarten's phrase, is aesthetic truth. The two modern derivations of the probable clearly diverge widely in their respective forms of elaboration. The probabilistic revolution—as Lorenz Krüger's research project was named with good reason[3]—appears as a machine that constantly reprograms its own applications anew and in ever more intensive and extensive ways. Mathematical probability is the colonization of reality through mathematical constructions that has continued to spread and become stronger from the eighteenth century to this day. The aesthetic experience, in contrast, has been justified from Baumgarten on as an alternative to the epistemic construction. Since the nineteenth and through the twentieth centuries, it has increasingly taken on the role of the custodian of a remnant of true experience, *Erfahrung,* that had to be won back from the construction of reality and preserved in its proper status. Mathematical probability and the aesthetic appearance of truth are connected not merely through their common derivation from the logical-poetological term "probability." In the eighteenth century, they still belonged to a common space of discussion and thinking. The fact that the calculable and construable world can never entirely free itself from perceivable reality was a fundamental determination of this discursive space. This characteristic was shared by the modern novel and the *logica probabilium* that counts among the core elements of the German Enlightenment from Leibniz to Kant. Reality, however—and this claim goes beyond the case studies of the eighteenth century presented here—is constituted by this very relation between perceivable and construable meaning; it is a name for the tension between probability, including mathematical probability, and verisimilitude, the semblance or appearance of truth.

We shall reach this theoretical level in our material explicitly, however, only in the later chapters of part II, starting in particular with the discussion of Johann Heinrich Lambert (chapter 12). This contemporary of Kant, known for his work as astronomer and theoretician of optics, produced a doctrine of phenomenology that embraces epistemological probability as well as an aesthetic doctrine of the appearance of truth in the Baumgartian tradition. Before we can turn to Lambert, however, we must explore the even deeper implications posed by the configuration of probability's representation in words or numbers. Semblance or appearance of truth takes place first of all in verbal representation, while mathematical probability requires numerical forms of expression. Both forms of representation can include diagrammatics, tables, and images—figuratively as well as materially. The question of verbal or numerical presentation is an internal relation and even a debate already within statistics, particularly in the way the discipline originated and was taught and developed in German academia.[4] As we shall see, some of the fundamental questions of any history of probability and probabilistic statistics return in this context: What set the stage for mathematical probability to use statistical methods and be applied to statistical data? Is this a development engendered by mathematics, and hence scientific thought in general, or was it rather propelled by the social and administrative complexes and problems that manifested in the mass of statistical data of population, wealth, and economic resources of the state? As far as these questions concern the forms of representing probabilities, they belong intrinsically to the historical contexts of our account of the formation of probability.

Two-Ply Statistics: Numerical (Political Arithmetic) and Topical (*status praesens* and *Notitia rerum publicarum*)

The proto-statistical literature that we find between the middle of the seventeenth and the beginning of the eighteenth centuries offers a confusing picture. Today's scholars most commonly read the documents to discover clues about mathematical probability, though it is difficult to determine the concrete relation between them. A number of different disciplines, moreover, lay claim to these texts as part of their own prehistory: statistics does so in a narrower sense; but demographics and political economy also look to these writings as documents of their beginnings.[5] We shall not be dealing with such questions of descent and genealogy here, however. Instead, this study aims at a typology of this literature's aptitude to be processed in the apparatus of probability calculation. In pursuit of this question, we shall deal with forms of representation in data, which only obtained the statistical sense we give it today toward the end of the seventeenth century.

John Graunt's *Natural and Political Observations upon the Bills of Mortality* (1662) has often been seen as anticipating modern demography or economic statistics.[6] But neither of these disciplines convincingly captures what Graunt was doing by correlating the London mortality numbers with the utmost hodgepodge of social and economic factors. Graunt first of all developed and defined the characteristics of data collection, the basic activity of a man of science as demanded by the Royal Society. Fittingly, his *Observations* are dedicated to Sir Robert Murray, one of its founding members. According to the Royal Society's charter,[7] it was not so much the case that there had hitherto been a dearth of observations or a lack of acumen in the ways to use them. It called instead for the implementation of new methods to raise and preserve those observations in such a way that they might be accumulated and compared. This proclamation was the harbinger of a change in the use of material observations: instead of combing them for direct messages and significations, an activity is fostered that we might describe as data collection.

Graunt's book is a striking case in point for this change. In it, there are messages and significations that over the seventy years of their existence in London, everyone had read the monthly *Bills of Mortality* without ever suspecting that they held in their hands a compendium of data. For this view to change, it was necessary that the Bills should be observed as raw material rather than read as a text for meaning. Graunt speaks about this in the preface and the dedication of the *Observations*. As Graunt notes in his foreword, the usual readers of *Bills* had long seen only the increase and decrease of burials. Doing so had provided them with conversation material and, in the case of approaching epidemics, had instilled fear and panic. "Now, I thought that the Wisdom of our City had certainly designed the laudable practice of taking, and distributing these Accompts, for other, and greater uses," he writes.[8] The attitude of reading and understanding had to be broken in order to extract the hidden data from the signs. Graunt claims not to know the particular higher purpose of the wise London city government in publishing these accounts. His professed ignorance provides room for a more important, general kind of purpose, which lies in the bare fact that it is numbers that are collected. Numbers can be processed in formalized steps. Numbers are data. A rewriting, or more precisely a graphic transformation, is the determining factor for Graunt's *Observations* from the point of view of the history of representation. He speaks of "having . . . reduced several great confused *Volumes* into a few perspicuous *Tables*."[9]

This is the necessary precondition under which Graunt claims in the preface he felt encouraged to dedicate his work to Sir Robert Murray; and it is conversely also its foundational epistemic gesture. Not burdening Lord Robert, who at the time of the dedication had just become Lord Privy Seal, with thick volumes meant breaking off the reading of a text and instead constructing *perspicuous tables*. The figure of a nonreading attention to the printed characters is thus invoked, an attitude that would have a great future in the sciences as well as in literature and aesthetics. This reading without a reading is characterized in rhetorical terms by *evidentia*, the figure of placing-before-the-eyes in tableaux, a reading that seems to work by transforming a text into a picture or a diagram.

At the beginning of Graunt's book, the reader finds an "index of the *Positions, Observations, and Questions* contained in this Discourse," with one hundred and six entries. The refinement of elaboration in these individual sections is beyond dispute with regard to their respectively chosen correlations as well as their mathematical treatment. Nevertheless, there is no unified conceptual or mathematical apparatus for the construction of those primary signs into data, nothing that would explain how they can function as that which is "given" (data) for further processing. The one hundred and six "observations" and "questions" elicit the character of being data for Graunt, not despite the invisibility of purpose in their choice and presentation, but precisely because of the lack or rather the erasure of any purpose. Lack of purpose is exactly what defines data. The impression of such lack of purpose cannot simply be ascribed to the preparatory character of Graunt's book. It rather belongs to the very nature of what would later, taking a hint from the title of an essay collection by William Petty, be known as *Political Arithmetic* (1690). Petty's *Political Arithmetic* similarly places various social and economic correlations alongside one another in the form of a balance, of bookkeeping. The higher purpose or "greater uses" of which Graunt speaks in this moment, is again marked as an empty space. Arranging fields of data is the use that begins first by extinguishing traditional semantic purposes.

Petty clearly seems to have based his *Anatomy of Ireland* (written in 1672, published in 1691) on the typical early modern concept of the body politic. But the concept takes a peculiar turn in this case. Ireland offered the monstrosity of a European state being framed in the laboratory situation characteristic of colonies. In a striking shift of political body metaphorics, Petty claims that Ireland is suitable for political-economic anatomy because it is only twenty years old and he has known it since its embryonic existence. With this, Petty is alluding to the 1652 Act of the Settling of Ireland, which regulated the settlement of Englishmen in Ireland after Cromwell had finally suppressed the revolt against English rule.[10] The *Anatomy of Ireland* is, we might say, a founding work of numerical statistics. But it is such only because it is does not apply to a state, but rather to individual fiscal, business, and economic policy problems within a state. The *Anatomy* is simultaneously the incision of a deceased body and the study of one that had just been formed. Petty can refer to neither a measurement nor a formula for whatever it is that makes the ensemble of his balanced materials into the presentation of a political body. If it is true that numbers present a colonized space of administration rather than a state, and if such numerical presentation is reached through a deliberate erasure of the usual "purposes" of verbal representation, we may conclude that the desemanticized measurements of Petty's *Political Arithmetic* in fact equate to the colonization of one's own land.

Political statistics, on the other hand, cannot do without the framework of a national state. Only the frame determines what random surveys and probabilistic calculations might signify—even and in particular if census takers are as obsessed with detailed rubrics as they would become, for instance, in Frederick the Great's Prussia (the charts of the administrative districts ordered by the king after the Seven Years' War consisted of ninety-seven folio pages with over ten statistical variables in the headings).[11] No statistics are possible without reference to a political unity, whose reflection it is supposed to provide in high resolution. In the traditional language of scholarship: the topics of the state are required for the numbers to speak. In the seventeenth century, however, numbers were still not expected to come up in these topics.

Two discourses served as starting points for numbers to find their place and meaning in statistics. One of them can be exemplified by Gottfried Achenwall's *Notitia rerum publicarum* (1748), the title of which refers back to the index of public offices in the late Roman Empire, or *notitia dignitatum*. The other can be seen in Edward Chamberlayne's *Angliae notitia, sive praesens Angliae status* (1686). This title distinguishes descriptions of the state of a nation from historiography, on the one hand (with statistics recording current as opposed to past data), and politics, on the other (with statistics treating the respective state of nations rather than a general theory of forms of government). Most prominently in France and England, but also in Holland, works began appearing around 1700 that provided general accounts of single nations, and usually for a single year. Those works would use the familiar topoi for describing the state of the nation as regards population growth, economic wealth, military power, and so on. Their authors were educated publishers. They often came from offices in the court or in the administration; but even so, they were not writing in line with their official courtly or administrative duties, or did at least not directly so. For example, Jean Pinson de La Martinière's *Estat de la France* undertook to portray "how France was governed in the year 1648," beginning with an account of the constitutional law and dynastic history of sovereignty in France and the current cast of officeholders in the government, from the monarch down to the parlements. In 1656, and with a new edition in 1661—hence in the year of Mazarin's death, in which Louis XIV assumed personal control of the reins of government—Nicolas Besongne brought out a follow-up publication entitled *Le parfait Estat de la France*.[12] William Aglionby's *The Present State of the United Provinces of the Low-Countries* appeared in in 1669 (2nd ed. 1671); also in 1669—between Graunt's *Observations* and Petty's *Anatomy of Ireland*—Edward Chamberlayne, a historical-political writer who was a founding member of the Royal Society, published the *Angliae notitia, or, The Present State of England*. Despite the title, it was less limited than the *Estat de la France* to the *notitia*, the list of offices. Chamberlayne instead embedded the administrative listings and the legal or dynastic legitimacy of the monarchy between two sides of the *status praesens*: between descriptions of the country, climate, inhabitants, and economic conditions, on the one hand, and descriptions of religious confession, coinage, legal institutions, and cultural customs, on the other. Abridged Latin versions or expanded English new editions continued to carry the *Angliae notitia* on into the later eighteenth century.[13]

The spaces in the lists and the subjects from the topics seem to provide a model for processing statistical data, except for the fact that they have little or no use for data of the kind Graunt and Petty constructed and handled. Traditionally, topoi had been seen as schemata for constructing arguments in dialectical probability but not as blank spaces to be filled with raw data. If such an expectation arose with the practices of the Royal Society and its call for a data-based science,[14] the change in attitude owed its existence to a certain pressure on the Aristotelian tradition to do scientific work. After a first attempt in the early seventeenth century (Keckermann), Hermann Conring and Andreas Bose, two scholars and courtly advisors in diplomatic and legal affairs,[15] redefined the *status prasentes* and the *notitiae* for such purposes. They outlined an academic discipline that was to embrace the world of nations in the European context and its understanding of the body politic.[16]

Not surprisingly, Arno Seifert's authoritative research on the early modern concept and practices of *historia* in the Aristotelian sense focuses in great detail on Conring and his invention of the *notitia rerum publicarum*.[17] According to Conring, the *notitia rerum publicarum* had to be squeezed in between history's knowledge of singularities (stating that something is what it is) and the general constitutional theory of political science (asking the question of why something is what it is). Many attempts at clarification and redefinition in early modernity tackled the opposition between *historiae* and *scientia* in Aristotle. The central point in this context was poetry, the realm of probability stretching between historiography in the narrower sense and philosophy as the quintessence of true, universal knowledge. The *notitia rerum publicarum*, the study of nations, do not, however, merely pose this problem as natural history, historiography, or practical philosophy had framed it; instead, Conring poses the problem of how to achieve such a point of convergence between *scientia* and *historiae* in the first place. This quasi-empirical turn of Aristotelianism in Conring achieves its most precise point with his understanding of statistics. Happily underlining the difficulties of his enterprise himself, Conring isolates the point of convergence from two sides. Explicitly, he deduces the necessity of a probable knowledge of the state and its functions out of the requirements of *prudentia civilis*, practical politics. The practical syllogism of the *politicus* connects the major premise from political philosophy with the singular circumstances in the minor premise in order to generate the political decision as the *conclusio*. Implicitly, however, Conring constructs a convergence of philosophical universality and "historical" singularity as they are characteristic of the topics of the state on the model of poetry in Aristotle's *Poetics*.[18]

The material for *notitia rerum publicarum* is thus collected and arranged in the blank spaces of the topoi: What is the extension of the state's territory? Is it whole or divided? Are there mountains, lakes, forests, sea coasts? Is the population great or small? What is the climate like? Are the inhabitants industrious workers? Are they suited to discipline and battle? In the terms of modern systems theory, the *notitia rerum publicarum* supplies a self-description of a nation by filling out these topoi; it constructs the state's historical semantics. The choice and organization of the topoi ensue hierarchically, as is the case with all well-composed compendia of topical knowledge in the seventeenth century. In the outline of the *notitia rerum publicarum*, the central perspective or topos is the *causa finalis rei publicae*, the ultimate purpose of the state. Conring and Johann

Bose disagreed about this fundamental decision right from the beginning. For Conring, who thought along the lines of a social order based on the three estates, the aim of the state lies in the *bonum* of political wholeness and in the *bonum* of the rulers. For Bose, a follower of Lipsius's doctrine of princely absolutism, the state's ultimate purpose lies in its strength and the growth of its *imperium*.[19] These final causes of the *notitia rerum publicarum* may remind us of what Michel Foucault famously called *gouvernementalité*, as distinguished from strictly constitutional thinking in early modernity. In constitutions, something like this summum bonum of government would find its place only much later. In the New World, it was called "the pursuit of happiness." Since that time, this element of statistical self-description has also entered declarations in (at least the republican) constitutional texts of Europe.

With Conring's and Bose's respective projects of *notitia rerum publicarum*, the state's self-description—required in the practical syllogism of the *politicus*, and settled in the place of poetry—still remains a text and not a tableau of numbers. Even Veit Ludwig von Seckendorff's 1656 *Teutscher Fürstenstaat* shows in its text a narrative account of what makes a nation into a nation, only the diagram of an empty form to be filled out both in words and in numbers in order to accomplish a "description of a country and principality as a whole."[20] The book contains a collection of tables for dividing and describing a given princely territory in districts in order to facilitate the survey of their demographic and economic situation. Depicted "as a whole," the principality embodies the Lipsian style of government with its anonymity of universal singularities ("offices of the principality of x"),[21] and thereby with the coincidence of enforcing standards and standardizing description. Embedded in political philosophy, statistics indicate the invention of the territorial state as an entity that has successfully colonized its own territory. Much can be said for the hypothesis that, some ten years after the Peace of Westphalia, when Seckendorff composed his hybrid of political philosophy and instructions for designing statistical reports, he was following a wave of actual surveys carried out by the respective principalities in the Empire after the Thirty Years' War.[22] Without these surveys, the undoubtedly justified tears over the ubiquitous destruction of the fatherland could not have become the tonic keynote of baroque poetry in Germany.

It was not solely due to arcane politics that after the questionnaires were filled out and compiled, they remained hidden in the secret archives of territorial states and their absolutist administering of reconstruction. Nothing between 1660 and 1740 indicated that publishing the numbers could ever acquire a political meaning for itself, as it did later in the Napoleonic state, in the Prussia of the Napoleonic wars, or even later in the citation of public statistics in Weidig's and Büchner's manifesto of the social revolution, the *Hessische Landbote* (Hessian Messenger, 1834). In the earlier eighteenth century, numbers remained disassociated from the semantics of self-description. They were—still—not yet ready to be processed as data of the state.

Table and Evidence Around 1700

We hardly find any intersections between political arithmetic and the semantic topics of the *notitia rerum publicarum* or *status praesens*. One exception turns out to be the manuscript of a numerical state budget composed by the diplomat and heraldist Gregory King (ca. 1696).[23] In it, numerical statistics did appear, but in the framework, not of the self-description of the state, but rather of the strategic situation of the Nine Years' War (1688–1697). Comparing the allies Holland and England with their opponent France motivated and determined this survey of population and military and economic strength.

We must not, however, retrospectively equate the division into "political arithmetic," on the one hand, and *status praesens* and *notitia rerum publicarum,* on the other, and their fusion with nineteenth-century statistics. In particular, the methodological debate around 1800, which targeted the relationship between textual and numerical statistics with great keenness, cannot be used to explain the process of fusion. At the beginning of the nineteenth century, a discussion raged about the possibility and meaning of statistical laws. Were there numerically statistical laws, which would then be merely laws of society, before or outside of the legal and constitutional laws, the normative laws of the state that required textual form?[24] This question was firmly anchored in the theory of statistics in nineteenth-century Germany. Prussian theoreticians especially insisted that the law of the state must never be reduced to the merely numerical laws of society.

They projected this debate back to the opposition between German academic statistics as defined by Bose and Conring and the English political arithmetic of Graunt and Petty. For the Prussians around 1800, an integral statistics meant a German description of the state enriched with English number-crunching techniques, while according to the Anglo-Saxon tradition up to this day, Conring and Bose contributed little more than headings and names for the numerical statistics of bookkeeping, which in fact developed out of political arithmetic. Both genealogies miss the fact that representation through tables that combine words and numbers resulted from an independent decision that can be traced neither from the history of topical statistics nor from political arithmetic. The decision on the form of representation was instead involved with the territorialization of the state and the nationalization of society.

The statistical table connects numbers with alphabetical and linguistic elements in a diagrammatic frame. Thus it forms a component of statistics that registers the numerical notation of political arithmetic within the topical frames from *notitia rerum publicarum* and *status praesens*. In the terms of John Graunt's "perspicuous tables," this configuration of words and numbers is tantamount to the aesthetic construction of diagrams, quasi-pictures that lend the embedded numbers and words a specific visual and cognitive force. To translate Graunt's phrase "perspicuous tables" into contemporary usage, we may also say that tables are evidential. Evidence—*evidentia*—in rhetoric means showing rather than telling. Put more precisely with respect to the table, the numbers embedded in the frame of the text can be said to show something that they do not express semantically. Or, in other words, diagrammatic configurations of number and words express something that they only show. In the times and circumstances of a Graunt or Petty, such tables belonged behind closed doors, however, in secluded spaces barred to the public. Administrators and members of learned societies employed them only for their internal modes of communication: in those realms of seclusion and controlled attention, the *evidentia* of tables exercised its specific force to function as immediate "evidence" in the narrower sense and, at the same time, as topical support for argumentation and "proof." Toward the end of the seventeenth century, such tables appeared in two contexts. First, tables exhibiting numbers were used in legal matters. The calculation of interest and similar contract-related situations belong to this mode of drawing tables, as do the actuarial charts of insurance companies, which later also employed particular tables when infinitesimal nesting was required. Prototypes of both sorts are found in Leibniz's pioneering 1683 essay, "Meditatio juridico-mathematica."[25] All of these mathematical considerations have a juridical character because of the underpinning relation of their operations to the formula of distributive justice. The second type of table use was for representing data collected under certain categories. Again, Leibniz is the primary example. He collected data of various kinds, encouraging the development of medical statistics and mortality tables, and supplied Caspar Neumann's Breslau mortality statistics to Halley, who based his calculations for life insurance on them. Leibniz continually connected his plans and work for a scientific academy with expanding these collections of data beyond merely private initiatives. In particular, he offered detailed suggestions about *Staatstafeln*—"state tables," or statistical surveys—for the Holy Roman Empire and the electorate of Hannover.[26]

In the terms of old European scholarship and its Aristotelian structure, all these instances of recording and presenting materials belong to the realm of *historiae*—whether this is the *historia* of a physical experiment or of the observation of rain in a specific area or the mortality in Breslau. These tables, *historical* in the traditional sense and *empirical* in modern parlance, convinced Leibniz of a case for calculation hidden in them. For Leibniz, such calculation could not be made visible in an explicit mathematical structure. Nonetheless, lawfulness sprang from historical-empirical tables. Both kinds of table—the table that arose from juridical semantics and the table that came out of "historical" singularities—raised the classical problem of evidence (evidence in the legal sense and evidence as proof) as characteristic for mathematics. For one thing, we have a numerical series that precludes our ability to grasp the formula behind them intuitively. On the other hand, we have a series of numbers that we size up with an intuitive view of its underlying law and pattern, but without attempting to determine a formula for its production at this juncture. Book 4 of Jacob Bernoulli's *Ars conjectandi* is located in the vanishing line of the convergence of both tables. Nevertheless, a distance was sustained between the two at least until into second half of the eighteenth century; "juridical" (and "economic") mathematics and "historical" statistics (of empirical data collection) remained separate undertakings.

Tables in the Eighteenth Century, 1: From Johann Friedrich Polack's *Mathesis forensis* to Applied Mathematics

In 1734, having just broken off his study of law, Johann Friedrich Polack became a professor of mathematics in Frankfurt on the Oder, which in the early eighteenth century was a bastion of Prussian police and administrative science. He would make a name for himself with the publication of *Mathesis forensis* (1740), a book that established an entire discipline despite its modest size.

Forensic mathematics, according to Polack, has to do with using geometry and arithmetic for surveying calculations or structural engineering, as well as with the theory of series formation as required in calculating interest. The first truly juridical chapter deals with the formula of commutative justice (an arithmetical proportion to be applied, for instance, in the division of an inheritance) and distributive justice (the geometrical proportion used to calculate profit and liability for corporations). Then Polack adds the interest calculation that Leibniz had placed under the title "Meditatio juridico-mathematica." The pool of mathematical-juridical themes in their first and narrow sense is thereby exhausted.

The mathematical nature of justice was clearly so evident for Polack that he scarcely felt a need to make it explicit: it is impossible to miss the correct understanding of what is just and fair, even without using outright calculation; and similarly one is already calculating in understanding the mere meaning of justice. On the other hand, Polack says, it is "entirely tasteless to measure the moral concepts, as legal studies have attempted to do, with mathematical yardsticks and measurements, the like of which length, width, and depth can well be used for bodies . . . , but cannot be applied to . . . legal concepts."[27]

The numerical representation of justice, which for Huygens and Leibniz had grounded the unity of mathematics and law, is veiled in this *Mathesis forensis* precisely because it seems so self-evident. Polack's argument hence culminates in a contradiction: justice presupposes calculation, although its concepts cannot be comprehended in numbers. For the Wittenberg professor Gebhard Christian Bastineller, who presided over the defense of a mathematical-juridical dissertation in 1741, only the implication of property and boundaries remains as a justification for juridical mathematics: "It is necessary to lay down boundaries because . . . nature is opposed to a common ownership of things."[28] In another example, the Leibnizian-Pufendorfian tradition of probabilistic conditional contracts found in Polack's *Mathesis forensis* undergoes a similar process. The work in question here, a juridical treatise from 1731, speaks of *aequitas in probabilibus*, that is, the right of one's expectations in purchasing things as expressed in mathematical terms. It performs such an application without paying any further attention to the fact that, methodologically, equity and probability already imply each other.[29] In a similar vein finally, many early German Enlightenment writers, such as Christian Thomasius at the beginning of the century,[30] would combine law and mathematics in the contexts of series and function theory, probability calculation and justice, without discussing the grounds on which such a configuration of law and calculation was possible in the first place.[31] *Mathesis forensis* builds on the implications of justice and probability in defining single areas of mathematical expertise in law but tends to forget why it could do so.

Polack came up with the title of *Mathesis forensis* based on the model of *medicina forensis,* forensic medicine. This appellation characterizes it as expert science in the service of the law, and no longer, as for Leibniz and Huygens, an immediate and necessary part of law in the representation of justice. It rather plays over into the law the way that medical knowledge of fatal wounds can do. It serves law. Polack gives a particular explanation for why jurists themselves desire to be trained in this complementary science. Applied mathematics is a requirement for the many jurists who do not become lawyers or councilors of justice and who do not pursue their careers in courts of law, but who find employment rather in the diverse commissions of the state for the administration of roads and real estate, for building inspection and financial supervision. Precisely in leaving behind corporate law, interest, and insurance, fields that are structurally related to probability, *mathesis forensis,* the mathematics of law, constitutes knowledge for jurists who work in "Policey," that is, in all administrative efforts for maintaining order within a given society.[32] And, as Polack emphasizes, the majority of jurists do just that.

The discourse of legal mathematics is interesting because it became applied mathematics at the point where it simultaneously named and forgot the example of games of chance. The dispersal of jurists into *policey*, which Polack identifies as his didactic challenge, is the pragmatic side of the displacement of theory. The former graduates of law schools, who scattered into the departments of the central state administration, or *policey*, from now on set themselves a new task outside of the law proper: they began bringing forth the "avalanches of numbers" that Ian Hacking speaks of.[33] They were thus the forerunners of an army of cameralistic mathematicians in the second half of the eighteenth century. The message of the earlier Enlightenment generation of Polack, Johann Ulrich Cramer, and Georg Bernhard Bilfinger in the 1730s and 1740s, who still closely followed Leibniz's juridical-mathematical writings, was conveyed and transformed, for example, by the famous Göttingen mathematician Abraham Gotthelf Kästner.[34] As in the generations of Huygens or Leibniz, he first studied law (1731) before dedicating himself to mathematics, though in his case the legal studies only served to compensate for the unfulfilled ambitions of his father, a Leipzig law professor.[35] Kästner took advantage of the legal education forced on him for bread-and-butter work throughout his career by translating a significant number of contemporary works in juridical mathematics.[36] In the system of his own later compendia, the objects of *mathesis forensis* were given a new name: *applied mathematics*.[37] The juridical-mathematical core was continued here in two ways: in supplementary mathematical courses for civil servants and in an applied mathematics subdivided according to physical and technical fields. Kästner's transition from *Mathesis forensis* to applied mathematics opened the floodgates:[38] in 1788, Friedrich Meinert, a professor of philosophy at the University of Halle, published the first of his countless textbooks of this kind, *Über das Studium der Mathematik für Juristen, Cammeralisten und Oekonomen auf Universitäten* (On the Study of Mathematics for Jurists, Cameralists, and Economists at Universities). Even though Polack's *Mathesis forensis* remained the model, jurists only made up part of the Halle audience, among whom cameralists and economists were now also present—precisely the aspiring bureaucrats into whose ranks Polack had sent the graduates of the law faculty from Frankfurt on the Oder some forty-five years previously.[39] Applied mathematics was now no longer even related in an essential manner to the question of law.[40] Instead it had an effect all the more concentrated "immediately on the well-being of the state and on the happiness of its citizens." *Mathesis applicata* was the cumbersome academic name for what Condorcet and Laplace referred to in France at this time as moral and political mathematics. They admittedly did so with the significant difference that mathematicians and philosophers of the French Revolution were above all interested in probabilistic calculations of the *volonté générale*. Election processes were at issue as well as voting procedures for sworn juries, a legal institution that the revolutionaries introduced according to the English model, though without the rule of unanimity.[41] Friedrich Meinert in Prussia applied mathematics for the well-being of the state and its citizens, not so much to questions of distribution and political welfare as to economic law and theory, and above all to a multitude of technical questions that fall under the jurisdiction of the government: structural engineering in construction inspections; ballistics for artillery; mechanics and machinery in mills, transport, manufacturing, and mining. Applied mathematics comprehended a neutral and higher knowledge that could recalculate and supervise the various forms of local knowledge. The universality of this administrative knowledge made it independent of traditional (often guild) skills and could, conversely, observe and supervise them. It was the implementation of universal knowledge about what is strictly individual and who is an individual.[42]

Tables in the Eighteenth Century, 2: Listing Population and Divine Order

Mathesis forensis and the mathematical evidence of its tables stood in contrast to the moral evidence of statistical representation in the eighteenth century.[43] Statistical representation for its part was again split in two: Conring and Bose's topical semantics furthering Lipsius's art of governing,[44] to which Gottfried Achenwall had assigned the name *Statistik* in 1748;[45] and the political arithmeticians' columns of numbers, which listed prices, population, and livestock figures, or tax incomes. The text of Achenwall's topical statistics was publicly accessible; the numbers were either the fragmentary work of individuals or inaccessible ministerial arcana.

Tables of births, marriages, and deaths made up one exception to this separation. They presented numbers in the evidence conveyed by semantic frames. This fact cannot be accounted for in the usual explanations that histories of statistics give for how political arithmetic and Gottfried Achenwall's statistics came together.[46] Neither does it quite merge with the modern question of how the avalanche of tables found its way to the concept of a statistical law.[47] Instead, a decision is at stake about representation, a decision molded by the institutional place where it first was made. Just as with the space where jurists forged the *mathesis forensics,* the exhibition of tables of births, marriages, and deaths again took place at the fringes of an institution.

In any case in the German lands, it was Protestant pastors, agents of the general semantics of God and the regional statistics of church registers, who took care of the publication and evaluation of these numbers; and they did so until well into the eighteenth century. Their parishes as a rule were invested with a series of sociopolitical administrative tasks: the oversight of schools, hospitals, and asylums, as well as of scholarships and fellowship funds of city governments; and the inspection of country parishes and theological seminaries. Caspar Neumann, for example, the son of a Breslau tax collector, became a member of the consistory of Breslau in 1689; in 1697, he took on the theological professorship in the high schools and the pastorate of Saint Elisabeth.[48] In a letter from 1689, he commended himself to Leibniz for his studies on language and natural sciences and particularly his expertise in population tables. In response, Leibniz would make him one of his first candidates for the Berlin Academy of the Sciences. For the 1689 letter, Neumann prepared a copy "of reflections made hitherto *about life and death* among those who were born and died in *Breßlau.*" "Reflection" refers principally to the preparation and representation of data, which Neumann managed to a high degree of refinement.[49] He first announces another sort of reflection, however: "At this time it is still not possible to see what the actual use of it will be."[50] This hesitation is all the more conspicuous since Neumann had been a member of an academic society founded by Bose back in Jena.

Mortality lists were included among the responsibilities belonging to Neumann's office. Since the Reformation, Protestant parishes in Germany had begun to set up death registers. Catholic congregations followed suit after the Council of Trent. In some cases, the numbers were collected again by state agencies. Thus, for example, the ministerial office of alms collected parish information in Breslau beginning around 1590. Working under consistory oversight, it put together data on marital status and the causes and dates of death in order to report to the council.[51] Neumann hence at first only expanded what the parish office required. As a congregational pastor, he entered notices in the church register; as a theologically motivated researcher, he made deductions from them. Neumann even recorded his basic principles as a theoretical reflection of these ascetic transfers: "Should God grant life so long that one might bring together the calculations of some number of years, or else someone in another city make similar *observationes* and would like to communicate them, then beautiful commentaries of divine providence in our life and death, preservation and multiplication in the world could be made."[52]

The *Providentia* Neumann is evoking may have appeared to him in the allegorical form penned by the Christian Stoic philosopher Justus Lipsius, and its dramatic appearance in the baroque mourning plays performed on stage in the famous school theaters of Breslau. Neumann no longer called for *experimenta* and *observationes* only in the *regno Naturae*, but now also in the "regno gratiae or in theology." This expansion emphasizes the corresponding nature of the two realms, as the baroque stagecraft of the Breslau diplomat and poet Daniel Casper von Lohenstein also did. In his choice of words, Neumann implicitly cited the so-called philosophy of experiment, as one said in the Royal Society. The numbers that he surveyed in virtue of the administrative share of his office, charged him as it were with the office of reading in them the physico-theological law of the long run of providence. By collecting numbers into tables and offering them to the planner of an academy such as Leibniz, the Breslau pastor only exceeded the requirements his office by a little. This little bit, however, was necessary in order to recognize providence's own record-keeping in the contribution of a corresponding member of the scientific society. Neumann clearly became famous because the English astronomer Edmond Halley later requested his numbers for the calculation of insurance statistics.[53] Neumann showed the so-called dying-out of a population in tabular form, and Halley calculated it for the first time according to probability theory.

It was another Protestant pastor in Germany who, some half a century later, finally read the great evidence of cosmological and universal-historical order together: the first edition of Johann Peter Süssmilch's *Die göttliche Ordnung in den Veränderungen des menschlichen Geschlechts, aus Geburt, Tod, und Fortpflanzung desselben* (The Divine Order in the Transformations of the Human Race, from Its Birth, Death and Reproduction) appeared in 1741. The divine order must again be seen as the neo-Stoic providence of Justus Lipsius, now complemented by the "physico-theology" of the English philosopher William Derham (1657–1735).

Süssmilch performed the theological-administrative office of keeping lists of marital status just as impressively as Neumann. In the year of the publication of the *Göttliche Ordnung*, he was appointed by Frederick the Great, against the will of ecclesiastical leaders, to the position of provost in Kölln and later to high consistory councilor in Berlin. The latter post brought the monitoring of Berlin hospitals and asylums, among other things, under his jurisdiction.[54] The difference from Neumann lay in Süssmilch's taking the step of actually publishing the tables. While Neumann remained in the shadow of his office, Süssmilch's publication depicts his office of collecting birth and death data literally and literarily. In composing his book, Süssmilch became a poet of his office, the curacy that had allowed him to keep accounts on the divine order of the movement of mankind. The finished work would not result in another case of descriptive semantics in the style of the old statistics of Conring and Bose.[55] But neither would it produce a mathematical text of the new statistics à la Laplace and Quetelet, which would provide the conceptual basis for and have to give a semantic account of the probabilistic calculation applied to statistical data. The diagrammatic appeal of Süssmilch's tables is to indicate an order that can only become obvious through the listing of numbers. The mathematical medium of this image-like visibility is proportion: the proportion of births to the whole population; of those who survive to those who die; of the growth of the number of survivors to the increase in food resources. In Süssmilch's large book, Protestant political arithmetic takes one half-step beyond the realm of the church by making visible its own providential presupposition in number and concept. Providence circumscribes here precisely the convergence of theoretical concept and pragmatic instruction characteristic of the Christian stoicism of the Lipsians. In his work, Süssmilch establishes an order whose evidence springs from his own pastoral records; and in order to give an account of the world's beauty and perfection, he bases the explanation of the work and its history on the design of its creator. From its origins with the humanist Justus Lipsius (1547–1606), the claim that the real is the rational has become a kind of official philosophy with Süssmilch.

While Süssmilch was working on the *Göttliche Ordnung*, he was however not yet master of the Berlin ecclesiastical numbers, but rather an educator, then a regimental pastor with Count von Kalckstein, who for his part was the tutor of the crown prince (1735–1736). During his time in education, Süssmilch seemed still to be following plans for a scholarly career, though he decided, by taking the army chaplain's post, to enter the clergy. This career decision then contained the initial impetus of the *Göttliche Ordnung*. It cast light in particular on the famous comparison of the tables and the order they show with that of a parading regiment.[56] A visible order in the image of military formations in marching motion anticipates the idea of providence. Providence appears in the image as a process that takes place and reaches completion in the ongoing march of history, and that the tables illustrate while waiting for the continued pace of time.[57]

When the book finally appeared in 1741, Frederick the Great had not only just appointed Süssmilch to Kölln, but he also fetched the Enlightenment philosopher Christian Wolff, who wrote the preface to *Göttliche Ordnung*,[58] back to Halle. This princely act of academic politics effectively decreed Wolff's doctrine—though not without resistance—to be Prussia's first state philosophy. In his preface to Süssmilch's book, Wolff makes energetic demands for the state to organize population lists itself, that is, to provide the basis for the experience required for any probabilistic science of the growth of population: "There would be a great deal to say for making these lists more complete, and one would at the same time find that private persons are not in a condition to undertake all that would be necessary for this enterprise."[59]

Wolff, as we see, analyzes Süssmilch's population statistics according to his concept of experimental knowledge. On the one hand, the probability of *historiae* characterizes its epistemic status in the sense of the traditional topical arts and sciences. The call to organize standardized and continuous surveys, the precondition for a regulated creation of data, on the other hand, draws its conclusion from the new concept of experimentation. With English experimental philosophy in mind, Wolff indeed rigorously distinguishes experimentation from observation and other forms of experience. With his double characterization drawn from the resources of traditional probability and modern experimentation, Wolff both supports and undermines Süssmilch's enterprise. His accolade once again celebrates the pastor who extrapolates the divine order from entries in his church register as if by way of a simple hobby suggested by the circumstances of his day job; at the same time, Wolff already presents him as a forerunner, a somewhat amateur private figure, who acts alone in an area that only a great coalition between state and science could successfully approach.

With the order in cabinet of March 19, 1747, six years after the first edition of *Göttliche Ordnung* and Wolff's preface, Frederick the Great arranged for the erection of the *Generaltabelle*—a complete and annually renewed statewide population count.[60] In this way, a kind of double bookkeeping began in Prussia. The population census, which remained secret, duplicated the lists of the church registers, which were published and available to the *res publica litteraria*. Süssmilch—whom scientific histories of demography like to apostrophize as the founder of the discipline—embodies the trigger point of this duplication. The first edition of *Göttliche Ordnung* in 1741 had clearly helped influence Frederick the Great's order of 1747. Its result, the general population table, however, made theological speculation and its simple mathematics of extracting mean values already obsolete; and it did so exactly according to what Wolff had suggested in his preface. Süssmilch nonetheless clearly sticks to the division of counting methods even in the later editions of his book. Although he now had access to the secret state archives, he never cites any numbers from the *Generaltabelle*.[61]

Süssmilch's decision is in keeping with the tradition of Prussian statistics. There had been continuous mortality lists in Prussia since the late 1680s.[62] The Great Elector had commissioned the Geistliches Department (Ministry for Ecclesiastical Affairs) to print lists for the provinces based on the church records and to submit a copy to the court.[63] Population surveys were provided by the Geistliches Department, despite some interruptions, throughout the eighteenth century. The Prussian royal Statistische Bureau was not founded until the beginning of the nineteenth century, when it was brought into being by Friedrich Wilhelm III. Its institutional origin was the General Directorate, Prussia's central administrative office since the 1720s. At that time, Friedrich Wilhelm I had ordered regular statistics in the most various areas—price statistics, agricultural and trade statistics, and particularly historical tables, an overview of inhabitants, buildings, economic conditions, and public institutions to be drawn up by municipal magistrates. When Frederick the Great ordered annual censuses in 1748, he conceived of them too as a generalized version of the historical municipal tables, not as population lists in the manner of the Geistliches Department. Later on (around 1770), though, the clergy was again relied upon for providing statistical material. Particularly in the countryside, where there was no *Policey*, and state commissions had to collect the numbers, pastors had to correct and complete the official lists. When Süssmilch, despite his access to the state archive, only made use of material obtained from church registers and through clergymen, he was thereby marking the exact systematic as well as chronological mean between the beginning of the population census by the Geistliches Department in 1688 and the establishment of the Königlich Preußisches Statistisches Bureau in 1805.[64] Immediately before its collapse in 1806, under the auspices of the Adam Smith enthusiast Leopold Krug and the sponsorship of Baron Heinrich Friedrich vom Stein, minister of state for trade, a Prussian state balance for the years 1804–1805 would be published. With this statistical publication, the Prussian state would place itself for the first time as a whole before the public's eyes in the *evidentia* of the tables.[65] The epoch of Neumann and Süssmilch had come to an end. A different law was now represented in the work of the compiled tables. With the law of large numbers, the state constructed its own counterpart, society, and, on the other hand, reflected its own image in it. It is not without irony that the Königlich Preußisches Statistisches Bureau developed from a rivalry and competition with the enemy. It was the Napoleonic state that first established a bureau for statistical services in Paris.

Just as the evidence of the *mathesis forensis* grew out of the side jobs of jurists, so too the evidence of population lists emerged from a side room of the parsonage. The evidence whose diagrammatics offered an elementary semantization of numbers derived from spheres where the law of the jurists and the law of the theologians are precisely one step beyond themselves. Asking for a statistical law—the concept that Quetelet would use in the nineteenth century when he applied probability to statistics and called the reference of the operation "society"—presupposes following up on the institutional conditions under which such lawfulness took form. That statistical law came about in the shadow of juridical and theological laws.

9

Novels and Tables

DEFOE'S *A JOURNAL OF THE PLAGUE YEAR* AND
SCHNABEL'S *DIE INSEL FELSENBURG*

Placing-before-the-eyes, *enumeratio*, and Tableau

In a number of works in which scholars have searched for the origins of the modern novel and the first developments of realistic probability, we meet with the contours of charts and graphs. Mortality tables and population statistics emerge that are interpolated into the text of narration. In Daniel Defoe's *A Journal of the Plague Year* (1722), for instance, tables make up an attraction in the foreground of the textual display. Their very prominence adds weight to a significant ambiguity that no reader of the work can evade: is the book a matter of poetic fiction or of journalistic research on London's Great Plague of 1664? In Johann Georg Schnabel's *Die Insel Felsenburg* (Felsenburg Island, 1731–1734), the first modern German novel, in contrast, tables merely constitute small insertions in the narrative. They appear as curious sideshows in the invention of the new genre, or as arabesque forms meant to supplement the amorphous nature of the novel.

The following pages focus on the graphics of text-embedded numbers and the images of number series that, by challenging the beholder's view, demand further reading. Here, we are concerned with the figural role assigned to the table through its insertion into the narrative text.[1] The figure of the *tabella*, of the quasi-pictorial placing-before-the-eyes, repeats the construction of its own graphic mode of appearance on the semantic level of the text. Writing itself is already infused with a nonliteral graphism displayed in the relation between letters and numbers, which is also simultaneously inherent in both aspects of writing.[2] The constellation of literal and transliteral graphism can be called the *diagrammatics* of writing. The table in the novel, as we then might say, restages the diagrammatics of its fundamental condition as a written text.

It is certainly true that such numerical images served for the most part as curiosities in early novelistic experiments. But it is no mere coincidence that the evidential placing-before-the-eyes, which the table embodies poetologically and consists in graphematically, led to decisions on the very status of the novel since its inception. Numbers in texts confronted readers with questions of how to assess the difference between fiction and nonfiction, the distinction between narration and description, and finally with the contouring of the novel's aesthetic form. The evidence of statistical probability was an epistemic category turned aesthetic form. By virtue of this intersection, then, it became a significant factor in the birth of the novel.

The epistemic and aesthetic modernism of the table has a prehistory in rhetoric and poetics. As indicated in the use of the French word *tableau*,[3] tables of statistical probabilities were connected in early modernity with the figure of the reversal of text and image. In other words, they were material instances of what rhetoricians called the placing-before-the-eyes of what is said as if it were a picture. John Graunt, for example, saw the accomplishment of his *Natural Observations* in the reduction of many volumes of text to the transparency of a few tables. In replacing alphabetical writing with numerical tables and dedicating this substitution to the book's inordinately busy addressee, Graunt heeded the rhetorical formula of *evidentia* in order to stage the history of the origin of his work: this was indeed showing instead of telling; image instead of text. At the latest with Leibniz—to name a prominent example among many others—the table or graph would become synonymous with the statistical enterprise in general.[4] Leibniz determined both the task and the method of *Staatstafeln* (state tables) as the process of placing-before-the-eyes, and he never tired of suggesting the use of such tables and the figurative method either for Prussia, Hannover, or even the Holy Roman Empire:

But to see everything . . . that belongs together as if in a single momentary glance is much more advantageous than what can generally be accomplished by *inventariis*. For this reason, I am entitling this work "Tables of the State," because the duty of a table is to represent the *connection* of things all at once . . . . Such advantages of tables can be found in maps and sea charts, in outlines, in the art of bookkeeping and well-composed calculations, all of which should exhibit their certain, as if mathematical, consistent model and form, so that everything is constricted and made visually evident [*augenscheinlich*] or graspable [*handgreiflich*].[5]

Thus reads the conclusion of the most complete analysis of tables in the history of early modern statistics. The metaphysics of the continuous connection of all things with each other, characteristic of Leibniz's *Monadology*, is brought into appearance in a kind of mathematical form. This aesthetics is brought about through processes of translation and projection that take place according to the rules of the rhetorical *ante oculos ponere* and of cartographic geometry. Two aspects of what it means to be an image intersect here. Rhetorically, the table offers a picture instead of a text; and graphematically, the construction, a process that itself cannot be taken in at a glance, produces a perspectival graphics in cartography. The placing-before-the-eyes [ante-oculos-ponere] of statistical tables of probability would become commonplace with the later Prussian theoretician of providence Johann Peter Süssmilch. In his case, it is the order of God and State that is placed before our eyes.

The placing-before-the-eyes of statistical tables belongs to a very specific brand of constructions within the family of rhetorical devices of evidence.[6] We find this type of figuration at the end of the speech, as a recapitulation, *anakephalaiōsis*, or, in Quintilian's Latin, an *enumeratio*, or enumeration. Its task, which methodologists of philosophical treatises declare to be the only function of the conclusion, is to repeat and compress what has come before.[7] In French rhetoric, this process is referred to as the *tableau*. The tableau of the conclusion repeats the argument in a brief and orderly summary; it condenses what was narrated into elements that it arranges quasi-simultaneously, hence pictorially, side by side. The repetition transforms narrative succession into virtually simultaneous enumeration. In this context, placing-before-the-eyes is graphic, not because of sensuality or vividness, but rather as a result of the spatializing effect of repetition that turns narration into simultaneity. In a law case in court, Quintilian asserts, the *enumeratio* at the end of a speech "places the whole Cause before [the judge's] eyes at once; even if this had not made much impression when the points were made individually, it is cumulatively powerful."⁸ Quintilian admittedly introduces the *enumeratio* as the type of the conclusion that is specifically related to the *res*, whereas another type, one used at least by orators, is related rather to the affects. It is not difficult to see, however, that the *enumeratio* in itself is already both a shortened recapitulation of a whole and, at the same time, an appeal to affects due to the quasi-visual nature of the image aimed for by *enumeratio*. Later on in the chapter about affects, in fact, Quintilian conversely gives an important role to the figure of evidence (*enargeia*: transparent representation).⁹ Hence the narration condensed in the enumeration is the affect of the optimization of representation; it is affect *as* the economy of representation. Regardless of how trivial Quintilian's concluding *tableau* may seem at first sight, by transforming narration into enumeration, it constitutes an epistemic device and an aesthetic design of a fundamental nature. The novel may not be a tableau in the technical sense, but it can only compensate for its poetological deficiency of form by revealing itself as the composed tableau of a world.

Res fictae and *res factae*: A Constitutive Distinction of Novelistic Form

Defoe's *A Journal of the Plague Year* is shot through with statistical tables. H.F., the writer of the diary, who, like the journalistic D.F. of the *Essay upon Projects*, is identified only by initials,¹⁰ cites London's municipal decrees and quotes from theological tractates of the seventeenth century, as well as medical-police treatises from the beginning of the eighteenth century.¹¹ But while he designates these writings as documents of the London plague and himself appears as an observer of contemporary social and moral events, the tables of the mortality registers play an exceptional role. They constitute document and representation, cited source and original observation, of the plague all in one. They provide a picture of the plague that can only be observed in statistical citations and only narrated in the graphism of the tableau. In other words, the cited numbers present a picture that cannot be traced back to any first-order observation, particularly wherever Defoe does not just scatter the numbers in the course of the text, but rather maintains the diagrammatical appearance of the table of mortality. These tables elude the opposition between the fact of cited documents and the fiction of individual experience. They give a picture of the plague that cannot be drawn except by the cited document of the table. The quintessence of the authorship of *A Journal of the Plague Year*, in which fact and fiction are inseparably superimposed, lies precisely in the tabular rows of numbers. This peculiar kind of visualization can be derived from the rhetoric of the table. It is not so much a matter of imitation or representation as of placing-before-the-eyes; and this device of placing-before-the-eyes allows the question about fact or fiction, the fundamental question of the novel, to develop out of the journalistic use of tables.

Like a series of articles that Defoe wrote for *Applebee's Journal* in August 1720,[12] *A Journal of the Plague Year* was inspired by the plague that broke out in 1720, spread until the following year in Southern France, and led to discussions about preventive measures by the City of London.[13] A few months before he composed the *Journal*, Defoe published his *Due Preparations for the Plague, as Well for Soul as Body*. Although *Due Preparations* did not appear in the newspapers, they continue the style and fabric of Defoe's journalistic work without a break. They provide suggestions and admonitions, which the author dresses up in little stories or scenarios. The first part, "Preparations against the Plague," relates to the body; the second, "Preparations for the Plague," deals with the soul.[14] Such warning and admonition aim for visualization in the vein of the most pristine pastoral rhetoric, which Defoe modifies only minimally for the sake of journalistic publication. The exemplary story at the heart of the first part—"Preparations against the Plague"—tells of a patriarch who, like Noah with his ark, defends his boarded-up house against the plague. What is put before our eyes in this almost biblical tale is meant as an appeal to fend off danger. In the second part—"Preparations for the Plague"—in contrast, the survivors of a fatherless family debate the advantages and disadvantages of spiritual practices in coming to terms with the plague. The two parts refer to each other in complementary ways: precautions to be taken; the regime of the body and the kingdom of the soul; actions of a father and debates in a fatherless family; the telling of an allegorical tale about the plague and the discussion of such tales and interpretive visualizations in meditative exercises. *Due Preparations*, it seems, is not interested in narrating the events "themselves" that are summed up in the name of the plague. With all their emphasis on the forms of presentation, they rather seem to circumvent the factual presence of those events. But then we should remind ourselves that the plague essentially consists of an ensemble of warnings and preparations, tales and practices, reports and numbers that make it present for people and let them visualize its approach, raging, and ebb.

In contrast to *Due Preparations, A Journal of the Plague Year* offers a certain representation of the plague itself. We can pinpoint the nature of this representation from the point of view of the two rhetorical strategies of journalistic visualization found in *Due Preparations*. The question implied here goes beyond the one discussed in Defoe scholarship as to the turn from realistic to allegorical interpretation. We should not only ask if there is a rhetorical strategy involved in the presentation of facts; the debate, in other words, is not only about the "reality effect" in the *Journal*.[15] Before we can pose such a question we should rather ask: How does the new rhetoric of fictive factuality arise at the intersection of the traditional rhetorical forms of visualization? In the precise place where they intersect with each other, we rediscover the rhetoric of the table, the figure of evidential visualization.[16]

In *Due Preparations*, the tables have only subordinate functions, but they are significant ones. In the first part, in the biblical ark-like house shut off from all contact with the infected citizenry, we read about the expectation and reception of numbers: messengers or watchmen bring the *Bills of Mortality* to the door in the beginning, and then they call them out to the housefather through the open, and later closed, window. The expressiveness of the numbers is dramatized in the story of their transmission. In the second part, in the discussion for and against visualizing practices of the catastrophe, the tables of the *Bills* become a signal for the beginning of the meditative exercises. With the mortality tables, God gives His people a sign to recognize that the time has come and the preparation for the reversal of things has to begin.[17] The evidence of numbers and their appeal to persuasion in both sections of *Preparations* is linked to the respective modes of their conveyance—the postal, news-like dispatch, and the divine, symbolic one. Numbers function in the rhetoric of texts and their transmission.

The *Journal* again begins at the intersection between a news-like transmission of texts that gives warning and a symbolic one that encourages spiritual exercise. At this zero point of persuasion, the idea of pure information arises, the existence of data that can only be read as a picture and only seen as text. An evidential placing-before-the-eyes emerges that no longer persuades rhetorically about any given matter, but that rather constructs its own object. From the beginning of the *Journal*, the plague has no other form of existence than its visualization: "It was about the Beginning of *September* 1664, that I, among the Rest of my Neighbours, heard in ordinary Discourse, that the Plague was return'd again in *Holland*."[18] From this first sentence on, the *Journal* reads like a small media history and critique of terror. The "ordinary Discourse" is its preliminary stage: the word-of-mouth circulation of news that reaches London through the private correspondence of merchants and spreads in the vicinity of the neighborhood. The first entry of a death by plague in the *Bills of Mortality*, meticulously portrayed in the *Journal*, occurs some time after the first wake-up call of receptivity through *ordinary discourse*: two strangers die in the parish of St. Giles-in-the-Fields. The family with whom they lived attempts in vain to keep their deaths a secret. A commission appointed by city government agrees that "evident Tokens" of the plague are present: "and it was printed in the weekly Bill of Mortality in the usual manner, thus, *Plague* 2. *Parishes infected* 1."[19]

Thus does the regime of the numerical tableaus begin. Among its consequences, we find the struggle between the city government, which attempts to record the plague fatalities, and the citizens, who want to prevent the entries of the deaths in the *Bills* because they fear that the city will close up their homes: "but from the Time that the Plague first began in St. *Gile*'s Parish, it was observ'd, that the ordinary Burials encreased in Number considerably."

From *Dec.* 27th to *Jan.* 3.	{	St. *Gile*'s	16
		St. *Andrew*'	17
Jan. 3. to 10.	{	St. *Gile*'s	12
		St. *Andrew*'s	25
Jan. 10 to 17.	{	St. *Gile*'s	18
		St. *Andrew*'s	18
Jan. 17 to 24.	{	St. *Gile*'s	23
		St. *Andrew*'s	16

. . .[20]

H.F. is the reader-observer of these tables, which endlessly crisscross the text of his *Journal* long before he can report any eyewitness account of sick people and deaths, or of the breakdown of moral and social order in the city. And even then it remains the *Bills* of the increases and later decreases in the fatality numbers that constitute the medico-political object of the plague "itself." This fact is all the more remarkable since Defoe sets H.F.'s data-reading in a kind of pre-medial world. He writes: "We had no such things as printed News Papers in those Days, to spread Rumours and Reports of Things; and to improve them by the Invention of Men, as I have liv'd to see practis'd since." H.F. also complains that the city government possessed "a true Account," but attempted to keep it "very private."[21] Between the narrative time of the *Journal* and the narrated time some sixty years before—Defoe is not overly precise in drawing the historical borderline[22]—lies the breakthrough of the modern newspaper in Britain. Defoe himself had become famous through this new medium of communication, for which he had written horrifying descriptions of and stories about the plague in Southern France in 1720 and 1721.[23] Defoe, the author of *A Journal*, thus determines H.F.'s observer situation very pointedly in a world before the publicity of journalistic evidence.[24] He does so certainly, not in order to fabricate a world of mere appearance, but rather to pinpoint in *ordinary discourse* and the *Bills of Mortality* precisely an irreducible mediality and construction of evidence that could be taken up by the journalistic enterprise.[25] H.F.'s *Journal*, Defoe's proto-novel, presents itself as media history and critique in the age of journalism: the reader's view of data is, paradoxically, a rhetoric of visualization purified of all pastoral-journalistic persuasions, mediality without the conventional use of media. The diary writer first and foremost witnesses the plague as a medial construction, but not in the sense of a distancing or of an "only medial" view. On the contrary, only through the frame of the tableaus, which are incessantly shown and discussed, can the horrors in H.F.'s eyewitness accounts appear as those of the plague.

The plague's statistical mode of existence determines the form and content of the *Journal*. Probability and its culmination in evidence thus not only retrospectively shape the question of genre and aesthetic form; they are already contained in the construction of the thematic object. The reader's view of the first plague case entry, and then of the rising and sinking mortality figures, sees the question about the probability of the text already solved in the tabular representation itself. In this context, the numbers and tables in the text have a double status. Introduced with "thus" or "for example," they are, on the one hand, documentary evidence of a diary keeper for his *Journal*; on the other hand, they appear as outtakes from the mortality lists glued into the text in the style of a collage. The tabular numbers are the plague in the text of the proto-novel.

Critics ever since Sir Walter Scott have been interested in the *Journal*'s genre and in the related question of its status as fact or fiction.[26] It is nevertheless not important in the final analysis what answer we settle on. It is much more significant that *A Journal of the Plague Year*, in the tabular visualization of its object, persistently confronts readers with the question of whether it is fact or fiction. What is probable in its very mode of existence already embodies in itself the old ontological question about the place of poetry between historical particularity and epistemic causality. Precisely this feature makes the *Journal* into a proto-novel. Only with the incipience of the novel does the differentiation between fact and fiction become the key gesture that, beyond any rhetorical and poetological schemata, determines literary form. Differentiating fiction from fact is the decision about form in the era of the novel at large.

The form-inducing nature of this differentiation does not, however, mean that fact and fiction are clearly separated in each instance in *A Journal* and organized in a hierarchical order. It can be characterized neither as predominantly fiction, into which facts are inserted as implicit citations, nor as a documentary compilation merely connected together by fictive elements. H.F.'s *Journal* unfolds its form solely out of the visualizing force of the tableau. *A Journal of the Plague Year* contains neither in its beginning nor at its end any indication about a fictive frame or a factual reference. We could instead speak of it as a work of fiction without boundaries or a textual compound of facts without references. H.F. neither raises himself to the status of a fictive first-person narrator—Defoe only made use of such devices in the conversion biographies of his later novels—nor is there any (as always, fictive) attempt to authenticate the writing "I"—this task is reserved for the journalistic works.[27] This double refusal coincides, however, with the peculiar legibility, or rather observability, of the mortality tables. Because it is the visualization of mortality and the plague, the tableau constitutes a space for the construction of an object that is neither fictive nor factual in itself. The tableau is proto-literary, just as the *Journal* is a proto-novel. We may recall in this context that John Graunt in his preface to the *Natural Observations*—from which Defoe took the tables used in the *Journal*—had connected the step from direct understanding to the inspection of data with a similar questioning of reading numbers. The tables are the core of a text without borders or reference, a circumstance that makes it irrelevant that the question of its fictional or factual status never receives an easy answer.

Reading the Novel by Reading Numbers

A statistical reading of the *Bills of Mortality* guarantees that the theme and the frame of the *Journal* are set from its first three pages. Of course, there are long narrative and descriptive passages: we have the circumstances of the diary keeper H.F.'s livelihood; the viewing of a mass grave at night; and the faces of families fleeing London, among many others. But the episodes of singular stories and private experiences have their place assigned to them only in the course of the increase, culmination, and decrease of the numbers. In the nineteenth century, the work of social statisticians such as Quetelet might lead us to expect a curve to portray such development. The graphical appearance of the statistical curve and above all its necessary prerequisite, the probabilistic treatment of raw data from the bills, did not, however, exist for Defoe. The connections drawn by textual elements are still necessary in order for the rise, apex, and fall of the numbers as presented by the tables of mortality to place before our eyes the epic construction of beginning, middle, and end. The evidence that Quetelet's curves would later claim in relation to the mathematical structure of probability had its roots in this pre-mathematical and pre-referential evidence of the tables. Defoe's tables already establish the fundamental epistemic implication of evidential representation. Listing numbers and drawing lists and diagrams for arranging them is a gesture of representation only. But when we look at those lists and diagrams, something like a picture of rise and fall emerges from the numbers and their arrangement. Knowing about rise and fall is an epistemic accomplishment. Thus epistemic elements are implied in the mere act of representation, an act that less depicts than generates a piece of knowledge.

If we turn our focus to the frame of the narrative provided by the tables of mortality, we recognize three stages in the *Journal* that largely coincide with the rise, peak, and fall of the mortality numbers. In the beginning, the *Bills* wield a kind of performative power. The news or the report is the event of the plague. The hearer or listener at the point remains one among many. Hence the above-quoted first sentence asserts "that I, among the Rest of my Neighbours, heard in ordinary Discourse, that the Plague was return'd again."[28] This mode of report continues up to the point where the composer of the *Journal* makes the decision not to leave London. His decision at the same time marks the crystallization of an individual reader of the *Bills of Mortality*. What does this mean? In the beginning, H.F. and his brother—both merchants—agree not to flee London. They at first only read in the tables that chance and risk exist. Chance and risk do not have any calculable values; they simply show that a situation of decision and probabilities is at hand. In the brother's words, "is it not as reasonable that you should trust God with the Chance or Risque of losing your Trade, as that you should stay in so imminent a Point of Danger, and trust him with your Life?"[29] A new increase in the mortality numbers then splits up the parallels between economic risk and the chances of survival in the name of God. The brother, who only sees the risk for life, finally decides to flee ("he would venture to stay no longer").[30] H.F., in contrast, randomly lights upon the Ninety-first Psalm— "I will say of the Lord, He is my refuge and my fortress: my God; in him will I trust. . . . There shall no evil befall thee, neither shall any plague come nigh thy dwelling."—and stays behind. In point of fact, his business suggests to him this alternative relation to risk. He runs a wholesale saddlery business that relies on warehouse stores and long-term business contacts. His "dealings were chiefly not by a Shop or Chance Trade."[31] It is therefore not a matter of an isolated choice between chance and risk. The long-term probabilism on which his decision is based predetermines his observer status, which he will assume from now on.[32] And it is in this context that he also reads the tables in his own way.

With this decision, the central and most extensive part of the *Journal* begins. H.F. is no longer one among many. He provides for his own protection; he watches the infected people on the streets maddened by fear of the pestilence's horrors; he hears about life in the houses closed down by London officials after the corpse of a plague victim was removed from them. H.F. cites and discusses the measures of the administration. He hashes over theories about the spread of the plague and draws medico-political conclusions. This individualization reaches a conclusion when H.F., the reader of statistical tables, becomes their agent: he is appointed for a while to the office of overseeing the closing of houses infested by the plague, and of counting the dead. Just as these tasks make inspection his duty, his growing certainty that he will not end up among the counted lists of the dead provides him with an outsider's view of the plague. In the precise sense of constructivist theory, H.F. becomes an observer.

As data collector and observer, H.F. is first of all a reader of the *Bills of Mortality*. His view of the tables, which accompanies all his perceptions and allows them to become observations in the first place, is increasingly a distance-taking reading in the style of Graunt. It is a reading of numbers as data that anticipates a still unknown mode of processing data according to the laws of probability. This begins with a harsh critique of the number. It is an established fact with H.F. that the *Bills of Mortality* are deceptive throughout half of the diary. His own observations in the nightly transports of bodies and their burial in mass graves convince him that the corpses cannot reliably be counted. The stories that he hears from all sides lead him to understand that families have every interest in keeping the plague as a cause of death secret. Corpses are hidden away, gravediggers and doctors are bribed. The numbers hence must be falsified. For the observer and reader of statistics H.F., the numbers therefore lose their primary evidential power.

For a while, what H.F. sees on the streets of London seems to step out of the frame of the tables. Accordingly, many modern readers have believed themselves to have seen, in H.F.'s criticism of the *Bills of Mortality*, their own discomfort with the regime of statistics and a salvation of the individual from numbers. This is a mistake. H.F.'s critique is a methodical process, the early form of an as yet nonexistent doctrine of error analysis. This fact becomes clear from a passage in the—again shorter—final part about the waning of the epidemic.

when the violent Rage of the Distemper in *September* came upon us, it drove us out of all Measures: Men did then no more die by Tale and Number, they might put out a Weekly Bill, and call them seven or eight Thousand, or what they pleas'd; 'tis certain they died by Heaps, and were buried by Heaps, that is to say without Account.[33]

This reflection goes beyond criticism of unreliable or falsified numbers. The criticism joins the power of a central allegorical scene in the *Journal*. At night, H.F. visits a mass grave, a scene that he experiences and reports with a nod to the story of Korah and his company rebelling against Moses and the earth's subsequent swallowing them up to join the dead (Num. 16).[34] This allusion highlights the horror of the uncounted nameless bodies. Yet H.F. goes on: even if much higher estimations would be thoroughly trustworthy, "yet I rather chuse to keep to the public Account."[35] He has by no means forgotten his methodological objections against the numbers and their reliability. His preference for official numbers over private estimations, despite these realizations, results from a simultaneously religious and methodical decision. Since "it is much to the Satisfaction of me that write, as well as those that read, to be able to say, that every thing is set down with Moderation, and rather within Compass than beyond it."[36]

The only thing that contents both readers and writer is a text that is written, so to speak, under the auspices of the numerical diagram. "Moderation" well expresses this double aspect of law-giving and methodical order. Jacob Bernoulli speaks, incidentally, of *moderatio* in introducing the analysis of events into equally probable contingencies, the methodical principle of probability theory. The word "moderation" originally also referred, in a sense now obsolescent in English, to governance, rule, or guiding leadership, which also implies divine providence. And, of course, it refers to what is moderate and does not go beyond the limits of what can be grasped and measured. Instituting order and providing the basis for measurement are different but related features of moderation. H.F.'s allegiance to the *Bills of Mortality* expresses their status as a methodological act of moderation appealing to the metaphysical order of nature and creation. Because of their confidence in counting and measurement as such, despite all errors and falsification, H.F. takes the *Bills* as an example for the universal moderation that is God's ordering of the world. The adherence to number is a this-worldly echo of faith in God and its epistemic implementation at the same time. The quoted passage thus does not contradict H.F.'s proto-methodical critique, but rather endows it with meaning.[37] In drawing this conclusion for the *Journal,* we may go even a step further and claim that the *Bills* and their numbers at this moment are the text's reference in the double meaning of the word. Through the numbers, the fictive journal refers factually to what happened in the Great Plague of London. But reference can also mean the source that provides force and legitimacy. In this sense, therefore, the passage can be called the moment of reference. In it, H.F. assures himself of the tableau of numbers as the source of moderation in life and death. We could even say that the remark corresponds to the *fondement* of the state and of knowledge at large that Leibniz always insisted on.

The passage, as incidentally as it may seem, also has a double significance for narration and narrator in the *Journal.* For one thing, H.F. here goes beyond his diary-keeping existence and becomes a narrator who addresses the reader. Then, however, it is also related to H.F.'s failed attempts to compile a private list of the officials, doctors, and pharmacists who remained at their posts during the plague and died of it. This list is clearly the one in which H.F. would like to have had himself entered had the pestilence taken him. *A Journal of the Plague Year* is written from here on as a candidacy for the impossible self-entry in this particular *Bill of Mortality.*

In a certain sense, the evidential power of the table from the beginning is reestablished with this consideration of H.F. It is certainly no longer the mere effect that the diary keeper experiences in viewing the rising numbers, while he himself is not yet counted. The decision for providence and the confession of faith in the official count, moreover, lays the ground for the founding statistical evidence as such. H.F., the observer, relates to this evidence, no longer as one of those who are not yet affected, but rather as a survivor. As a survivor, he has already been subject to the statistical effect of the plague; he is one who would be also counted. This is a first step toward becoming a first-person narrator like those in Defoe's later novels, such as Crusoe and Roxana, figures that look back at their lives from some point of confirmed ending. But it is only the first step, since no allegorical conversion seals any such conclusive point of view for the survivor. A noteworthy passage toward the end of the *Journal* is connected to this curious structure of figuring as survivor in one's own narrative. In the course of a digression about London cemeteries that had to be set up in the plague year, some anonymous authority interrupts to add a parenthetical reference to the site of H.F.'s grave. The diary, which seems to be that of a survivor, revised without the viewpoint of a retrospective on his whole life, suddenly proves to be the document of a dead man. It is like the final self-entry in the *Bills of Mortality* of those who remained at their posts during the plague and died as survivors.[38] This act, we might say, is the final price for subscribing to the moderation of the *Bills of Mortality*.

Interpreters have usually regarded this detail as an oversight on the part of the author, and they have assumed it to be the forgotten incipience of a planned frame or a documentary reference in the *Journal*. A frame of fiction is missing, however, just as much as the attempt to give the text documentary references; there are hardly any traces that would indicate a difference between a written and a published *Journal*. The remark about H.F.'s grave site is stuck like the splinter of a missing fictional frame or dropped documentary reference in *A Journal of the Plague Year*. H.F. managed to escape the *Bills of Mortality* in 1665, but the data collector and reader of tables nevertheless already stands in the diagram of the dead.[39]

Population Lists on Felsenburg Island

Johann Gottfried Schnabel's *Die Insel Felsenburg* (Felsenburg Island) is not affected by any uncertainties of genre or fact and fiction.[40] Riding the coattails of *Robinson Crusoe*, which had been translated into German in 1720, the first three volumes of Schnabel's novel appeared in 1731, 1732, and 1736, some ten years after Defoe's *Journal of the Plague Year*.[41] Actually, *Die Insel Felsenburg* belonged to the even older body of biographical travel literature. Convoluted narratives about men and women stranded on desert islands had been appearing in western Europe since the close of the seventeenth century.[42]

In the second volume, annual tables of marital status among the Felsenburg population appear in two places. (It is noteworthy here that Prussia had begun conducting statistical surveys during the 1720s and 1730s under Frederick Wilhelm I.)[43] In the novel, the first-person narrator, Eberhard Julius, has been a guest on the island for quite a while when this survey takes place. He has already toured and viewed all the various departments of the island with its sovereign ruler, Altvater ("Old Father")

Albert Julius. As we shall see, such visitation, data collection, and surveying are the basis on which this first modern German novel develops its contents and its form, the *enumeratio* or comprehensive picture of the island state. The reprinting of the annual tables of those baptized, married, and deceased "from the first Sunday in Advent 1725 through the same time in 1726"—that is, in the course of the Protestant church year[44]—is announced shortly before the beginning of the life story of the surgeon and barber Kramer.[45] As readers have learned some hundred pages previously in a detail not without bearing on the announcement, Monsieur Kramer makes his appearance on Felsenburg as a "great lover of garden work and stock breeding." Also, the general narrator Eberhard Julius is on the point of "incidentally" reporting "how profusely the [domesticated animals of Felsenburg] have multiplied within the first year of our presence." Certainly: "By rights, I should have thought about the multiplication among men first, but I am saving that not without reason until the end of the church year, since Magister Schmeltzer . . . publicly proclaimed from the pulpit the *specification* of those who were born, died, married, and confirmed."[46]

"Be fruitful, and multiply, and replenish the earth," the first human couple are admonished by God, and Eberhard Julius does not let anything stop him from introducing stock breeding in obedience to this "divine commandment." The implications of Genesis 1:28 for population statistics had been noted long before Malthus—as early as the 1680s, a Berlin court chaplain denounced the Elector's government's plan to conduct continuous population censuses as a sin.[47] Nine years after publication of the second volume of *Die Insel Felsenburg*, Johann Peter Süssmilch's *Göttliche Ordnung* (Divine Order) interpreted Genesis 1:28 in an equally drastic manner. Süssmilch believed that replenishing the earth was the true intention of the order to multiply.[48] However, in his opinion, men differed from animals in how such replenishing takes place: whereas animals fill their place in a homeostatic balance of destruction and reproduction, men replenish the earth through discovery and colonization.[49] *Die Insel Felsenburg* at least alludes to the distinction: the livestock get only a simple, informal listing, while human statistics are ritually announced from the pulpit.

At first, the lists of Felsenburg's livestock exhibit fertility and procreation as commanded and promised: "(1) the young breeding mare bore two foals; (2) four cows had just as many calves; (3) breeding sows gave birth in all to thirty-three young pigs," and so on, up through "(13) the rabbits," which could not "really be noted, since they were all white and never came into view all at once."[50] As far as the representation of their breeding goes, the domesticated animals must be satisfied with this form of list, which often appears throughout the text of *Die Insel Felsenburg*. The novel contains calculations and records of goods of all kinds: foodstuffs and building materials; trade goods and money; animals and humans, insofar as they are brought in to replenish the island.[51] The representation of human multiplication, in contrast, receives a tabular form. When the narrator tells about the advent sermon from 1726, he returns to his promise "to append a *specification* of the multiplication of the Felsenburg people through the year 1726." He now desires, "in order perhaps to satisfy the curiosity of a few, though certainly not all readers," to fulfill his word "by means of a table, which I have extracted from Master Schmeltzer's church registry."[52]

The table of the population list is richly staged when it makes its appearance in *Die Insel Felsenburg*: it is a printed reproduction of a handwritten copy from the church archives; and its publication follows a ritualistic process. Finally, it is a bonus for the notoriously modern drives of curiosity and inquisitiveness, but it is said to be so only for the few. These three characteristics of the table's appearance in the text are confirmed when Eberhard Julius inserts into the text for 1727 "a renewed catalogue of those who were married, born, confirmed, and buried, by means of the following table."[53] The table in the text is once again a transcription (of a handwritten document), the substitution for the ritual event of information (in the first sermon of the church year),[54] and a colorless enumeration (which excites *curiositas*, but remains difficult to read). In precise opposition to the common legibility of the tabular image in Defoe's London, where the *Bills of Mortality* could be bought weekly by anyone since the late seventeenth century, the tables of Schnabel, a government official in a small German principality, constitute an evidential image that, as a piece of writing, is inaccessible, rare, and challenging to readers' hermeneutic skills.

Describing and Narrating: An Internal Distinction of Novelistic Form

Die Insel Felsenburg does not present its tables of official municipal statistics until the second volume, which fittingly opens with the description of a large wedding celebration on New Year's Day in 1726. The tables here have their carefully chosen place in the simple but momentous architectonics of the novel.

The tables demand to be contrasted with the adventures recounted by the characters in their life stories, their seeming counterpart. In order to make this claim plausible, the partly epistemic and partly aesthetic architectonics of the novel—which have evaded criticism since the novel's publication—must be sketched briefly.[55] The classical symmetry we find in *Die Insel Felsenburg* lies neither in its colorful stories nor in the sprouting subjectivity of the narration. The relation between the life stories and the political utopia of Felsenburg Island, too, only has a mediated bearing on our focus. The basic structure of the novel has much more to do with the differentiation between narration and description.[56] Narration and description denote the fundamental difference of historiography, and they thus refer to the difference according to which the early modern theory of historiography modeled the intrinsic form of the new literary genre of the novel. In the words of Giovanni Botero from the foreword to his 1591 *Relationi universali*: "It is fitting for a history not only to tell what happened, but we must also know where and under what circumstances any given event took place;[57] and particular attention must thereby be devoted to the *descriptiones locorum, urbium,* &c."[58]

Botero's words are of particular importance, since the *Relationi universali*, which he composed for the head of the Counter-Reformation, Cardinal Carlo Borromeo, while papal Rome was reorganizing itself into a modern territorial state, was an archetype of early modern statistics. Thus when Schnabel inserts the—in his case, rare—evidence of the table to supplement the description in his running text, he is following a *dispositif* of old in the formal construction of his novel. In the unique employment of this *dispositif,* the three first volumes of the *Die Insel Felsenburg* present above all the multilayered configuration of narration and description.[59]

Narration and description are strictly separated from one another in Schnabel's work. Narration is confined almost exclusively to life stories told by individual first-person narrators, while descriptions, delivered by the one general narrator, are devoted almost entirely to Felsenburg Island. The individual narrators report their stories in uninterrupted chains of events. Conditions to be described, on the other hand—institutions, customs, buildings, and landscapes—are portrayed as they exist on the island almost without any intervention of events that could change their nature. All in all, description focuses on the location where narration takes place.

Just as only the island becomes an object of portrayal, it is also the site of narration for almost all of the autobiographies describing the various winding paths taken to the island by its inhabitants. One exception is Captain Wolffgang, who brings Eberhard Julius, the general narrator, to the island. He already fills the empty time of the journey by beginning the story of his own first landing on Felsenburg. The main narrator, as a passenger on the captain's ship, is necessarily the second exception to the rule. Eberhard Julius narrates parts of his biography at the beginning of the first volume until, conducted by Captain Wolffgang, he lands on Felsenburg; at the end of the second volume and the beginning of the third, we learn further details pertaining to his father and sister. Apart from these fragments from his own life, Eberhard Julius only ever lends his own voice to descriptions. It is conspicuous—and hence the exception does indeed confirm the rule—that his descriptions only apply to Felsenburg Island. Until he arrives on the island, he is just another individual who narrates biographically, and hence without paying attention to circumstances. His journey to the island remains just as void of images of place and time as all the trips and journeys of the other autobiographers who, having arrived on the island, narrate the lives that led them to the site of their narrating. Felsenburg Island, in an entirely structural sense, is therefore indeed a U-topia—if by that word we understand a place from which past lives can be narrated, but at which no reversal from good to bad fortune, no contingent series of events can take place. The statistical tables and the ritual of their promulgation are the sharp counterimage to the narration of the contingent reversal of things. The evidence of the numerical pictures stands in contrast to the probability of narration.

Novels and Tables 239

The first three volumes of *Die Insel Felsenburg* differ in how each of them arranges the basic relation between description and narration. The first volume organizes the narratives according to a pattern of symmetrical frames. Eberhard Julius establishes himself here as the principal narrator. Within this main narrative, Captain Wolffgang recounts his life in three installments. Between these, the autobiography of Albert Julius, the discoverer and ruler of the island, extends over eleven installments, whereby in the final installments are inserted brief autobiographies of his sons- and daughters-in-law, who had contributed with their arrivals to the hermit's transformation into the founder of a whole people. Near the volume's exact middle, we find the Latin autobiography of a Spaniard who had come to the island before the German, Dutch, and English discoverers, and whose body Albert—as he reports in the course of his own inserted autobiography—manages to uncover. This life story, left blank here at the book's center, is then added to the first volume as an appendix, almost as a prehistoric document.[60] The symmetrical framing lets readers listen to initial journeys to Felsenburg Island in four chronological stages, which reflect the great eras of colonial history in the Americas. The successive framing offers a pyramid of founding stories.

The description of Felsenburg in the first volume is cast in an entirely institutional form. Albert Julius takes the new arrivals along on a "general inspection" through the settlements on the island,[61] the departments laid out in a circular formation around the "Albert Citadel," each of which bears the name of the head of another family. Inspections, here undertaken by the ruler in person, but most commonly conducted by commissions in the territorial states of the first half of the eighteenth century, are enterprises for the production of statistics. When Eberhard Julius, between the segments of his life story as the founding father, describes what he sees, he devotes himself to "description" in the manner of seventeenth- and eighteenth-century descriptions of lands and nation-states.[62] The eleven alternations between narrative and description in the first volume thus represent the precise program, in novelistic form, of the twofold *historia reipublicae* as it was cultivated since the Renaissance. Narrations of a founding act that form a pyramid of repetitions of that first foundation alternate with descriptions of the area whose progress is measured by an inspection trip by an official or the sovereign himself. Consequently, the ruler's inspection on Felsenburg always stumbles over monuments of what has recently been narrated: the graves of those whose death has just been reported by the sovereign; or the manuscripts and memorials they have left behind. In this series of life narratives and serial statistical reports, the possibility of a history of Felsenburg emerges. We can therefore say that the distinction between narrative and description in the first volume is organized around a virtual, greater narrative—history.

Instead of three or four framed biographies, the second volume offers nine life reports in a row. The framed autobiographies of the first volume originate with discoverers and founders; in the autobiographies strung together in the second, it is the people newly arrived with Captain Wolffgang and Eberhard Julius who tell their stories. In their life accounts, the skills and occupation of the respective narrator play a decisive role. Starting with a pastor and a learned academic, we come to a surgeon and pharmacist, then to cabinetmakers, millers, threshers, lacemakers, and carpenters. In short, even if they all repeat the gesture of the first founders and discoverers, the lives of the second generation of autobiographers are determined by social typology. The passages of island description scattered between these stories of the sampled careers are arranged not spatially but chronologically. The text moves through ceremonial events, the days of the church calendar and the times of agricultural work. The volume begins, for example, with a wedding, which is followed by descriptions of a shooting competition, the gathering of the harvest with its attendant celebration, and several church ceremonies with their detailed programs, songs and sermons. Beyond this, the general narrator reports on customs and habits, *vita et mores*, as the rhetorical topic of national descriptions is called. He describes the distribution of work, sartorial details, rites of birth and death, and festival customs. In the second volume, which for the first time makes room for a statistical table, descriptive conditions dominate in the configuration narrative/description. The stories exhaust themselves, while the form of serialization and the occupational exemplarity render them elements of the larger description that extends around them. Instead of history, as in the first volume, the dominant feature in the second volume is the quasi-ethnographical mode of *mores*.

The third volume clearly corresponds to the first again. Plans are being laid on Felsenburg Island "to provide for a new inspection in all colonies in order to have a close look in particular at the situation of skilled tradesmen and artists, and to find out how they might still be helped or served."⁶³

During the years when Schnabel was writing *Die Insel Felsenburg*, Frederick Wilhelm I similarly ordered his officials to carry out surveys, which in fact were intended to be repeated on a regular basis. In the novel, the insular inspection trip has to be broken off shortly after it begins, however, because earthquakes and volcanic eruptions shake the island. These events are interpreted as foreshadowing the impending death of the sovereign, Albert Julius, which prevents the second inspection, just as the general inspection in the first volume is justified by the very fact of his ability to undertake it.

Thus firmly linked to the life and death of the first ruler, the whole undertaking of the inspection revolves around the founder's view of what he has founded. Only the sovereign's life story and the description of the state he has founded can satisfy the structure of this history book (*Geschicht-Buch*). With the first ruler's death, this specific parallel of narratable life and territorial realm to be described comes to a stop. This ending then extinguishes the need for every story told also to be the history of the territory to be described. Now, for the first time, stories, and hence the novel, can take place on Felsenburg Island. First, we read about regulating the succession and providing the state with a constitution;⁶⁴ then about the introduction of administration and bureaucracy, and the establishment of more permanent commissions charged with surveying the country. As in the first volume, a certain parallelism of narrative and description remains the guiding principle of the text. Now descriptions no longer cover what has resulted from the stories of the autobiographies; instead, the narrated stories supply keys to what had remained unexplained and mysterious in the context of description alone.

The third volume corresponds to the first in many striking ways. Not the least interesting thing about it is how the longest story in the third volume, told by a certain Monsieur van Blac, responds to the essential story of the founding father in the first. Van Blac is no important person in the sense of a political novel; he is no founder or sovereign like Albert Julius. Monsieur Blac's story nevertheless defines the third volume as much as the eleven installments of the founder's tale determine the first—though in a very different way.

Van Blac is the first and only true novelistic hero in *Die Insel Felsenburg*. Right at the beginning of the third volume, before the return to Felsenburg, he makes his appearance, heralded by a conspicuous proclamation by the main narrator: Eberhard Julius introduces the episode with the telling note, "There is one other thing I nearly forgot!" What follows is the only instance in the book of a portrait of a miserable, half-naked person, an Odysseus arriving on the Phaeacian coast, the quintessential adventure hero. After the close of the report about van Blac's experiences, Eberhard Julius explicitly refers to them as an "Avanture" [sic] and announces that van Blac "became a very noteworthy person . . . over the course of this story, and one thereby has cause to observe the peculiar direction of fate."[65] Conversely, it is only with van Blac that we are finally given a structure that can be called a "course of the story" in the first place.

Van Blac constantly threads his way in and out of the entire third volume as a dramatis persona and as the narrator of his life story. After arriving on Felsenburg, he sees the grave of the traveling companion with whom the island's founder had landed, whom he identifies as a long lost relative. This is only the first of a considerable number of scenes of recognition for van Blac, another staple of the adventure story. The novelistic adventurer, we might say, is no longer a discoverer but a rediscoverer. In this vein, the volume ends with the rediscovery of a heathen temple on Felsenburg. Entirely in harmony with historiographical speculation in the sixteenth and seventeenth centuries about the American colonies, a further origin turns up behind the pyramid of discoverers who always find another predecessor—from Captain Wolffgang to Eberhard Julius to the Spaniard. This temple finally sets the colony into the frame of a pre-European prehistory. As Eberhard Julius puts it later on, van Blac, a novelistic rediscoverer well-versed in identification and *anagnorisis*, is "the main character [literally, the "central figure," *Haupt-Person*] of this discovery."[66]

This is the one adventure that can be experienced on Felsenburg Island in which narration and description run in tandem: the adventure of recognition and the discoveries of origins. No longer is the founder looking out over his territory, whose founding his life story narrates. Instead, the central figure (*Haupt-Person*) of the adventure knows how to narrate the appropriate story of the origin of what can be observed and described. The stories of finding the temple remain doubly mysterious: for one thing, Eberhard Julius and his friends cannot agree on how to decipher the inscriptions or on a hypothesis of the origin. And the publisher of the manuscript shows up for once behind the main narrator with the supposition that Julius may have kept quiet about finding treasure in the temple. In which case, the conspicuously emphasized adventure would become quite different and more picaresque than the pure adventure of an archeological finding by itself.

In the first volume, space and time, description and narration correspond, each in its own way referring to the establishment of the same territory or colony. Hence within the framework of their common object, the territorial state, narration and description relate to each other. This structure implies a dominance of the spatial side. In the end, it is the existing territorial state where the nation's history and its statistical status praesens come together. In the third volume, time returns in the constituted space of the territorial state. The adventure of rediscovering the origin of the state uses the spatial remnants and testimonies in order to go back in time. But whatever they find, such a rediscovery can only lead to other new stories about the state of affairs that lies before the eyes ready for description. In other words, the reinscription of time mirrors and potentially modifies the configuration of description and narration. But it can never unravel the fundamental relatedness of narration and description within the state and its novel. Two corresponding constellations of narrative and description thus come to light. The two outer flanks of the novel contain the history of the state: we find the story of the state's territory in the first volume, and stories that take place and play out on the territory of the state in the third. In the second, middle volume, we see the present state of things—the *status praesens*—the epitome of statistics. Here, narration and description are nearly identical to each other in providing one and the same thing: the portrayal of types of being or *mores*. In this sense, the second volume can truly be called the center of the novel. Description and narration are not only coordinated but nearly identified with each other. The brief instances of population statistics that turn up occasionally in the text thus structurally form the novel's core. They present in a nutshell the scenario of how what can be narrated turns into a describable condition, and vice versa.

With its emphasis on the symmetrical configuration of narrating and describing, *Die Insel Felsenburg* is epistemologically and aesthetically an epitome of the so-called *Staatsbeschreibung* (description of the state of a nation) of seventeenth-century statistics. This kind of *Staatsbeschreibung*, which included the corresponding relationship not only of narrative and descriptive elements but also of text and numbers, remained valid at least through the second half of the eighteenth century, up until the tenure of the Göttingen successors of the famous professor of history Gottfried Achenwall. The representation of the *status praesens* of a state, in the definitions proposed by neo-Aristotelians from Conring to Achenwall, had always led to the conceptual vicinity of what characterizes poetry in Aristotle's *Poetics*. Johann Andreas Bose, at the beginning of his 1676 *Introductio generalis in notitiam rerumpublicarum orbis universi*, analyzes the differences between the universality of politics (constitutional doctrine) and the singularities of history. Statistics, Bose reasons, is therefore not part of political philosophy, since it does not manifest itself in the discussion of constitutional law as applied to a given state. Nor can statistics be classed as a genus of history writing, which "represents individual state actions with the details of time, place, social condition, character and other circumstances," even if it includes further causes and consequences, pretexts and judgments. Instead, statistics bestows a "more general treatment" (*paullo generaliore tractione*) on the details of history (*particularia vel singularia*). Statistics represents the "concerns of individual states," and does so "principally from a universal point of view and not tied to this or that point in time or these or those specific persons."[67] In accordance with the famous formula from Aristotle's *Poetics*, which claims that poetry, in the medium of probability, brings the singularities of history closer to the universality of philosophy, statistics is poetical. Statistics is the poetry of the state. The "respective individual state" (*singularis res publica*) should be represented in its particularity as a whole of its own. In this task of constructing the unity of a single object, the characteristic mode of representation for statistics remains hidden. This observation leads Bose finally to the well-known metaphorics of the image and the evidential tableau. Bose continues that a "more general [that is, probable and poetic] mode of representation" in statistics places "the entire form and constitution of every kingdom before the spirit of the reader or listener as if in one single view."[68] *Die Insel Felsenburg* makes the poetical and aesthetic implications of statistics explicit in terms of the modern genre of the novel.

Lists, Tables, Maps: Evidential Graphics in the Labyrinths of the *Staatsroman* and the Adventure Novel

Whereas in Defoe's *A Journal of the Plague Year*, the reading of numbers leads straight to the constitutive moment of a decision for measures and measurement in general, the small population lists in *Die Insel Felsenburg* have their place on the surface. Located in the middle volume, with its dedication to statistically portraying *mores*, they are clearly separated from the founding stories of the first and the third volumes. They are only linked to the founding acts of narration by means of the various forms of ritual announcement (the church calendar, the agricultural year, and the inspections).

The fact that numbers are located on the surface of the novel is also true in a typographical and grammatological sense. Numbers in *Die Insel Felsenburg* share this status with the maps, outlines, and "genealogical table" found in the three volumes. Territory, population, and the genealogy of the ruling family: these three elements of political arithmetic are the three types of reference in the novel. The reader can also read or view them in the graphic mode. The tables or tableaus, the maps constructed to scale, and the visually organized tables are all inserted into the text in the same way, and they belong to the same type of diagrammatic visuality according to their rhetorical effect of *evidentia*.[69]

A map of Felsenburg Island appears in the first volume. It is proleptically inserted into the text, immediately after the narrator and grandson, Eberhard Julius, sets foot on the island. The accompanying parson in the narrative has just "presented our first offering on this island" when the ruler's narrating descendant announces the map, which embodies the gesture of a second land acquisition in narration and as description. The map thus represents once again the architecture of descriptions and stories of origin: "Layout of Felsenburg Island, discovered in the Year 1646 by Albert Julio . . . sketched by Monsieur Eberhard Julio to his best abilities, Anno 1726."[70]

The map has the peculiarly pictorial character found in the art and cartography of the seventeenth century, as Svetlana Alpers has authoritatively shown.[71] Just as in the city views and maps from the famous collection of Georgius Braun and Franz Hogenbert, which appeared between 1587 and 1617,[72] the projection of the island topography on Schnabel's map is simultaneously an object in the space that the reader-viewer perceives from a raised point of view. The mapped island is presented in a perspectival world with a foreground and a suggested horizon line. Throughout the novel, the reader is constantly referred back to this map. The text describes what the map shows, but we read it differently on the map than in the text: "It is," says the narrator in inserting the map, "impossible to describe everything thoroughly at once to the gracious reader, as it conveniently appeared all around to our eyes, and for that reason I wanted to insert a small layout of the Island."[73]

To show "everything at once" is precisely what the figure of evidence in rhetoric promises. In this case, however, the verbal figure is applied to an actually visual object. The rhetorical figure of *evidentia* means that, in speaking, something is pictorially shown instead of told. The map in the text of the novel exhibits this figure of medial substitution, but it does so in a reverse sense. Eberhard Julius's map is not the image that at once shows what could only be described successively in the text. Only by means of its explicit introduction as the particular image that it is does the map take on the task and gain the possibility of carrying out this function in the text that it is meant to pull together into the simultaneity of an image. We clearly recognize this process just by observing the map and the way it is inserted in the text. The legend refers the map back to the status of the text that it pulls together into an image: the scale of length and the compass rose let us read their fundamental subtext in the geometrical construction of a map; differences in shading symbolizing land or sea, schematic trees, and houses are iconic signs that the viewer must read. The map only provides an evidential picture for the text because it is not merely a picture, but rather already a text even within the picture itself. This text refers the picture in which it is inscribed symbolically to the text of the novel that the map summarizes as a picture.

In similar ways, maps and construction outlines continue to duplicate and embody the text's descriptive and narrative vividness. Recognizing the common structure of maps, numerical tables, and other diagrams, however, has only recently become so simple a task. The reunion of map and quantitative information developed around the end of the seventeenth century from the old fund of maps that offered symbolic representations of a qualitative nature—of flora and fauna, of populations and their habitual customs. In his "Entwurf gewisser Staatstafeln" (Draft of Certain Political Tables), Leibniz had made the synthesis of statistics, tabular representation, cartography, and mathematical formulae explicit as a concept.[74] A first example of a so-called thematic map bringing together topography and quantitative information, on the other hand, was the world map by the astronomer and statistician Edmond Halley, which appeared in the 1686 *Philosophical Transactions*. Halley marked the direction and strength of winds on the oceans by directed bundles of strokes.[75] In the modern terminology of E. R. Tufte, the map is an example of information design, or the visual display of quantitative information.[76] This term does not refer to this or that form of familiar knowledge or recognition, but rather to a diagrammatic presentation of data. The two parts of the definition imply each other. We can only call "data" that which can be entered in a constructive frame of representation and can be further worked out therein without changing this frame. The reverse is also true: data are the kind of elements of knowledge that we enter into diagrams and can work out according to the standards of those diagrams.

Such is the future of the figure of *evidentia*, beginning with the test forms printed by Veit Ludwig von Seckendorff in his 1665 *Deutscher Fürstenstaat*, and carried out further in the tables for which the provincial civil servant Schnabel made room in his novel.[77] The architectonic precision of his enormous novelistic work is demonstrated not least by the way in which it incorporated within itself this evidential-graphic constellation of narration and description, the diagrammatics of alphanumeric writing itself.[78]

10

The Theory of Probability and the Form of the Novel

DANIEL BERNOULLI ON UTILITY VALUE,
THE ANTHROPOLOGY OF RISK, AND GELLERT'S
EPISTOLARY FICTION

Probability Beyond Poetics and the Theory of the Novel

Graphical presentations of *Bills of Mortality* and pastoral registers only show up scattered throughout works that are considered experiments in modern novel writing in the early eighteenth century. But wherever tableaus of statistics appear in Defoe or Schnabel, they are linked to decisive distinctions for the form of the novel beyond traditional poetological patterns. One of these distinctions concerns the question of *res fictae* or *res factae*: the semblance of truth or truth without further respect to appearance. Another case is the opposition between narration and description: the contingency of events in the world or the world that provides events with a field for coming into appearance. These alternatives open up the fundamental question about form or nonform (in the case of fiction or fact) or propose the internal distinction between devices that bestow form and those that presuppose it (in the case of description versus narration). These questions are not, of course, answered once and for all; nor does one side of the divide gain permanent preference over the other. The important thing to note is that, in the text and for the text, the tableaus mark the possibility and the necessity of posing these questions of external and internal evidence of form. They define the existence and the scope of a theory of the novel.

Traditionally, probability, the semblance of truth, determined the efficacy of devices in rhetoric and poetics. In contrast, tableaus and their defining figurative quality of placing-before-the-eyes exhibit the appearance of truth characteristic for any theory of the novel. Semblance of truth is thus transposed into the phenomenology of reality and the aesthetics of form. The theory of the novel in this context refers to a very particular but pervasive circumstance in literary history. The novel had no guaranteed place in traditional poetics until deep into the eighteenth century, and its discussion was therefore relegated to the prefaces of novels, digressions in novels, reviews of novels and metaphorical uses of the word *novel* in the context of philosophy.[1] (Leibniz loved doing this.)[2] Moreover, as a genre without existing poetological definition, the novel required in every instance the explicit confirmation of its own constitution in itself in order to assert itself as a literary form. Referring back to Defoe and Schnabel, we might say that novels needed to reinscribe an allegory of probability, the theory of the semblance and appearance of truth, within their own composition. The evidence of the tableaus amounted to such allegories of probability in early examples of novelistic writing. In these cases, the allegory was even visually realized in the layout of the printed page. Such allegories of novelistic reality fundamentally redefined the traditional probability of rhetorical persuasion and old European poetological verisimilitude, which were cast according to "what usually happens." They were not only reframed in terms of *evidentia* and evident appearance of truth, but such devices of evidence also linked them back to what part I of this study calls the institutional logic underlying such forms of evidence. This institutional anchoring can, for example, consist in the methodical and moral decision that the plague fatalities should be counted in the first place, or it can coincide with acknowledging the state as the agent and purpose of counting a population.

When used as allegories of probability, evidential representations thus serve both to justify and endanger it. These seemingly contrary tendencies become clear in Jonathan Swift's *Gulliver's Travels*. To begin with, Gulliver does not provide any description of Lilliput in the account of his travels. The reader learns that this is the case because such a national description, a Lilliputian history, along with its *status praesens*, is supposedly already in press. When it comes to Brobdingnag, the land of the giants, on the other hand, we do read a description of the country in the novel. This description turns out to consist mainly of geographical statistics. In composing it, however, Gulliver does not place the country itself before his eyes; instead a huge map is spread out for him, which he traverses with measuring paces. Further on, the description of Laputa, the kingdom of speculative philosophy and science, is entirely contained in the astronomical and physical conditions that let the island float in space. Gulliver finally postpones the description of the land of the Houyhnhnms for the announcement of a further and independent statistical work, since the portrayal of the wise horses, their virtues, and their customs, would go beyond the limits of the novel.[3] The only regular description of a country that we get to read in *Gulliver's Travels* is of England. Gulliver presents it to the king of Brobdingnag, who, after listening to the descriptive account, proceeds to laugh at the "odd kind of arithmetic... in reckoning the numbers of our people by a computation drawn from the several sects among us, in religion and politics."[4] The statistical descriptions of the strange countries are either left out of the novel completely or are representations of other representations in the novel, portrayals of maps or astronomical models. Only his own country, which the author and the reader have before their own eyes and whose description they can look up in Chamberlayne's *Angliae notitia*, receives a first-order representation in the text. This distribution of missing and offered statistics amounts to an obvious and conservative satire: the novel, which seems to lead readers into exotic, distant lands, leads them in truth only into the familiar. The reality in which we live always contains our confrontation and occupation with other possible and impossible worlds; in studying all other worlds, we are really just dealing with our own. But the conservative satire is foiled by contrary irony: the fact that the descriptions that could lend the novel probability are either missing, because they should become their own independent books, or are second-order observations, because they rely on maps and models, also undermines the only remaining report of Gulliver's own land. Even the description of England, where author and reader are already sitting, becomes the discourse of a character in the novel, Gulliver himself, who is here narrating his narration. The probability in which we already live as in our reality, in which we write and read, that is, not only does not contain the other probabilities; it itself is only one of the transportable representations and contingent models of the semblance of truth always already modeling reality. The satire that insists on the one constant reality and the irony that exploits techniques of the semblance of truth make up the two sides of evidence as an allegory of probability and probabilities.[5]

Much of the remaining second part of this study is concerned with a reading of the modern novel that allows for both interpretations. In order to accomplish such a reading, we should presuppose neither the uniqueness of a single reality nor the arbitrariness of many probabilities as given. In other words, the intention is not to switch from the poetological *narratio verisimilis* into a rigidly constructivist theory of the novel and its reality all at once. The aim is rather to unravel the very process that brings the allegory of probability into the position of a theory of the novel. As the first part of this book shows, such aestheticization of probability is only possible, however, because the interpretation of the game of chance as probability enabled the play of contingency to turn into the theory of probability.

Beyond the ephemeral cases of citing numerical tableaus in the text of the novel, the process of turning poetological probability into the appearance of truth can be approached from two sides. On the one hand, methodological problems in measuring narratable situations of human life by the game of chance had already resurfaced in the eighteenth century. From inside probability theory, the question came up of whether its modeling of probability suitably resembled truth. On the other hand, these questions found resonance in thematic complexes of novels that dealt with contingency, probability, and providence. Probability, which took into account the novel's narrator and dramatis personae, was conceived of according to probabilistic criteria. We can contextualize this discussion around 1750 with two examples. One of them is the Berlin Academy's prize question for the year 1751 on a possible philosophical account of chance and its many contributions, which comprised a reference to mathematical probability. The other example is Christian Fürchtegott Gellert's epistolary novel *Leben der schwedischen Gräfin von G.* (The Life of the Swedish Countess de G***) which imported the new novelistic genre into German literature based on French and English models. More than probably any other work of the century, this novel is obsessed with the representation and critique of providence and chance.

The Prussian Academy of Sciences' Prize Question for 1751: Ontology and the Morality of Chance

Classical probability gets close to the question of a theory of the novel when it is not concerned merely with the affairs of Bernoulli's *vita civilis*, but rather when it touches upon calculation of chance and fortune in the context of providence and morality. In 1751, the Royal Academy of Sciences in Berlin, to which Frederick the Great had appointed the French philosopher and mathematician Pierre-Louis Moreau de Maupertuis as president, in order to carry out his wide ranging plans of reform, announced the following contest:

> Assuming that the events of fortunate and unfortunate chance depend alone on the will or at least on the allowance of God, the question becomes: whether these events demand the observance of specific duties from men, and what the nature and extent of these duties might be.[6]

The question was a matter of Leibnizian metaphysics in an environment of Newtonian science. The theme of contingency and freedom aimed for ethical or pragmatic commentary. In fact, the published solutions are hardly impressive for their metaphysical distinctions. Their versions of determinism remain so unspecific that we can scarcely distinguish between Wolffians, English deists, representatives of physicotheology, and disciples of Maupertuis's cosmology. Instead, the question about duties brings the modes of social behavior and practical knowledge implied in the various cosmologies to the foreground. The entries read like a tutorial in moral and social overcoming of contingency in the transition between the baroque period and the Enlightenment, or more comprehensively, between early and modern modernity.

The mathematician and poet Abraham Gotthelf Kästner was the final recipient of the prize. Born in 1719, Kästner lived until the very end of the century as the leading German authority in mathematics, head of the famous Göttingen observatory, and a well-known writer of epigrams and essays of Enlightenment-style *Populärphilosophie*. Forgotten though he is today, Kästner was the central German representative of the European breed of intellectuals who claimed to master words and numbers, mathematics and literature alike, much as d'Alembert or Condorcet did in France. Kästner's was in fact the only entry to introduce mathematical probability theory into the literary-philosophical field of the prize question. The academy's leadership had meant the question to address Leibniz's principle of sufficient grounds—the principle according to which everything that exists has a cause—and thereby to give an account of the metaphysics of human freedom in a thoroughly deterministic world. In his own, first Latin and then German, translation of the French prize essay, Kästner gives the question itself a slight twist.[7] By paraphrasing the original wording, Kästner claims that the Academy desired "to show: *Whether the recognition of this principle* [i.e., the principle according to which everything we attribute to chance takes place because God permits it] *obligates people to certain duties, and what such duties might be.*"[8] It is one thing to ask about duties laid on us by the events of a fortunate or unfortunate chance, when otherwise we still know that a providentially ordered world is at work behind them. But it is a different challenge to inquire into the duties that result from the knowledge that the "adverse and advantageous circumstances" are actually not contingencies at all. In both queries, an impenetrable curtain divides the phenomenal world of contingency from the true world in which laws according to providence are in effect. In both, we must answer out of our knowledge that the world on the other side exists. But in each case the answer comes from a different side of the divide. The French original asks about the duties in the world of contingency—under the supposition that a rationally and providentially ordered world can be detected behind the contingencies. Kästner asks instead about the duties arising from the realization of an actually rational and providential world—under the supposition that we have to act under circumstances that are not transparent to such a rational order. In these two different ways of interpreting the academy's prize question, we might trace the conflict between *philosophes* dedicated to the study of Locke, on the one side, and the German students of Wolff who swear by Leibniz, on the other. But even if we assume that such political wars in the theories are at play beneath the surface of the paraphrase, the important point here is the possibility of understanding the academy's assignment in either of the two ways. This double possibility has to do with the fact that we *must* calculate with contingencies, although we know that divine providence is at work behind them—but that we can only *calculate* with them because we start off from this assumption.

Already in the introduction, Kästner explains that we can only speak of the obligation of duty—as it is called in the language of natural law—when we presuppose fundamental order; but the precarious situation of a duty to follow only comes about where such order does not simply manifest itself openly and accessibly.[9] The double view that the question presupposes and for which it seeks a solution clearly no longer aims to close the hiatus between the constant laws and the events that can turn out in this or that way. Nor does it have to do with the one game that has the same extension as human life (Pascal), nor with the system of games according to the measure of the one fair game (Huygens). And, finally, it also is not directly connected to the mathematical-metaphysical solution of Leibniz, who makes contingency the moment of order itself, which can only be the best one because it could not be any other of those possible. To acknowledge a particular duty in a world that presents itself as marked by contingency, although in fact there is no contingency, means to no longer seek a solution for the problem of an order of contingency. Instead, the split view allowing two worlds to coexist with each other establishes a practical solution of how to act and plan under such circumstances; and it does so precisely because the illusion in all supposed contingency has been completely seen through.

While Kästner wrote his answer, he was preparing to leave behind the Leipzig of Gottschedian reason and move on either to the Berlin Academy of Science or, as he actually did in 1755, to the University of Göttingen.[10] These happened to be the two centers for scientific thinking in the German-speaking world, with Berlin closer to French philosophy and mathematics and Göttingen forcefully developing a British agenda in the natural sciences. Kästner's prize-winning essay was thus a career-making text; and it seems as if every decisive scientific and literary issue in the north and northeast of the Holy Roman Empire was at stake in it.[11] The Academy had only been advertising prizes for seven years, but the presence of Maupertuis and the Swiss mathematician Leonhard Euler on the award committee and d'Alembert's victory in the second contest guaranteed European attention. The emphasis lay on mathematics and scientific research. The Academy supplemented the scientific focus in the first ten years between 1745 and 1755 by launching three philosophical topics at issue between Leibnizian metaphysics and what was called *Philosophia Britannica*.[12] The tenability of Leibniz's metaphysics in the Enlightenment was to be discussed, particularly in the situation of the sciences, which had recently been dominated by English influence. Thus, the question for 1747 was on the doctrine of monads; for 1755, it addressed the cosmological doctrine of perfection. In 1751, the midpoint between these two years, it dealt with the peculiar natural law theme of duties regarding chance as conceived of in the doctrine of human freedom.[13] Kästner was well prepared to come out victorious in this topic: he had repeatedly studied juridical mathematics in the Leibnizian tradition.[14] In addition, he had been working on developing binomial coefficients, which Jacob Bernoulli and Pascal had already in 1744 hit upon as a model of a limit theory in connection with probability. A small publication on probability calculation from 1749 had a clear-cut purpose in method and intellectual polemics: in it, Kästner defends Jacob Bernoulli's interpretation of gaming calculation against the polemics of an eminent German historian. In 1748, Johann Martin Chladenius, theoretician of historiography and of hermeneutics, published *Vernünftige Gedanken von dem Wahrscheinlichen* (Rational Thoughts on Probability), a book meant to reclaim probability for the humanities. The wide acceptance of probability calculation among important German Enlightenment philosophers discomfited Chladenius so greatly that in 1748 he referred to probability as the idol of his time.[15] And indeed, the idea of grading probability had already shown up tentatively around the turn of the century with the philosopher Andreas Rüdiger.[16] In his *Logic*, Christian Wolff pursued Leibniz's old desideratum, the *logica probabilium*.[17] This branch of logic developed even more definitively in the 1730s and 1740s with professors of philosophy such as Friedrich Christian Baumeister and Ludwig Martin Kahle.[18] Not to forget that in 1741, Christian Wolff gave the enterprise of probabilistic logic the final stamp of approval with his foreword to Süssmilch's *Göttliche Ordnung,* which was brilliant at least in its programmatic ambitions.[19]

By admitting the gradation of probability, these philosophers, with the exception of certain followers of Rüdiger, finally broke up the Aristotelian order that made a fundamental distinction between knowledge of singularities—in *historiae*—and knowledge of science—in the laws of *causae*. If in fact graded probability as defined through the game of chance is finally and fully accepted, we are faced with a space of continuous increase and decrease of probabilities, a continuum between the zero (0) of negative certainty and the one (1) of positive certainty. Chladenius in contrast wished to reestablish the old probability of the topics. The topical idea of probability, however, presupposed a qualitative and irreducible distinction from certainty. To insist on the qualitative distinction between probability and truth thus meant to demarcate the realm of historical and topical probability as its own hermeneutic space,[20] as opposed to truth (including the mathematically probable). In his answer to Chladenius, "That Degrees and Measure of Probability Exist," Kästner gives the most concise summary of probability theory that had appeared in German-speaking lands up to that time. Against the conservative Aristotelian polemics that also anticipated the separation of the humanities from the natural sciences in the style of the nineteenth century, Kästner once again defends the Leibnizian concept of an integral *logica probabilium*. He does so, however, with the help, not only of the newest mathematics, but also of a new twist on the notion of probability itself that is indicative of its own theoretical development in the middle of the century.

A Utility of Chance: Daniel Bernoulli's Theory of Risk Taking

The idea distinguishing Kästner's entry from all the others is expressed in a subdivision he adds to the assignment. He follows up his treatise on the *duties* of chance with a supplement on its *use*. Through this new correspondence of terms, Kästner, who had studied law in Leipzig,[21] makes use of the legal philosophical vocabulary in which Enlightenment thinkers since Thomasius and Wolff had treated moral problems.

With *utility*, Kästner is also taking up Daniel Bernoulli's further development of probability theory. In a reflection on the history of probability theory, Kästner starts out with a remark on the ideal nature of the classical game of chance: "Math experts originally assumed that one is to the other as the sum of the possible cases favorable to us is to the sum of unfavorable cases." Such a game only takes place in exceptional cases. For the game to follow this pure ratio, we have to assume that "what we hope or fear does not contribute anything noticeable in relation to those goods that we own." Kästner refers with this critical remark to Daniel Bernoulli's *Specimen theoriae novae de mensura sortis* (1738; Exposition of a New Theory on the Measurement of Risk).[22] In order to capture ordinary situations in life, we cannot take for granted that they should follow the ratio of favorable and unfavorable events as modeled in games of chance. Instead, as Kästner explains with Bernoulli, we must "not only gain a view of the magnitude of what we can win or lose in itself, but rather simultaneously estimate how severe the loss would be to us and how happy the gain would make us."[23] What Kästner designates as a measure of how much a given gain or loss means to us (the measure of our *Empfindlichkeit*, or sensitivity) is the attempt to solve a crucial problem of mathematical probability. By considering differences in the experience—the sensibility or *Empfindlichkeit*—of gain or loss, Daniel Bernoulli marks a point where the structure of the game and the semantics of probability seem to run counter to one another. In the moment when that happens, however, the interpretation of the game as the theory of probability threatens to break apart.

Daniel Bernoulli's *Specimen theoriae novae* and its key term of risk were inspired by a problem of probability known as the St. Petersburg paradox, after the solution that he had published in 1738 in the records of the St. Petersburg Academy.[24] This problem exercised a lasting impact on probability theory up through the 1760s and 1770s. Its chance setup implied a situation in which the winning sums should rise with the number of the attempted throws toward infinity, assuming the agreed winning toss is infinitely postponed. In this scenario, the value of the expectation remains over ½ (with the game thus proving to be advantageous mathematically), even though it is intuitively (according to the semantics of everyday life) foolish to stake any larger sums upon the expectation of an infinite win. There is clearly no possibility of playing the game of the St. Petersburg paradox in the sense of the primary exemplarity of the game as a fair one. Although the analysis of strict calculation indicates that the stake is advantageous, putting down any large sum of money upon the expectation of infinite gain would only be sensible for the extremely rich and only fair in an infinitely continuous game. D'Alembert and Diderot, Lichtenberg and Laplace, all puzzled over the paradox. The solution that Daniel Bernoulli bequeathed, and that most of them, with the exception of d'Alembert, accepted, remained the essential problem for them to discuss.[25] Georg Christoph Lichtenberg, Kästner's later colleague in Göttingen, was the first to see this solution as the archetype of a different sort of applied mathematics, a mathematics that produced derivative models that did justice to everyday situations.[26]

The solution that remained a challenge for Diderot and Lichtenberg, but that provided Kästner with a formula for the *utility of chance*, had started out with Daniel Bernoulli as a fundamental rethinking of mathematical probability. It amounted to correcting the basic hypothesis of gaming theory interpreted as probability theory. The mathematical theory of games of chance always represents equal expectations of two players with the same value. It is assumed that "*the risks anticipated by each [player] must be deemed equal in value.* No characteristic of the persons themselves ought to be taken into consideration; only those matters should be weighed carefully that pertain to the terms of the risk."[27]

Behind the axiomatically assumed *aequalitas* according to which an event is equally likely to occur as not to occur, Daniel Bernoulli thus discovered a different kind of equality. According to Bernoulli, this second equality had previously been assumed without reflection, and only his own theory was to promote it to the rank of an empirical hypothesis. The axiomatic or methodical *aequalitas* in the game is based for Daniel Bernoulli on an arbitrarily assumed *aequalitas* of the players in their condition to play the game. His theory of value now models the hypothetical equality with mathematical means so that they can be factored into the calculation of chances.

We can say that Daniel Bernoulli finally takes the example of the game seriously or that he forgets its methodical meaning. He takes it seriously as a single and empirical process in a real world—the prerequisites under which the players play finally enter the grid of methodical investigation. He forgets about the game, however, insofar as the game is the model according to which events in the world can be formed as strictly contingent in the first place—he no longer asks, as did Pascal and Huygens, from what point of view the game can be called a model of life (in its finitude) or of society (in its systematic nature). With Daniel Bernoulli, the (axiomatic) justice of the game is admitted into the (hypothetical) continuum of situations spanning from the probable inequality to the improbable equality of players. The mathematical structure of probability, the game of chance, is based on a semantics of social relations that renders its exemplarity improbable. This shift, as Lorraine Daston shows, is linked with a transition from juridical to economic thinking.[28] Daston's interpretation becomes immediately evident in Bernoulli's classic example of hypothetical inequalities: a lottery ticket that has equal chances of winning or losing in a game whose prize amounts to 20,000 ducats has a different value for a rich man than for a pauper, if we understand value to mean the advantage or disadvantage that a win or loss would bring to one or the other. For the poor man, it is advantageous to sell the lottery ticket for less than 10,000 ducats—the quotient of probability; for the rich man that would not be the case. But the theoretical shift is not fully explained with the reference to law or economy alone. We are not merely dealing with two different models, but rather with two different ways of operating the formation of models. In the first case, which is probability according to Huygens and Jacob Bernoulli, the primary level at which to construct an example for calculation consists in its legal meaning. The legal example semanticizes calculation; and calculation finds its application in the legal sphere of the example. In order to exhaust the potential of this primary semantics or application, human life or the world themselves would have to be mustered or modeled in light of the example. From what point of view, by what justification, and under what aspect, we must ask in this first scenario, is life or the world a game of chance, a game and a play of equally probable events? In the second—Bernoullian—case, the example is framed in economic terms. Starting out from the difference between the utility or disadvantage to the poor and to the rich gives the game of equiprobable events a place and meaning in social life. In such a context, the task is to detect potential differences between the structural example and the empirical situations to which it is applied, and to add new elements to the theoretical formula that help to do justice to such deviations.

This shift from the legal to the economic reading of the example follows a direction that had already been implied in the defining act of identifying gaming theory with the epistemology of probability. If a metatheoretical problem in all probability theory is the question of how to adapt the rigid structure of the game of chance to situations in social life and in the natural world, then the semantics of probability can be seen as the medium through which such adaptation is performed and carried out. A more legal or economic interpretation of the notion of probability leads to different modes of such adaptations, with legal probability closer to the structure of the game and its rule of fairness, and economic probability more responsive to empirical situations in people's lives.

Daniel Bernoulli's new theorem of risk—his supplement to the interpretation of the game as probability—thus models the value of gain and loss depending on given data, for instance, the initial financial situation of the players.[29] In the nineteenth century, the theorem of economic value would be used to describe the model of the *homo oeconomicus*, or participant in the market. Kästner, in his supplement to the duties of chance, gives a less consequential but even more ambitious reading. In Daniel Bernoulli's addendum, he first of all sees the modeling of man in general as that of an animal whose form of life is dependent on and specialized in economic dealings with chance and fortune, that is, its utility in general. Kästner is not exclusively impressed with man as shaped by the special sphere of economy, which is a sphere of chance management. Rather, he recognizes that humans need to handle chance events in economic ways in life at large.

With the natural law term of a duty regarding chance, Kästner is thinking of the same things other entries also offered in many variations: duty is accordingly immunization against both good and bad luck. Through spiritual exercises in equanimity in the face of fortune and misfortune—whose basic psychological techniques had been taught since the Hellenistic philosophers of the good life preached *askesis*, and had become part of Protestant and Jesuit meditation since humanism—we are able to reassert the view of ourselves *sub specie aeternitatis*. Duty regarding chance means the permanent establishment of a double view: by practicing our capacity to experience events as fortune or misfortune with an ordering, more or less invisible hand behind them, we place ourselves in a position to confront every new event we face in this same way. The religious or moral *exercitium* absorbs the possibilities of the unforeseen so as to prepare practitioners of spiritual exercises in the contingent world for all seeming contingencies.[30]

The Protestant theologian Johann Gottlieb Töllner, whose entry took second prize, summed up what this deterministic metaphysics meant as follows: one should always follow the voice of one's conscience, thank God for everything, and entrust one's salvation to Him in prayer.[31] Phrased as spiritual exercises, we must achieve "such a degree of spiritual strength that fortune and misfortune make no particular impression on us." In misfortune, we must "practice total surrender to providence," realizing that whatever happens is either a matter of deserved punishment or inscrutable grace. And, finally, we must work on our own perfection and thereby our worthiness for fortune.[32] We should strive simultaneously and with the same engagement to expose ourselves to contingencies and withdraw from the world. Both are united in the paradoxical strategy of foreseeing the unforseeableness of an event and fathoming the unfathomableness of its providential meaning. The negation of chance as an instance of providence is already implicit in attributing the event to chance. This paradox does not have any logical solution, but rather stabilizes itself as a paradox in the self-affection of those performing the spiritual exercises.

As another contributor to the contest remarks, by aiming at its neutralization, confronting chance activates one's "cognitive faculty," although nothing specific can easily be said about one's "powers to act."[33] This comment is not meant to be an objection, but rather refers to the purpose of the exercise. It is a matter of continuously transforming the acting person into an observer, a process that has no real end. It is readily apparent how closely the poetics of the baroque theater and novel, the evocation of admiration and terror, are related to this.

In contrast to these meditative exercises, which some entries in fact still present in a baroque tenor, Kästner focuses rather on their cognitive counterpart, Leibniz's and Maupertuis's cosmologies. A proper view of the world's wholeness thus allows seeming contingency to be recognized as a moment in the causal chain of events. The installation of the double view of the chance event is then removed into a perspective of the philosophy of history: "My laborious work is lost if a *blind chance* hinders the execution of my plan; it is not lost if providence determines my exertions for purposes other than the ones I had directed them toward."³⁴

In other words, if I shape my intentions dutifully, that is, in accordance with justice, then I paradoxically no longer need to pursue them as my intentions. I can and must, as Kästner says, using the appropriate term, "leave [them] to providence."³⁵ Kästner does not distinguish himself substantially from his fellow competitors, however, until the *supplément* on the "use" or "utility" of chance according to Bernoulli's theory of value. If, regarding contingency, knowledge and morality go in tandem under the heading of duty, then, regarding utility, the two aspects are separate again: "In some of these incidents we behave as mere onlookers; others influence the course of our life such that it becomes different than it would have been without them."³⁶ We may speak of an aesthetics and an economy of chance events, which are sundered from each other here. Men split up into actors and observers of chance.

In the economy of life, we deal with the contingency of chance. As onlookers, in contrast, we recognize the whole in which chance is revealed as an element of order. Alone among the contestants, Kästner thus attempts in the *supplément* to move behind the façade of spiritual exercises against the vicissitudes of chance and their baroque theatricality. He installs a complementary procedure that simply gives up on the unity of worldview and divides into economic behavior, on the one hand, and an aesthetic view, on the other. With economic behavior vis-à-vis chance distinct from aesthetic observation, the probabilistic calculation of chance finally comes into focus. At this point, the calculation is no longer concerned with the game and its methodical justice, but rather with a theory of utility that sees the model of fair play disappear behind the hypothetical *aequalitas* of the players. As first glimpsed with Leibniz's reading of Huygens, the notion of probability in the mathematical theory of chance is finally anthropologized. We are no longer constructing a difficult coextension of game and life. In Kästner's reading of Daniel Bernoulli, human reason leads to the game of probability as though of its own accord, as Lorraine Daston argues is the case with classical probability in general.³⁷

In contrast to the baroque theatrics of spiritual exercises, it makes a clear difference to the utility of chance and its goal of rational action whether one is facing fortune or misfortune. In the case of good fortune, there is in fact nothing to calculate. Calculation only becomes an issue when misfortune repeatedly strikes. Through calculation, we attempt to transcend chance and to understand the underlying law governing contingent events. This is not done merely contemplatively, however. We even cannot wish to "avoid all danger, since then we would not have to undertake anything." It is much more important to calculate the imminent danger, and then "to leave just the right amount up to luck so that its loss cannot ruin us completely."[38] This is precisely where Kästner introduces Daniel Bernoulli's value theory into the provisions supposedly built into the fabric of human nature. Our hopes, Kästner argues, correspond to the mathematical probability of the occurrence of a favorable event, if we only take into account where we come from as players of this game; our desire in a given case expresses the value the favorable event has for our specific starting situation in the game. Thus the incentive to make this or that choice is an emotional product of hope and desire that coincides with the rational outcome of calculation. "I think I hear the voice of nature," Kästner writes, a voice that advises us as to whether to take a risk or not.[39] This is what the utility of chance installs in the human psyche: in contrast to all dutiful preparedness vis-à-vis chance through spiritual exercises, the play of chance realizes itself in the prevailing or succumbing energy of taking certain risks. In terms of risk, the utility of chance, rational choice is for once in sync with emotion and sensibility (*Empfindlichkeit*).

While a theory of risk hence already exists for those in whose lives chance interferes, the theory of the spectator still needs further elaboration according to Kästner's prize essay. It is true that the tragedies of Sophocles or Voltaire provide examples for the lessons of ethics; and history is known to be a great collection of examples of God's providence. But Kästner is concerned with clearly distinguishing the new aesthetics of spectatorship from the old poetics of fear and wonder. The spectator should be presented with spectacles of physico-theology in the vein of William Derham, representations of a nature in which even the Creator can no longer intervene with special providence. Whereas such teleological histories exist in philosophy for the realm of nature, Kästner argues that a similar view is still missing with regard to human history. For this he could have pointed out Süssmilch's *Göttliche Ordnung*, which had appeared exactly ten years previously. Süssmilch's population statistics—which fittingly kick off with a reference to Derham—fit the bill precisely: in their recourse to statistics, they amount to a physico-theology of man.⁴⁰

Exercitium, Calculation, and Aesthetics: The Economic Idyll in C. F. Gellert's *Schwedische Gräfin*

In Kästner's winning essay in the Royal Academy of Science's 1751 prize contest on the ontology and morality of chance, and the whole corpus of entries, we have unique access to the state of thinking on chance in mid-eighteenth-century Europe and a fundamental epistemic and social shift in what today we would call risk management, or the politics of how to use the future.⁴¹ It may well be fruitful to complement the analysis of Kästner's use of probability theory in this context by looking at a book published only a few years earlier by Christian Fürchtegott Gellert (1715–1769), who had lectured on morality, natural law, and poetics at the University of Leipzig since 1745. In his novel *Leben der schwedischen Gräfin von G.* (1747–1748; The Life of the Swedish Countess de G***), his greatest literary success aside from some sentimental comedies, Gellert adapted Samuel Richardson's epistolary style of narration to the German context. Although it may seem a poor imitation if viewed exclusively in terms of its adherence to Richardson's model, Gellert's *Schwedische Gräfin* is interesting because he adapted the British model by reinscribing its new aesthetics into an older, "baroque" genre of novelistic narration, enabling us to study the transformation of the theme of chance at the threshold between early modern poetics and the aesthetics of the modern novel.

Beginning with the title, chance takes on a complex central position in the novel.⁴² Like "history" elsewhere, the word *Leben*, "life," posits a distinction here from the older early modern and baroque novel. It had already figured in titles by the likes of Defoe and Marivaux—*The Life and Adventures of Robinson Crusoe* and *La vie de Marianne ou les aventures de Madame la Comtesse de ****, to name the most popular. Gellert's title omits the "adventures," however, and precisely this deletion of chance and of fortunate or unfortunate reversals reveals the latent center of his (anti-) novel, that is, "life."

Gellert's splitting of the aesthetics and the calculation of contingency, which Kästner treats under the dual utility of chance, had already been prepared for in his lectures on morality at Leipzig, which he proceeded to incorporate into his novel's message. According to these lectures, a "living trust in divine providence and governance" is the "great thought" and "divine reassurance of the heart" that "often transforms into conviction and feeling."[43] The contemplation of divine providence, conveyed to us by revelation, eventually becomes part of our interior nature by spiritual exercises, and thus concludes the list of the "noblest duties of man," duties against his own heart and mind. This contemplation has the same relation to the merely moral techniques of acquiring spiritual strength or equanimity as poetry does to rhetoric.[44] We learn here to see events as an inscrutable play of chance in which we finally recognize the mask of providence. But the inverse is equally important. Not only does the reassurance of the illusory nature of chance have moral value, but so too does the disquiet that emerges from the resistance of the illusion to clear sight: "Ignorance in view of future events seems to be a failure of our mind, but in fact it is its good fortune. [This ignorance] preserves [the mind] in fortunate circumstances from pride and assurance, and in unfortunate circumstances from idleness and despair."[45]

All of this can be summed up as confirming once again the old position of immunization against the vicissitudes of chance. In the novel, we see the pair of necessary reassurance and its conducive failure as the position of individual characters. The first-person narrator, the Swedish countess, says of a mother who just lost her daughter: "She took everything as a decree of fate whose causes were inscrutable. . . . It is certain that the support of religion has an unbelievable power in the face of misfortune. Take but the hope in a better world away from the misfortunate, and I know not how I could ever help them back onto their feet."[46]

During his long captivity in Russia, her husband, the count, makes a remark that shows on the other hand how important the natural failure of such immunization is. "When we really tried to be calm, we grew more dissatisfied than ever." The "heart" however—that is, the desire to help oneself—replies to "reason's" theory of providence with dissatisfaction; and by doing so, it is finally able to "befog" reason's "light" and alleviate the prisoner's condition.[47] These reflections give a preliminary idea of the subtlety and complication to which Gellert leads the techniques of spiritual exercises in the age of their psychologization. But even if the failure of self-inoculation against excessive misfortune acquires a functional meaning in Gellert's psychology, this meaning does not yet convince the individual to take a risk.

The case is different in the central passage of the novel, which has justifiably always provided interpreters with fodder for new commentaries. It is the stretch spanned approximately by the second half of the first book. The Swedish countess of G*** here neither bears this name, nor does she live in Sweden. After her husband's sentencing and presumptive execution, and following the confiscation of her property, she flees from the sovereign territory of the Swedish crown and arrives in Amsterdam together with Mr. R**, her (supposedly) executed husband's middle-class traveling companion and family friend. A kind of bourgeois and economic idyll intervenes at this point in the narrative. The idyll is performed—as is fitting to a pastoral parody—by masquerading and switching places. The "undisclosed social position" of the countess allows her marriage to Mr. R**, a "young man who wasn't good for anything in a large company other than filling an empty seat."[48] Aristocratic incognito and the bourgeois destiny of being a place-holder taken together create the paradox of the private under the protection of the public life: "We lived in the most populous place in the greatest quiet."[49] This part of the novel is determined by two sets of events. The first set has only one element: the marriage between the masquerading countess and the essentially place-holding middle-class commoner. The second set consists of a larger number of catastrophic events that befall the un-family of the count's former mistress: for example, the incest between her children, Mariane and Carlsson; the latter's murder by a friend; and Mariane's final marriage to the murderer of her brother and husband. The important point here is how the view of the catastrophes of the mistress's family is built into the steady idyll of marriage. The mistress's family represents a perverted picture of social and familial relations, colorful actions, and great passion. It is a baroque world of events in catastrophic density. In their marriage, the couple has not exactly perverted, but so to say suspended, their social conditions and relations: the countess hides her rank in society, the bourgeois has none to begin with. Accordingly, their own life has no room for any action and passion, reversal and recognition. The baroque novel of the mistress's family reverberates, as it were, in the perfect echo chamber of the quiet marriage. Through sympathy, the couple take part in the world of deeds and events, which is the world of the others; the countess's and her bourgeois husband's sympathy reflects the ostentatious and perverse world of passions and actions in states of emotion and moral reflection. In short, through the mirror of sympathy, a modern novel originates in the relationship between the two sides of this passage.[50]

Kästner's dual utility of chance is the material condition of this peculiar arrangement. On one side of the marital life of countess and bourgeois, we find the risk of capital investment. During their Amsterdam idyll, the married couple make a living by participating in a commercial company. A relative of Mr. R**'s, who provides the couple with an abode, is the head of this company. The countess and her husband manage to salvage some money from the ruin of their Swedish livelihoods, which they proceed to deliver to their landlord to invest in his unspecified business. The form of this investment is discussed in great detail. Though they themselves would like to have as little to do as possible with the management of their money—they propose lending it at interest—the host offers them an active share in his business. The Countess agrees, but insists "that he never show me an account, but rather maintain myself and my two travelling companions instead of paying me the interest."[51] Near the halfway point of the Amsterdam economic idyll, when the Countess and Mr. R** resolve to marry, their "small capital" has exactly doubled.[52] They want to live on the money they have, but the owner of the house and the company insists again on continuing their silent partnership in exchange for providing for their living expenses. These business investments multiply further through inheritances from diverse accidents. The businessman-landlord manages the money and employs it for the couple's discreet accommodation and financial support, and his death shortly before the return of the count, who has been presumed dead, precisely marks the end of the Amsterdam period: "We had to take over the management of our capital."[53]

One bookend limit of the Amsterdam idyll, that is, coincides with the utility of chance in the risk of business shareholding. Their distance from the silent partnership is so great, however, that the couple do not seem touched by it at all. At the other end of the idyll, in contrast, terrible accidents are piling up, apparently recapitulating possible disasters from the baroque novels in the picaresque tradition. Sibling incest, murder, and treachery become interlinked in a comparatively narrow space to a sequence of tragic fates waiting to be worked through, in which the countess and her bourgeois husband must share. But here, too, they are a kind of silent partner. None of the chances and accidents, horrors and deceptions, touch them or the family of the count of G***, but rather that of his former mistress, with whom the Swedish countess cultivates friendly relations. The countess and Mr. R** provide commentary, offer advice and tears. But they are only concerned with comprehending the chance that affects others as a tool of providence, and with observing how they struggle in coping with their misfortunes and seeing them as providence. At the beginning of their marriage, the countess announces: "We were both able to occupy ourselves with the noblest pastime, with reading and thinking."[54] Thus also does the report of the misfortunes end: "Our misfortune now seemed pacified. . . . We returned to our books . . . . My husband wrote a book during this time: The Steadfast Wise Man in Misfortune."[55]

Gellert's novel as a whole is wrapped, as it were, in the theme of duty apropos of chance. This means, in effect, that the characters' actions and behavior remain connected by the paradoxical continuity between acting and observing. It is the aristocratic world, hence that of the Swedish count, that is determined by this unified but internally graduated structure. The intrigues that lead to the count's fall from grace with the prince, as well as the subsequent adventures of his exile and captivity in Russia, find him and the countess in the paradoxical situation of acting in a world of fate's fortunate and mostly unfortunate vicissitudes, while at the same time recognizing and confessing that providence reigns over all of these events. Only in the central interval of the marriage between the countess and Mr. R** can we recognize the contours of Kästner's two ways of treating chance that resolve the paradox into two strictly separate halves. The separation becomes possible through the two forms of silent partnership performed by the married couple: their partnership in the business company for their own sake, and the play of providential destinies experienced by others. This particular arrangement prevents the two ways of treating chance corresponding to Kästner's probabilistic argumentation from coming in contact with each other.

By Implication: A Theory of the Novel and Its Form

In all its specificity, the idyllic segment in Gellert's epistolary novel is significant for understanding the internal need for a theory of the novel in the eighteenth century in general. Chance and providence, or more precisely, acting and observing in the realm of contingency and onto-theological order, have a peculiarly ambiguous character between the theme and narrative structure of the novel.[56] Negotiations between trusting providence or resilient desperation characterize different characters in different situations, and people sometimes do rebel against the cruelty of misfortune or set out to compose a treatise of man's resignation to God's will. At the same time, risk taking and the story's teleology are the basic devices by which the narrator gives the novel its start and its form. When the quiet risk taking of the quasi-bourgeois couple in its observant sympathy mirrors the baroque chain of catastrophic contingencies, the configuration of events and ideologies amounts to an implicit statement on form. The result is a transition from a rhetorical poetology (of probability) to a (probabilistic) theory of the novel.

It is well known that in the seventeenth and eighteenth centuries, the novel remained marginal in terms of the norms and rules of poetics. What has come to be referred to as theory of the novel since the late eighteenth century, since Christian Friedrich von Blanckenburg and above all since Friedrich Schlegel, clearly had its genesis in distinct discourses. One source was the critique of fiction whose roots can be traced to the seventeenth century and even further back to the fathers of the church; another was the differentiation of history and philosophy going all the way back through the Renaissance to Aristotle. But such questions and discussions either had no place in traditional poetics at all (the category of fiction) or they comprised an issue preceding poetics proper and reflecting back on it from the outside (the status of poetry between history and philosophy). From the late baroque *Staatsroman* to Fielding, Voltaire, and Christoph Martin Wieland, an immanent discussion on the novel becomes evident. This usually occurs in novels, but sometimes also in critical reviews or small treatises in which providence and contingency are approached in such a way that it becomes difficult to decide whether we are dealing with moral and theological or poetological and formal debates.

Pierre-Daniel Huet (1630–1721) emblematizes this discussion in his *Traité de l'origine des romans* (1670). Juxtaposing the prose novel and the characteristic features of poetry in general and the metrical epic in particular, he writes that epic "poems" are "more regular and subject to order," containing "less subject matter, events and episodes." "The novel," in contrast, contains "more subject matter because its stylistic level [niveau] is lower and its stylistic execution is narrower, and it therefore puts less strain on the mind, making it possible to occupy it with a larger number of diverse fantasies."[57]

Two possible conclusions can be drawn from this remark regarding rhetoric and poetics. First, the novel is the less regulated and regulatable genre; because excessive in terms of its subject matter, it stands out from poetical rules to a great extent. Second, the formality and regulation of the novel takes place on the level of its content, its invention, and, more precisely, the critique of such invention. Such a critique, however, necessitates criteria of narrative figures and devices that had hardly been developed in traditional rhetoric and poetics. Theoreticians of the novel would nonetheless long be unable to say more about this than Huet manages to formulate: the novel must, though in prose, still be composed "with art"—otherwise it would be "a chaotic mass without order and beauty." And under the pretense of titillating love stories, it must provide the reader, "to whom it must always display the triumph of virtue and the punishment of vice," with a single and continuously repeated lesson.[58] Because it cannot be regulated in terms of its poetical form, the novel is determined by regularity in the aspect of invention, which in the traditional distinction had been the aspect of content and subject matter. This aspect, however, is thereby exposed to a further distinction between form that can be regulated and formable material. In the novel, distinguishing between the order of narrative form and the finality of what has been narrated—or between the narrative rhetoric that makes the novel possible and the teleology of the story communicated through the novel—is no longer a matter of simple evidence. Such ambiguities affect literary texts for the first time with the modern novel: is it rhetoric or reference that we are dealing with? And for the novel it is in fact true that the medium of the form *is* its message!

Huet solves the ambiguity specific to the novel with great elegance. After having strictly differentiated the novel from the epic in the preceding poetological discussion, he brings the epic back again in order to provide the novel with some formal rule through a historical account of the genre. According to Huet, the origin of the novel can be found in the ancient Egyptians' predilection for mysteries, the Arabs' love of fables, and the fantasy characteristic of the Persians. The Greek reception of this material complex after Alexander the Great then subjected these wild, exotic, non-European narratives taken over from Asia Minor to a secondary regularity of sorts, which applied above all to the construction of the plot and the interlinking of the episodes. This regularity, however, had been handed down to novelistic narration from the metrical epic of Greece. The epic regulation of Oriental narration then made the Roman art of the novel seem to take a step backward, since the Romans inherited the narratives independently from the Milesian fables, without the Greek modifications in terms of formal discipline.[59]

The fundamental ambiguity in Huet's construction of the novel can still be found in modern theories as well, in which it only becomes stronger. They suggest that in the genre of the novel, which counts as hardly regulated by rhetorical and poetological standards, the criteria for order or chaos, form or formlessness, are located in an entirely different place, on a level fundamentally different from the poetological one. Form in narration is provided by theoretical reflections on the narrative process inserted within the course of the novel itself.

For this reason, felicitous or disastrous contingencies and providence, as they manifest themselves in Gellert's novel, play such a decisive role. They are crossovers in a unique way between the thematic and formal categories of narration. The providential steering of events or its probabilistic evaluation are present on the level of the characters and their consciousness, as well as with the novel's model of the world and its own form of narration. The questions of the story's probability, deriving from rhetorical *narratio verisimilis*, and the questions of its unity, deriving from poetical verisimilitude, emerge from these traditional orders of literary discourse. With mathematically modeled probability in the background, they make their appearance on the novel's surface as implied theories. Anything that can fall under the category of form—such as the handling of probabilities from the point of view of the characters or in the competence of narration—is not simply and not with certainty distinguished from a worldview, or *Weltbild*, a nineteenth-century term that has its philosophical source in Leibniz and literary followers of his like the Swiss critic J. J. Breitinger. History itself may precisely also be a novel, Leibniz observed, insofar as it goes according to God's plan.[60] As Werner Frick's seminal study of *Providenz und Kontingenz* (Providence and Contingency) in eighteenth-century Europe shows, it is important to search out "the connection and interaction between the erosion of a traditional worldview based in theology and metaphysics," on the one hand, and "processes of crisis and innovation in the genre of the 'novel'," on the other.[61] The link unifying the two sides in Frick's account is the self-description of society, or "historical semantics," a term borrowed from Reinhard Koselleck's history of historical concepts and Niklas Luhmann's sociology. Because it is marked by a lack of poetical form, narration in the novel is a gateway for such historical semantics.[62] In the narration of the novel, social semantics becomes form and form becomes a bearer of social meaning. Probability, in its development from the semblance to the appearance of truth, is the medium of this transformation.

11

"Improbable Probability"

THE THEORY OF THE NOVEL AND ITS TROPE—
FIELDING'S *TOM JONES* AND WIELAND'S
AGATHON

Improbable Probability: Between Sophism and Framing Device

In the incipient stages of the modern novel's development, probability only ever attained the status of the appearance of truth in particular passages or for a very specific reason. Here and there, with occasional proto-aesthetic gestures, *Wahrscheinlichkeit*, the semblance of narration, developed into the aesthetic *Schein of Wahrheit*, the appearance of truth. This tendency could reveal itself through the formal detail of a statistical table and its diagrammatic evidence, or, as in Gellert, with a particular situation of the novel's characters in which risk taking and an aesthetic of contingency replaced baroque meditation on good and bad luck. Such instances offered particular and sometimes even extravagant solutions for the problem of the novel. In a coherent and structurally sustainable fashion, however, probabilistics and poetological verisimilitude were brought together for the first time in one clever trope from the history of rhetoric and poetology—in the figure of *improbable probability*. Improbable probability is deeply connected with the development of the novel, or more precisely: with the theory that compensates for the novel's formlessness. The trope also carries with it an epistemological commentary on probability theory in a way that science was hardly able to formulate before Laplace.[1]

Ancient authors were quite familiar with the "improbability of the probable." For them, it was a favorite paradox to be brought up at the end of a symposium when no participant could offer anything effective to argue or possible to say beyond what had already been brought forth. Naturally, it was regarded as a *sophisma,* a spurious argument or one that plays foul with appearances in probability and improbability.

In the topical-logical analysis, the sophism remains rather innocuous. To begin with, the logical investigation only makes the rather uncontroversial claim that the improbable can sometimes be probable, and, conversely, the probable also improbable. Judgments containing the predicates "probable" or "improbable" belong to the larger class of arguments that could be used "as the result of considering a thing first absolutely, and then not absolutely, but only in a particular case."[2] This operation can then be applied to probability in a narrower sense. Whereas, in its general use, a certain argument would have to be called "improbable," we can sometimes speak of "probability" or just "improbable probability" in a particular use—taking account, that is, of additional complex conditions. This is, at least, how the relevant passage in Aristotle's *Rhetoric* is usually interpreted. The wording (*para to haplōs kai mē haplōs, alla ti*) could in fact also be understood to be juxtaposing a simple and honest use of arguments to a complex and dishonest one.[3] In any event, that which is absolutely probable (when meant in the normal sense, speaking honestly and without reservations) can be improbable in particular cases (when one takes particular circumstances into consideration and tacitly presupposes them in one's arguments for strategic reasons or out of ignorance). Probability and improbability in this respect are concepts on the same logical plane. They are valid in one and the same world. But in this single world, they can be applied to a variably large or limited number of cases.

The consequences are even more far-reaching when we apply the opposing concepts to each other. This mutually retroactive potential of the sophism's terms always seems to have been built in: if what is otherwise normally improbable suddenly appears probable in a particular case, we can understand that what had seemed probable in itself turned into an improbability when seen from the point of view of normality. This old, familiar turn of the argument paves the way for the framing and modeling of probability by an artificial arrangement. No longer do the predicates "improbable" and "probable" contradictorily oppose each other on a logical plane or in a world of facts. Instead, we have framed off a state of the world into which we peer from without—from an outside that represents a different state of the world embracing and dominating the first. In other words, we differentiate two planes that are hierarchically arranged on different levels. The predicates of the lower plane can be judged from the point of view of the higher plane, but not vice versa. Epistemologically, this logical argument moves us into the realm of probability in the modern sense. Mathematical probability always and inevitably presupposes the differentiation of planes and their arrangement in a logical hierarchy. As soon as the probable has been tied to the model of the game, we can no longer think our way back behind such a framing figure. Gradable and calculable probability is always and fundamentally possible only within the framework of a game and, as such, is subject to the framing probability or improbability. Probability (of the game) is then always and fundamentally probable or improbable (in view of the establishment of the game).[4]

The modern logic of modeling realities, with their meta- and meta-meta-levels, can thus be derived from the special effect of a sophism in ancient topics. This fact deserves our attention all the more inasmuch as we cannot help but rely on rhetorical topics for argumentation in the one and only reality we live in. The final transformation of improbable probability into a reality-constructing figure was never treated conclusively and robustly in antiquity. In other words, it was not applied for the purposes of experimentation and the differentiation of logical planes, as was the case in modernity. Nevertheless, we can still find particular passages in which the juxtaposition of the *sophisma* seems to cross over into the figure of framing and hierarchy of tiers. Ever since Aristotle, for example, poetics has tied the paradox of improbable probability to dramatic peripeteia. Out of the simultaneity of the probable and the improbable, poetics unfurled a figure of what in the eighteenth century would be called wit or esprit. Through potentially redefining the one side with the other, this figure grew more important than any other poetological device for the theory of the modern novel.[5] Through the specification of peripeteia as the unexpected turn from fortune to misfortune, the paradox receives its particular methodological significance. Aristotle considers the change from fortune to misfortune to be improbably probable, because he sees in it a dual perspective. The clever man, with his wickedness, becomes the victim of an act of deception; the heroic man who is unjust is defeated in battle. First there is the simple paradox of the probable and the improbable. The clever man, who otherwise never misjudges, is mistaken this once for particular reasons; the heroic man, whose defeat is never expected, loses because of certain circumstances. For the spectator viewing the action, however, such juxtaposition of general improbability and particular probability means something completely different. For him, probability and improbability enter into a hierarchical relation with each other, and he recognizes the particular improbability of probability. He thus sees the specific improbability of the probability of a world in which the bad and the unjust lose; he sees the improbability of the probability of his own moral or political order.[6] Peripeteia, though its events may well be connected with each other in a probable manner on the level of the plot, in both cases derives its improbability from an act of authorial meta-directive or poetic justice. The hiatus between plot-defined connections and the connectivity established by the narrative discourse, a meta-factual connectivity,[7] characterizes no other poetological concept as thoroughly as it does that of peripeteia.

The improbable probability exemplified for Aristotle by peripeteia is precisely the principle of the dual view of poetic action. The absolute probability of an event appears as improbable in view of its particular embedding. This understanding indicates that, in the firm linkage of the paradox to peripeteia, the particular case does not come about through a simple addition of circumstances and determinations. Instead, it is constituted by the fact that the (general) pragmatic context of an event simultaneously forms its (particular) frame. It is the frame, then, that semantically overdetermines the primary event and its context. Improbable probability proves to be the poetological figure in which a methodical stance is introduced and practiced as a posture toward the sophistical juxtaposition of improbability in general and probability in particular. This stance is assumed by any spectator before the scene on the stage or beholder in front of the framed picture.

The paradox of probability's coexistence with improbability introduces the assumption of a boundary marking off the area in which an event has a certain value of probability assigned to it. The value cannot easily be deduced from the autonomous world of those arguing. In its hierarchically ordered way of reading, improbable probability characterizes the limits of a probabilistic test arrangement. The view through the frame that thus becomes possible is referred to in today's cybernetic theories as the view of the observer (von Foerster, Luhmann). The figural heritage and the transformation of the *sophisma* into the observer's position do not, however, find expression in the hermeneutic power of distinction invested in this suggestive concept. Wherever the figuration of improbable probability carries out the task of modeling a probabilistic situation, it can be understood as the figuration of observation. Such figuration is a device that only installs the initial spaces and limits of any possible observation. In this context, we can make use of the concept of the "device" (*priem*) of framing from Viktor Shklovsky's theory of the novel. With this term, Shklovsky discusses the familiar phenomenon that novels share with collections of novellas: a narrative sequence of the first order frames one or more sequences of the second order. In novella cycles, the framing part is usually a thin, rudimentary plot, while in novels it makes up the central thread. In novella cycles, the second-order sequences are formed by the actual narrative unities; in the novel, by more or less independent interludes that are illuminated and assessed by the main strand.

Precisely in this case, Shklovsky's concept of a figural device (*priem*) becomes productive.[8] Otherwise, the phenomena in question are just shuffled about between thematic considerations, genre theory, and media or sociocultural factors. In the perspective of the Shklovskyian device, however, they display, not only simple commonalities, but also at the same time a quasi-epistemological character. Such an extension to methodical relevance renders them figurations of observation in the first place. This is precisely what Shklovsky intends to accomplish. Every novel narrated after *Don Quixote*—presupposing the perspectives from which Bodmer and Breitinger, and above all Fielding and Wieland read Cervantes's novel[9]—adheres, according to Shklovsky, to the framing device. Framing figures emblematically demonstrate what Shklovsky's notion of device is all about. The framing device makes itself conspicuous and identifiable as an artistic technique or figuration, and it offers itself to analysis as an instrument of cognition without deriving from any conventional inventory of tropes.[10] Devices in general work themselves out in the course of their application as the rules that they are already following. Improbable probability, we may say, is a traditional figure that translates itself by means of its own recursivity into such a Shklovskyan device, an artistic figure of epistemological relevance.

Figures of Probability: A Short History of the Paradox of Probabilities in Neoclassical Poetics and the Probabilistic Theory of the Novel

Even before the paradox of improbable probability became a hallmark of the theory of the novel, it began to affect the poetics of probability as a whole. This fact becomes apparent in the literary discussions in German-speaking lands between 1720 and 1740. The debate between literary critics Bodmer and Breitinger in Zurich, on the one hand, and Johann Christoph Gottsched in Leipzig, on the other, demonstrates the point best.

The Aristotelian subtext of this discussion, surprisingly, is not defined in terms of the paradox of improbable probability in the narrower sense, but rather in the milder version that opposes probability to wonder. This formula became one of the core principles in classical poetics of the epic: "One ought to take likely but impossible things," Aristotle asserts in chapter 24 of the *Poetics,* "in preference to possible but unbelievable things."[11] Possibility and probability are arranged together in a significant chiasmus in this bit of advice. They do not form a paradox, however, because the distinction between ontology (possibility) and epistemology (probability) keeps the terms at a distance from each other.

Commentaries on this sentence proceed between two poles without ever reaching either extreme. One pole is the strict separation between possibility and probability; the other, their absolute equation. If we come close to strictly separating the possible from the probable, we avoid the paradox, but run the risk of getting tangled up in a scandalously wild divergence between knowledge and being. If we deny any essential difference between possibility and probability, we accept the paradox of improbable probability (or of probable impossibility), but avoid the scandal of the divergence of ontology and epistemology. André Dacier's straightforward comment, "This passage is very important," wielded influence all the way up through European Romanticism.[12] He introduces a further distinction that allows him to remove a potentially scandalous if unparadoxical case from a version that was suspected of paradox but free of scandal. Probable impossibility, according to Dacier, can come about first in an absolute and second in a specifically human sense. For poets, the first case—the scandal in which ontology and *epistēmē* fall asunder—is forbidden. The second version, however, is allowed. Things deemed impossible by human measure are nevertheless probable if, for example, we assume that they result from the intervention of the gods. In this case, improbable possibility is allowed to poets, though they should use it sparingly. Things impossible to humans can also be probable in a human way. Examples include the deeds of heroes and demigods that go beyond human capabilities, but still technically remain human actions; or the supernatural stories humans tell one another, which they credit with belief despite certain extravagancies. This last variation is that of poetical *vraisemblance*, which allows for the widest application and causes the fewest difficulties. Following the familiar attachment of classical probability to moderation and moral norm,[13] Dacier's reference to man converts the poetological principle of the probable impossibility into a hierarchical arrangement of being and knowledge. A specific realm of possibilities and probabilities emerges, a space that allows an observer to look in from the outside and to which he ascribes a certain mode of being in the name of his assumptions of what is human.

Classicism transforms the old discussion about wonder and plausibility into the more or less strict paradox of improbable probability. Johann Jacob Breitinger says so literally in his 1740 *Critische Dichtkunst* (Critical Poetry). "Wonder [*das Wunderbare*] is... nothing other than disguised probability."[14] Truth for the Enlightenment thinker Breitinger, no less than for his adversary Gottsched, is the only legitimate object of poetical representation. It is only possible to show that which is or can be. The lie, however, is not. Certainly, the poetical devices of "making things present" or putting them "before the reader's eyes" also need their own category. Before Baumgarten and Kant, however, the concept of aesthetic appearance (*Schein*) was not simply available. Breitinger distills it from two forms of improbable probability: from what is strikingly new and from what is wondrous. The "new" or the "wondrous" for Breitinger is the "semblance [*Schein*] of truth."[15] It does occur in the world, but because it deviates from our common perception, an element of subjective falsity is attached to it. This semblance of truth consists in the impression something makes on us, affecting its true being, when we encounter it for the first time, but for the same reason, the semblance is also the way through which such true being must be brought into appearance for us in each instance. A "wonder," in contrast, has the *Schein des Falschen* (semblance of falsehood). It does not occur in the world, but serves to introduce something that is true. It is mere seeming, through which, however, a certain truth is brought onto the path of appearing.[16] Seeming true and false are thus types of poetical improbable probability in transition to epistemic and aesthetic bringing-into-appearance. What is new refers to the view into a world that seems foreign, but that is in fact our own; the wonder evokes the perspective onto a foreign world in whose semblance we recognize our own. In newness and wonder, the improbability that affects a certain realm of probability distills an element of semblance from that probability. This semblance, then, is the very essence of what renders a certain experience of probability improbable (whether the objects in question exist truly or not). This is the case because, either way, the element of semblance intervenes between what we knew as being probable before experiencing newness or wonder and the probability through which we reconstruct our world after such experiences.

Ten years before, in his *Versuch einer critischen Dichtkunst* (Essay on a Critical Poetics), Gottsched had prepared the way for this argument, even if he, as always, seems to say the opposite and avoids all puns and ambiguity. At first sight, Gottsched appears to demand for poetry the unlimited rule of the probable. He admits variations according to geography and history, however, and thus provides space for three kinds of poetically sanctioned wonders. First, whatever we might think about the fantasy of mythological stories, we know that the Greeks told them. Second, it might seem ridiculous that in the Trojan War, two peoples would slaughter each other for the sake of a single woman, but this was in accordance with the habits and customs of the time. Finally, the idea of talking animals offends against the laws of nature but is based on a convention in literature.[17] In all three respects, Gottsched argues in the crosshairs of the developing historical-philological method. Yet in another passage, without speaking of the wondrous or the probable, he had already handled the paradox in a much more radical way than Breitinger ever would. Gottsched defines the so-called Fabel (fable), the story or subject of a poetic work, as "the narration of an event that might be possible under certain conditions, but that never really took place. . . . Philosophically, we might say that it is a story from another world."[18] The probability of poetry therefore cannot only vary according to determined criteria and thus take on wonder in itself. Instead, the probability of what can become poetry's object is always and fundamentally only probable "under certain conditions," and that means, with a certain improbability. When the debate about wonder and probability finally coincides with the figure of improbable probability, it no longer does so on the plane of a specific poetological form. At issue, rather, is aesthetic form, the fundamental requisite and characteristic of the literary work after rhetoric and poetics. With this argument, Gottsched points beyond the limits of poetry in two respects: first, he cites the authority of a philosopher; second, he refers to the theory of the novel, the nonpoetological genre par excellence: "Mr. Wolff himself . . . has said that a well-written novel is one that contains nothing contradictory and that can be seen as a history from another world. What he says about novels can with equal justification be applied to all fables."[19] For the first time the novel—precisely insofar as it seems not to fit into poetics but requires form—sets an example that should be able to stand for all of literature.[20]

Gottsched's reference to Christian Wolff provokes the question of why it is precisely the theory of the novel that figures so prominently in the philosopher's discourse and allows him to transform poetological probability into aesthetic *Wahrscheinlichkeit,* the appearance of truth. Historically, the transformative role of the novel is connected to the rhetorization of poetics under way since the Renaissance, a long process familiar to students of early modern European literature. In Renaissance treatises, poetological probability is often dealt with in terms of rhetorical *narratio verisimilis.* This rapprochement is particularly pursued by the champions of classicism among the interpreters of Aristotle in Renaissance poetics. Antonio Riccoboni's commentary on the *Poetics* (1587)[21] precedes Dacier in this regard.[22] For this branch of commentators, the rhetorical device of increasing or decreasing the probability of an event by manipulating its circumstances implements technically what probability or wonder mean in poetics. The rhetorical technique turns the poetological choice between probable and wondrous into the question of the more or less probable. Hence it is a proto-probabilistic gesture. The wondrously probable, a fundamental figure that goes beyond the special case of peripeteia, thus also marks poetology's self-transgression into aesthetic form. The transgression was already laid out in Aristotle's *Poetics*.[23]

At first glance, Huet's *Traité de l'origine des romans* is an odd candidate to be the model for the theory of the novel in the eighteenth century, since it was originally meant to apply to the courtly novels of a Madame de Scudéry or a Gautier de Coste La Calprenède. Its wide application was indeed not due to the descriptive elements and contents of Huet's theory. Instead, it achieved its relevance because of the way a theory of the novel appears as possible and necessary over and against the poetics of the epos. This theory marks the contours of an aesthetical form by indicating the very lack of poetical figures and genres; in doing so, the theory can be comprehended as the prerequisite element of a new, no longer courtly, novel. The novel has a definition for Huet—after Friedrich Schlegel it would be called a "theory"—insofar as it can be distinguished from the epos and hence from traditional poetics. The first criterion applies to prosody and style—epic verse and elevated style are opposed to prose and the low style of the novel. This is the criterion of the words, *verba*. The second criterion applies to the material, *res*. The epic is characterized by wonder; the novel, by the probable. Epic narration follows a clear order; the novel is organized rather by episodes, events, and the demands of the material content.[24] In all of this, we recognize the schema of being more and less artful: the epic displays artifice in a higher measure than the novel. The emphases on the side of the *res*, however, are distributed differently than for the *verba*. On the *verba* side, at least at first glance, it looks more like a simple easing up of strictness and order in the novel; on the *res* side it becomes clear that this decrease corresponds to a restructuring, a kind of compensation for artifice and order. The novel, as Huet explains before opposing it to the epos, is a fiction in prose with art, and that means: a work written in prose, but still with art.

Lennard J. Davis introduced the idea of a phantom theory in his study *Factual Fictions: The Origins of the English Novel* (1983). The poetological vocabulary of the probable is no longer appropriate to expound upon the procedure of the novel after Defoe or even Aphra Behn. A discrepancy emerges between the moral probability heralded in the prefaces and theorizing remarks of novels and the "actual" or "realistic" improbability of probability that constitutes the material of a novel. According to Davis, this new type of probability, a probability specific to the novel, belongs to the source from which novels are not so much derived in terms of content as in terms of form—in terms of the way in which novels have form at all. Davis points to the daily news, the balladesque sensational report, and the journalistic announcement, realities whose determining characteristic is the improbability (the unexpected or unusual character) of their probability. In fact, something phantomlike seems to hover about the discussions of the probable and the miraculous, the expectable and the unexpected in the digressions of eighteenth-century novels. They come across sometimes as pedantic and old-fashioned; sometimes as playful and arabesque. Nevertheless, the phantomlike aspect Davis mentions with regard to the concept of the probable cannot be resolved and laid to rest as simply as he suggests.[25] The paradoxical constellations of probability and improbability, the ordinary and the unexpected, are concentrated precisely in the figure of improbable probability, a figure that can be observed again and again in the theoretical asides in novels of the 1740s and 1750s. The realistic probability that Davis, like Ian Watt and others before him, extracts from the English novel of the eighteenth century is in fact inseparable from the paradox formed out of the old rhetorical-poetological notions of probability and improbability. Unless, that is, one were to move directly over to a mathematical concept of probability for calculating the information value of the news from which Davis seeks to derive the structure of modern novel writing.

Improbable Probability in the "Prefatory Account" to Wieland's *Geschichte des Agathon*— and Fielding's *Tom Jones* as Its Subtext

When scholars today put together collections of texts from the seventeenth and eighteenth centuries on the "theory of the novel," they tend to include Puritan polemics against fiction such as Gotthard Heidegger's *Mythoscopia romantica* (1698), the prefaces of novels (like the tracts by

Huet and Birken), critical reviews (Christian Thomasius's famous piece on Lohenstein's *Arminius* and Leibniz's discussion of the *Mythoscopia*), or treatises on the history of the novel.[26] They cannot offer any straightforward poetics of the novel or rhetoric about how to narrate it. If the novel was discussed at all, it was either in a novel or in direct connection with a novel. Treatises on poetics and rhetoric of the time concentrated on circumscribing the space where the literary work and its form can originate. The novel did not receive or allow any such regulation. Instead, it developed those forms of theory that would dominate the discourse on literature after the period of traditional poetics and rhetoric, among them most notably literary criticism, history, and the philosophy of history.

Around the middle of the eighteenth century (in England and France somewhat earlier, in Germany somewhat later) such a theory of the novel was decidedly marked by poetological probability, that is, by the neo-Aristotelian paradox of probabilities and their improbability. This was particularly the case in novels that would not fall under the category of realism as understood in the nineteenth century. In all of these novels, the paradox of improbable probability is developed in ways that would have literary form emerge from it.

Christoph Martin Wieland's (1733–1813) *Geschichte des Agathon* (The History of Agathon), which appeared in its first version in 1766 and 1767, is a prominent example for staging the paradox of probability as a theory of the novel. Set in ancient Greece, *Agathon* obviously echoes Fielding's *History of Tom Jones*, beginning with the emblematic title that avoids any allusion to fortunate and unfortunate adventures familiar to the older generation of novel readers. In highlighting the paradox of probability, the novel is a rather late specimen by western European standards, but an early one in the German context, and the most thorough example perhaps for the theorizing effect of the paradox anywhere. It is in particular the novel's beginning and ending that make probability a theme. With their characteristic irony accompanying Wieland's points on literary theory, the "Prefatory Account" [*Vorbericht*], the first book, and the last book frame the three major parts of the narrative proper of the protagonist's development or education: Agathon's life between the philosopher Hippias and the courtesan Danae in Smyrna; his youth and early political career in Athens told in retrospect; and finally his frustrating experiences as a political reformer at the court of Taranto.

It is as if the editor who speaks in the preface would like to exhaust the term "probability" utterly with one fell swoop. His first sentence reads: "The editor of the present history sees so little probability of persuading his audience that it was in fact culled from an ancient manuscript that he thinks it best to say nothing at all in this matter."[27] He does not, however, wish to rule out the possibility of "proving in a court of law" that a corresponding manuscript had been "found in the archives of ancient Athens." Thus, the editor immediately gives up any claim to the probability that could lend him the rhetorical and poetical arts of persuasion. Instead, he yields to the calculation of probability that pertains to events and their narration on the basis of documents. This practice was the first field in philology and history on which mathematical probability had been applied. From Thomas Craig through the juridical probability of the Bernoullis and numerous theologians and philologists down to Laplace, formulas were invented for calculating how the probability value of a testimony, for instance, depends on the number of witnesses or years it ran through in a chain of transmission from the first through the last phase.[28] In Wieland's "Prefatory Account," this probability amounts to the probability of the voice speaking as editor actually belonging to an editor at all. The same publisher-editor who gives up any attempt to prove, by means of rhetorical and poetical probability, the actual existence of the document that he purports to be publishing in the first place, also refrains from convincing his audience that he is the one who has the right to speak in the "Prefatory Account." He himself and his status as editor-publisher depend upon the mathematical probability of the events of which he speaks.

The editor sets out by leaving the discursive realms of rhetoric and poetics in favor of mathematical probability. In the end, though, it will be improbable probability, the old rhetorical-poetological figure and the new formula of the theory of the novel, that again offsets the implications of the game of probability. Following the introduction, however, the editor first of all becomes entangled in more rudimentary paradoxes, paradoxes that function according to traditional poetological probability. This happens under the heading of truth, the very notion that the editor has just defined in terms of mathematical probability. He first introduces the argument of poetological probability, that is, verisimilitude, as narrative truth: "so that everything harmonizes with the course of the world." This demand evokes the probability of Aristotle's *Poetics*, according to which poetry is closer to the sphere of natural laws and stable causes than to singular events as reported by historians. Later on, he appeals to the traditional formula according to which true facts and that which to history has testified can nevertheless be called improbable. Such historical facts, however, reintroduce the singularity of the singular event that the poetic probability means to transcend. In the briefest space, the editor thus enlists the two contrary positions of the Aristotelian probability debate and ascribes them to his own cause. Accordingly, both the probable and the improbable are true. This result of the editor's figurative move so far does not amount to a clever and significant paradox, but rather to plain and simple inconsistency. Verisimilitude recurs contradictorily but stubbornly under the code notion of truth, which leads back to mathematical probabilistic and its implications. In an aside in his "Prefatory Account," the editor touches on an intricate argument that Huet had already discussed. He calls "Agathon and most of the other characters" "real persons," which means such as "the likes of which have existed since time immemorial." *Die Vielen*, hoi polloi, whose stories no historical document mentions, are those who still remain to be counted. They do not belong to the world of politics and the state and have no place in Plutarchian biographies. Their mode of being is not that of truth or verisimilitude, but rather of mathematical probability.

Agathon is their emblem. As Wieland explains, the name of the tragedian Agathon as he appears in Plato's *Symposium* is well-known from Greek literary history. The editor furthermore suggests that the novel's hero could be identical with this historical figure. In the novel, however, nothing intimates such identity—except perhaps for Agathon's dealings with Plato, the circumstances of which, however, do not accord with the specific facts of the poet's life. In other words, Agathon is no historical figure with additional biographical traits invented according to verisimilitude, but, under the code name of a historically testified personage, he is rather an element from the class of the many. Under the notion of verisimilitude, the novel is dealing with mathematical probability, the very probability that stems from the order of the archive and implies calculation instead of narration. At the time of the novel's first edition, it had been just twenty years since Christian Wolff had declared the divine order of statistics in Süssmilch to be the first test of a universal theory of probability. The fact that the hoi polloi are strictly private persons is thus connected with their mode of representation in the novel. They are not subjects of historical narrative, but their bare private existence is recognized by calculation and probability.

The editor thus traverses the entire course of the classical argument that opposes verisimilitude with the truth of history, and he enlists both poles of the continuum simultaneously for support. Under the title of truth, he lays claim to both the probable—that which approaches universality and does not require testimony—as well as the improbable—the singularity of events and occurrences that need not fear any discrepancy with the more universal nature of probability. The outright contradiction refers to the middle passage and the calculation with large numbers hidden therein. A technical solution of the semantic squabble is suggested in this passage. Every probability proves simultaneously improbable in the measure assumed for such probability. When compared to the traditional semantics of verisimilitude, calculated probabilities highlight the question of where their implied measure comes from and how such a measure is able to determine degrees of probability, alias improbability.

With this opening, the novel distances itself, as a form requiring a theory, from its own poetics and poetics as such. In doing so, Wieland references the tradition of Aristotelian poetics in general. But more precisely he refers to its contemporary adaptation in the "prefacing chapters that serve as a kind of poetics" and that introduce each of the eighteen books of Henry Fielding's *History of Tom Jones* (1749).[29] The reference here is to the introductory chapter of the eighth book in particular. The chapter comes at a moment when Tom Jones leaves his boyhood home, Mr. Allworthy's paradise in the country, and makes the transition to the adventurous and dangerous journey to London, from whence he will finally return to take up his role in Mr. Allworthy's realm. By then, he will have become the permanently adopted bastard looking forward to marry Miss Sophia. The final chapters of the preceding book have already announced the contingencies of "adventure" and "dread" in their titles; and they do so for the hero as well as for the reader. The introductory chapter is then called "A wonderful long chapter concerning the marvellous; being much the longest of all our introductory chapters."[30] The treatise on the marvelous becomes so wonderfully long because it attempts to determine "certain bounds" of the marvelous.[31] Boundaries of the wondrous, however, as we have known since Dacier, lead in the end to the improbability of probability. It is this consequence that finally turns the poetological preface into a declaration on the theory of the novel.

Fielding, like Wieland after him, begins his "introductory prattle," in William Empson's phrase,[32] by establishing narration in the epoch of the novel on both sides of the poetological distinction: on the side of the particular historical event that can be and always is improbable due to its singularity; and on the side of verisimilitude, which is nearer to philosophical universality. Fielding's narrator proffers this consideration in the same breath as he equivocates about its provenance in Aristotle. With this turn, the narrator raises the question of probability, not about the factual status of the novel and its documentation, as Wieland's editor will do, but rather about its theory. Again like Wieland, Fielding also discusses in this context the difference between the realm of history as conveyed by literature and writing and the details of private life that can only be ascertained by probabilistic methods. While Wieland sees *Agathon* as already composed in an irreducible intersection of both parts, however, we can watch Fielding simply putting them together for his novel.

As Fielding's narrator explains, the historian must report unusual events when they are testified. But the license for the wondrous extends only so far as the "essential parts" of the historical narrative:[33] Henry V's victory at Agincourt and Charles XII of Sweden's in the battle of Narva provide two examples of improbable facts that the historian is obliged to report. His license does not extend, however, to incidents grouped around the "essential parts"; it does not apply to the ghost that appeared to Mrs. Veale, for example, a circumstance reported in Defoe's famous narrative. Regardless of how well an improbable occurrence is documented, so long as it lies outside of the "essential parts," the historian may do better to consign it to oblivion in order to secure the readers' trust. The poet on the other hand deals from the beginning in "private character" and thus with circumstances and incidents outside of the "essential parts."[34] *Tom Jones* keeps brushing up against the heights of historical events, for instance, without ever making them the focus of the narration. Hence the novel has "no public notoriety, no concurrent testimony, no records to support and corroborate" its narrative.[35] According to its mode of being, everything that happens to private characters is based in probability.

This discussion of documented or undocumented marvels is not just a matter of opposing history and poetry, however. It leads rather to recognizing a delicate intertwinement of the two in the case of the novel. The legitimate historical wonder is here involved with the probability of inner, narrative circumstances that, as such, border on private life. And the probability of private characters is arranged around the core of an event that, at least in *Tom Jones,* indeed draws upon the traces of official history. To state the case even more strongly: the inessential parts of history are nothing other than the novels of private persons. And novelistic probability works through the circumstances of historical events. In this complementary relation of probability and improbability within the novel, history and poetry are precisely reproduced as mutually different from each other. By being reinscribed as a defining formula in the novel's text and theory, the distinction of history and poetry gives shape and boundary to the new genre without poetical form.

In the course of the poetological discussion, we rediscover the very constellation we first glimpsed when interpreting game theory as probability theory: the singular event (now invested with the pomp of history) is equated with the contingent event that can occur or not occur (this time emerging in the proto-statistical probability of private characters). Thus a problem is posed that Bernoulli's probability theory implies but does not discuss and that reaches significantly beyond rhetoric and poetics. While the singular event means an unmistakable particularity in a given world, the contingent event that occurs or does not occur takes place in a prearranged setup, a framed field of possible events that is observed from without. Traditional rhetoric and poetics, however, assume a single order of things; they argue on the basis of the one and only one world in which things happen and statements are made about them. The world's simple and unique status, moreover, implies the impossibility of integrating events and statements about them into one and the same level of consideration and calculation. In the one world, there is no vantage point for observations from without; there is no way of applying the probability of events to the viewpoint of the one making statements about them and that in turn makes such statements observable as events. The simple and unique world is the complex one.

Fielding's and Wieland's theoretical prefaces continue to use the language of this traditional rhetoric and poetics. In the course of speaking, however, Fielding's narrator and Wieland's editor transform the grammar of their language. For this reason, the theory inserted in the novel is indispensable for its narrating what it narrates. The prefaces and insertions, which have been understood as the struggling self-interpretations of a new genre, instead implant a framing device in the text of the novel necessary for its own functioning. In fact, we are not concerned here with a matter of poetological commentary, but rather with the passages that, although they form a meta-level of narration, precisely therefore belong to the construction of the novel. The meta-discourse belongs to the discourse of the novel; the theory of the novel is part of its text.

The introductory chapter to the third book of *Tom Jones* is famous for its "introductory prattle" on the "vacant spaces," the undescribed expanses or blank spaces. As such, they imply a reader's view from without. This view would then shape the novel at the intersection of history and poetry, improbability and probability, and give it the form of a paradox.

The literary critic Wolfgang Iser refers to Fielding's passage on "vacant spaces" and ties it to the concept of the blank space [*Leerstelle*] as Roman Ingarden, a disciple of Husserl's, had developed it in *The Literary Work of Art* (in German, 1931).[36] For Ingarden, the blank space, or *Leerstelle*, is a virtual hole in a text that, through the very activity of reading and understanding, the reader produces and, at the same time, fills in. Texts often and structurally do not contain every semantic element that presents itself in and for the act of understanding. In his version of the reception theory of literature, Iser develops this argument into the concept of the "implied reader," the virtual reading conscience that we reconstruct when interpreting the meaning of a text. Iser's comment on Fielding's passage is of primary significance in this context. By taking up the introduction, specific to Fielding, of the general literary-theoretical figure of an "implied reader's" *Leerstelle*, Iser identifies the framing figure of novelistic narration par excellence, and he does so at a most significant juncture of the history of the novel: "for besides that by these means we prevent the reader from throwing away his time, in reading without either pleasure or emolument, we give him, at all such seasons, an opportunity of employing that wonderful sagacity of which he is master, by filling up these vacant spaces of time with his own conjectures."[37]

"The vacant spaces in the text," Iser comments, "are offered to the reader as pauses in which to reflect. They give him the chance to enter into the proceedings in such a way that he can construct their meaning."[38] The comment of the author, to which Iser attributes significance for the theory of reading and the construction of meaning, clearly anticipates the discussion of probability at the beginning of the eighth book. It jumps ahead to credit the reader of the novel with what will be attributed at this later occasion to the historian. The historian, as the introduction to the eighth book explains, should stick to the essential events reported in the sources. As for the rest, he should judge for himself and supplement facts according to the laws of probability. Just as the historian fills in the gaps and emends the improbabilities of historical documents with his conjectures, so too the reader should fill out the blanks left in the novel by the narrator of *Tom Jones*. The novel *Tom Jones* is to the reader what history is to the historian. The historian relates to history in its own implied narrative as the reader of the novel reads the narrative as an implied history.

The reader of *Tom Jones*, according to the "introductory prattle," is constructed according to the figure of improbable probabilities. This reader will have proved himself as the historian of the private characters of Mr. Allworthy, Squire Western, Sophia, and Tom Jones. If, however, standards are lowered as to what counts as "essential parts," and the novel can pass as history, then such history is possibly also merely a novel. This would at least be the case under the supposition that the standards for what counts as historical and hence suitable for being an essential part are raised to considerable heights. The reader, who fills out the blank spaces [*Leerstellen*] of accessory circumstances around the core of inaccessible events, reads history as a novel and the novel as history. He reads both under the implied figure of improbable probability, that is, as a world whose probability he can supply because he understands the improbability of its constitution. Such mutual implication of "history" and "novel," it might be argued, defines their meaning in modern times.

The conjecture about singular occurrences necessary for this operation is exactly what is dealt with in probability theory and Leibniz's logics of contingency. After *Tom Jones,* the modern novel no longer requires character to be deduced from the action, but rather "to foretell the actions of men, in any circumstance, from their characters."[39] The beginning of the second book shows that the theory of the novel as implied in the vacant spaces really comes about with mathematical probability and statistics. The introductory chapter had announced that the novel would only report extraordinary scenes, and would leave unremarked entire years without relevant events. As if anticipating the formula of the vacant spaces from the beginning of the third book, the narrator refers here to "blanks in the grand lottery of time."[40] The vacant spaces of time from narrative theory are blanks in the lottery of time. Deriving from the early modern idiom of lotteries (from the Italian *bianche [charte]*), blanks still refer in English today to worthless lottery tickets (as in the expression, for example, "to draw a blank"); in the seventeenth century, "blanks" could still stand metonymically for the lottery at large.[41] History then divides its lots into the winners and losers. Winners are the great deeds, as reported in the newspapers, of which one day history will consist. The losers, "blanks in the grand lottery of time," are the many events that never trouble or even reach the public.[42] The blanks, which the narrator presents in the theoretical preface of the third book, are chance events of the first-order things happening. They are furthermore not merely chance events of everything that ever takes place; above all, they are subject to the specific contingency of being or not being historically relevant. The "grand lottery of time," to which we owe the metaphor of the blanks, which for its part becomes the metaphor of the hermeneutic blank spaces [*Leerstellen*], thus plays out the statistical relation of event and circumstance. In this lottery, all theoretical relations of narrating, reading, novel and history are at the mercy of contingency.

To understand fully the importance of this double irony (noted by Empson), we have to go back once more to book 8's introductory chapter, in which Fielding discusses the limits of probability for the marvelous. It will be apparent here just how central it is for *Tom Jones* that the reader be able to distinguish between history or (historical) event and the novel or the circumstances, and thus differentiate between context and what is to be contextualized.

The theory of the marvelous and its transformation into improbable probability sets the great trek or journey into relief that will finally bring

Tom Jones from Mr. Allworthy's country estate to London. Tom Jones, due to the false accusations of young Blifil, has been expelled from a paradise that will become his legitimate home only in the course of his exile in London. The wonderful and the marvelous as discussed at the beginning of the book are thus in the first place a world of adventures, which take place in the streets, in impassable terrain, in public houses, wherever Tom meets with strangers, travelers, soldiers, or the mysterious Man of the Hill. While Tom thus gets caught up in adventurous stories, he is also entangled in history. When he crosses paths with soldiers who are marching against the Jacobite uprising, the reader does not become aware of the fact that this event belongs in the context of political events at the time of the novel's composition until the moment when Tom actually encounters the soldiers. The chapter that first introduces the category of adventure ("Book VII. Chapter XI. The adventure of a company of soldiers") also introduces historical names and the name of the historical for the first time: "The sergeant had informed Mr Jones that they were marching against the rebels, and expected to be commanded by the glorious Duke of Cumberland. By which the reader may perceive (a circumstance which we have not thought necessary to communicate before) that this was the very time when the late rebellion was at the highest."[43]

With the soldiers, Tom encounters history. What's more, he encounters a history that was still current for Henry Fielding and his readers: the novel appeared in 1749, and Bonnie Prince Charlie had been defeated at Culloden just a few years earlier, in 1746. The extraordinary nature of the adventure is simultaneously the extraordinary nature of the historical event, albeit located in its nonextraordinary and inessential complications. This intertwining of history and adventure does not mean, however, that the private story of Tom Jones is admitted into the historical history of the Jacobite rebellion and the conclusive pacification of Britain by the House of Hanover. There is no supplementary play between foreground and background, historical event and further context here. The distinction between the public matter of political actions and what happens to private persons is no longer identical with the fundamental distinction between the event that takes place, and can be documented, and its filling out according to the standard of circumstances and their probability. The grand lottery of time rediscovers the play of contingency at the very roots of the distinction between the singular event and its probable circumstances.

Who Narrates *Agathon*?

Wieland—to come back to *Agathon*—follows Fielding when he transforms the juxtaposition of probability and improbability into improbable probability, the figure of novelistic theory. But Wieland accomplishes the transformation not through the implied reader, but rather by refining the position of the narrator.[44] This transformative process is certainly true for the first edition of *Agathon* (1766–1767). But even the revisions that were meant to convert the novel of probability into one of Goethean *Bildung* were never able to wipe away this original narrative logic entirely.

In the first book of the novel, this logic is revealed through the parody of a baroque hero's story; in the final book, it is manifest in the ironic presentation of a fairy tale. They are the two ends of a figuration of improbable probability that has its origin in the preface. Who is narrating the novel? This famous question was posed with regard to the twentieth-century novel by the German literary critic Wolfgang Kayser in 1955 in his book *Entstehung und Krise des modernen Romans* (Crisis and the Origin of the Modern Novel). The question already arises, however, for the first time in German literary history, and with full rhetorical precision and narrative consistency, in Wieland's *Agathon*. In the first book of this novel, a triumvirate of writers begins to emerge. First of all, Agathon, the novel's protagonist, had composed "a kind of diary." This journal of events "from Agathon's own hand" was available to the "anonymous author [*Verfasser*]," the "history writer," or, as he is called in the eleventh (and final) book, the "author [*Verfasser*] of the Greek manuscript," who based "his narrative" on it.[45] Responsibility for the final text of the novel is then taken on by the editor of this manuscript, who emerged in the preliminary report and who remains present throughout the text as the first-person voice. The exact nature and extent of his interventions are never explained. At the two ends, therefore, we find opposite each other the *I* of the statements in the text (the *I* of the publishing/editing writer) and the *I* whose story gets told (the *I* of the diary-writing Agathon).

"Improbable Probability"   295

These two first-person authorities at the extreme ends of the novel locate themselves in the first book by taking a stance on the paradox of improbable probability. At first we hear from the editor. His intervention occurs in the second chapter, "Something Entirely Unexpected." Agathon lands amid the preeminent scenery of mythological wonder and disorder, the dance of the Bacchae, and the editor introduces the chapter by saying: "If it is true that all things in the world have the most precise relation to each other, then it is no less certain that this relation is often entirely imperceptible among some singular things; and thus it seems to be the case that history sometimes relates much more strange events than the novelist would ever dare to compose."[46]

Within the course of a few chapters, Agathon manages to escape the Bacchae, only to find himself in the clutches of pirates. On this occasion he is reunited with Psyche, whom he had thought lost, and listens to her equally wondrously probable story. He then makes the following remark to himself (and precisely in this connection, the textual fabric and the three writers' cooperation are explained, thus rescuing the probability of the wondrous fact of the remark's existence): "If order and coherence are the characteristics of truth, then, oh! how very much do the coincidences of my whole life resemble the erratic game of dreaming fantasy!"[47] With this exclamation, the first brief autobiographical account of Agathon begins: a report of his earlier life that takes the reader back to the beginning of the novel.

A noteworthy inversion of the paradox of the wondrous and the probable intervenes between the comment of the editorial *I* and that of the autobiographical *I*. Both statements agree on the deterministic understanding of the efficient cause. Every event has its reason; the whole encapsulates the order of things. The rebellion of an individual event or of chance, in contrast, is mere illusion. The ways the two writerly authorities articulate this illusionary discrepancy of the event, however, are quite distinct. For the editor, the unconnectedness of individual events makes itself conspicuous in the reality of things. The order of the whole remains invisible and can only be obtained as poetry and through the poetics of narration. For Agathon, the protagonist, in contrast—in the tradition of Leibnizian doctrine[48]—the unconnected individual occurrence under the mythological name of Chance is characteristic of dreams and poetry.

Order, the coherency of things, is for him the mark of true reality revealing itself to quiet contemplation. The editor argues sensually, one might say, whereas Agathon argues rationally. Based on Agathon's and the editor's views of improbable probability, poetry and truth appear to form a chiasmus.[49]

This becomes a particularly striking observation when we realize that the editor, mediated by the Greek author, is working from Agathon's diary. The editing/publishing *I* in the text of the novel construes the wondrous and the probable in view of poetry and true knowledge to be quite at odds with the way the *I*, in whose name it speaks, does so.

The empirical *I* of Agathon—the one whose story the novel tells, but who never speaks himself—judges from the point of view of an enthusiastic affirmation of connection and order. The editor's scholarly *I*—who speaks in the text, but who is never the subject of the novel's storytelling—judges from the point of view of the skepticism of the unconnected individual. The narrative's text—or *Geist der Erzählung* (spirit of the narrative), as Wieland put it long before Thomas Mann and Wolfgang Kayser[50]—speaks, however, from a perspective well beyond this opposition. Its logic, the contextualization and probabilization of the individual event, is poetical-epistemological. It cannot be reduced to an emphatic understanding of the whole compared to which all singular occurrences have the questionable interest of poesy and dream. But neither can the logic of a narrative that integrates both the *I* function of the editor and the personal *I* of Agathon be identified with the dry knowledge of singularities, a knowledge for which all connectivity exists only through fiction.

The order of writing in *Agathon* does not only thus consist in the division between an unknown autobiographical journal that serves as its basis and a text dominated by an editorial *I*. Instead, there is a mediating authority between the two, the unknown-known of the narrative process that only defines the rule that it is already following in the course of its application. Critics have hitherto overlooked this decisive moment: in the Greek author's manuscript—in the narrative—the lines of Agathon's diary and of the editor's novelistic text intersect. The writing of the anonymous and formless narrator lies in the zero point of the crosshairs formed by Agathon's and the editor's inverse versions of improbable probability.[51] This writing of a narration beyond autobiographical source and edited text is located where the distinction between the rule-abiding probable and the unrelated singularity, as the editor and Agathon formulate it, has not yet been made. It is a mere technique of narrative representation, an improbable framing of probability and a spontaneous calculation of contingency. The writing of the third authority, the Greek author, is improbable probability *in actu*.

The final book of *Agathon* (in its first version) comes into action at this precise point. Taking a hint from some of Wieland's own remarks, readers have usually seen the so-called fairy-tale ending of the novel as an artificial solution. It has been criticized as a conjurer's trick or appreciated as playful irony. The problem of the ending, like so much in *Agathon*, was inherited from Fielding's *Tom Jones*. Just as this hero's notorious "good heart" made it improbable that the story could have a good ending for him, it is improbable that Agathon's "enthusiasm" could ever be reconciled with a reality that allows a happy ending, notwithstanding his ability to learn. In the concluding fairy-tale, the program of the neoclassical novel is spun out of the poetological figure of probable improbability. The figure is represented through the experiment of letting the moral subject follow its internal goal in the context of external circumstances, an experiment that foreshadows its inevitable failure.[52] The circumstances—the *circumstantiae* of the *narratio verisimilis*—stand in each case against the morality of the subject—against the good heart or the enthusiasm, indeed, against any will to self-determination. Karl Philipp Moritz's *Anton Reiser* (1785–1786), which has been called the first sociopsychological novel, at least in the German tradition, makes this condition the impetus of persistent failure with a resolution as stubborn as it was impressive.[53] The "good heart" and "enthusiasm" are individual resistances against circumstances as such, at least before the invention of a subjectivity like that of Wilhelm Meister, which works through dialectical mediation. But because in this concept individuals are devices of restless resistance that can never come to a conclusion, they will render the (good) ending of a novel always an exercise in the improbability of probability. Because the "story of our hero . . . corresponded to the orderly course of nature and the strictest laws of probability," as Wieland's editor remarks, the fairy-tale nature of its ending is all the more conspicuous.[54]

In the "apology" of its "Greek author," the novel's functional narrator, the final book of *Agathon* thus delivers the very thing that had been intended in the first book with the chiasmus between the Agathon of the diary and the first-person editor.⁵⁵ The utter casuistry of the improbability of a probable good ending concludes *Agathon*, one of whose very first words is *Wahrscheinlichkeit*—"probability." In the apology, the editor lays all the wonders of the end to the account of the narrator, explaining them as consequences of the latter's attempt to adhere to the law of probability in narrating the ending. Prominent in this respect is Agathon's enthusiasm, which can never be entirely stifled, and that could only allow an ending by means of the improbable probability of all the peripeteias in the novel (the fortunate chances, the recognitions, the reconciliations, etc.). Other considerations apply to Agathon's character (the enthusiast can never turn away from the world into misanthropy or bitterness) and the customs of the time (ancient Greece did not offer the possibility of withdrawing as a hermit from society).⁵⁶ There would have been only one possibility of escaping the drive of probable improbability characteristic of the novel, but it is certainly understandable that the editor does not find it appropriate to take this recourse, as he explains in the "Apology of the Greek Author." The Greek author would simply have had to do without an ending (and correspondingly a beginning) of the novel—in which case he would not have fulfilled precisely the function that he exerted between diary keeper and editor: the function of narrating.⁵⁷ The result of the apology, therefore, is that with the improbable probability exhibited in the ending, the "Greek author" is simply performing his function.

The fairy-tale conclusion of the first version makes improbable probability unmistakably the figuration of framing.

> Under such versatile circumstances, and because (as said) the author's intention had been to make his hero into a virtuous, wise man, and in fact in such a way that one might easily comprehend how such a man—born and bred in such a manner—with such abilities and dispositions—with such a determination of the same—according to such a series of experiences, developments, and transformations—in such circumstances of fortune—at such a place and in such a time—in such a society—in such a clime—with such nourishments (because even these have a greater influence on wisdom and virtue than some moralists imagine)—on such a diet—in short, under just such given conditions as constituted by precisely those circumstances in which the author had so far placed Agathon—could have been so wise and virtuous a man, and (those who are not accustomed

to thinking may believe it or not), how under identical or rather very similar circumstances, he could become the same even in these days and times.[58]

The excessive enumeration of circumstances introduces the framing effect of conditional logic, the "(always) if-then" [*(immer) wenn-dann*] into the course of the narrative. It does so at the novel's ending, where ending as such is at stake. The ending of the novel, to refer to Arnauld's argument, is thus composed out of singularity and contingency and forms a probabilistic event. The singular moment in which endings necessarily consist is guaranteed by the repeatability of a contingent event, a repeatability that comes into play with the given conditions of the circumstances. The point of a singular event, seemingly lacking any extension, still offers room for the scenario of the repeating event that occurs "(always) if-then." This is a kind of irony that is not identical with the obvious one exhibited by the editor and first-person narrator. Instead, it originates with the nameless author of the manuscript. Precisely the moment of the ending virtually disappears beneath the piling-on of circumstances and conditions that are meant to demonstrate the repeatability of the moment up through the present moment of the reader's involvement with the text. In the moment of the ending, what we might call a poetics of the contingent event expressly becomes the repeatability of the singular and the singularity of the repetition. The single event appears at a place already marked for repetition and seriality, in a place of marked contingency. The simultaneously singular and iterative event no longer dominates the sequence of the narrative; but neither is the hermeneutics of observing this series and its corresponding space of contingency yet established and concluded.

In antiquity, as has been noted, improbable probability was to start with a paradox, under the supposition of a single world of validity and argumentation. In poetics, then, it referred to an initial resolution of the paradox by introducing the separation between two worlds, a world of the event and one of observation. With Wieland's lesson in mind, we may however understand improbable probability as the very kind of figuration that does not refer either to the paradox or to the hermeneutics of observation, but rather to the shift from one to the other. In that case, it corresponds precisely to what, as regards events, can be called singular repetition and the repetition of the singular.[59]

The trope of improbable probability marks Wieland's novel throughout in the name of its titular hero, Agathon. For Aristotle, in his *Poetics* as well as in the *Rhetoric,* it is tied to a line in a tragedy by the poet Agathon: "One might perhaps say that this very thing is probable, that many things happen to men that are not probable."[60] Wieland's *Agathon* can be read as a continued and relentlessly advancing citation of this citation. In pursuing the figural concept of improbable probability, which Aristotle firmly connects with Agathon's name, *Agathon* incessantly but inconclusively converts the paradox of probability and improbability into the hierarchical figuration of framing. Framing, however, is as relevant for the theory of the novel as it is for probability.

The Device of Narrative Framing and a Short History of Academic Statistics

In providing a narrative framing, the paradox of improbable probability functions as the theory of the novel within the novel. Derived from rhetoric and poetics and developed into a narrative device (in Shklovsky's meaning of the word), the paradox is also at work in the tableaus in the novel and in the novelistic aesthetics of contingency and risk calculation. What is implicit in the constellation of Daniel Bernoulli's value theory with Gellert's novelistic critique of providence becomes explicit in the framing figure of improbable probability: functioning as framing figuration in novels, probability and its improbability form the semblance of truth, or *Wahr-Scheinlichkeit*. Although the theoretical explication of improbable probability in Fielding's and above all in Wieland's novels points toward the problem of statistical probability, no manifest discursive link connects novelistic writing with the constellation of number and text in statistics until the middle of the eighteenth century.

This lack has its roots in the constitution of eighteenth-century statistics. Political arithmetic and topical statistics continued to refer to each other without ever forming a compact unity. The methodical implication of a frame in the tableau of numerical statisticians remained without any direct connection to the narrative semantics as elaborated in the topical statistics of the German statistical school. Johann Peter Süssmilch's 1741 *Göttliche Ordnung* and Gottfried Achenwall's 1748 Göttingen dissertation on the *Notitia rerum publicarum* did nothing to change this fact. Any attempt to construe a deep discursive turning point here would lead us astray.[61] Süssmilch converts Petty's political arithmetic into universal laws of nature. Achenwall reproduces the familiar seventeenth-century academic description of a state.

Nevertheless, beneath the threshold of programmatic writings and explanations, minor discursive changes work their way up to a statistically probabilistic text and a semantically saturated framing of the tableau. To begin with an example from political arithmetic, a return to Süssmilch may be in order. In 1749 and 1750, Süssmilch held two lectures in the Berlin Academy of Science which he then turned into a combined publication on the statistics and history of the city of Berlin. The first part deals with the growth of population in the vein of political arithmetic. The temporal dimension of this part is defined by *historia* opened into the future; its mode of representation is statistical. Süssmilch compares the number of houses, for instance, between a municipal inspection from 1645 (as conducted throughout Prussia at the end of the Thirty Years' War) and the current numbers from 1747. He obtains a growth rate between these dates of 1 to 4.5.[62] The rapid growth of the population then turns out to be even more meaningful when calculated by an average relation between inhabitants and deaths through the church registries (for single congregations since 1585, for all of Berlin since 1692). Süssmilch comes up with a ratio of about 1 : 4 for a period of fifty years, and calculates a ratio of 1 : 7 for the time between 1590 and 1748.[63] As the divine order would have it, the acceleration of Berlin's growth, and not its size, makes the Prussian capital comparable to Paris or London. The other part of Süssmilch's Berlin publication deals with the "age" and "construction" of the city. Here, the author is occupied with the past and above all the founding events. The method is narrative and amounts to a historical critique of sources. Again, Süssmilch's aim is not so much to prove the city's age. Quite the contrary, it is precisely the city's relatively late founding date—Süssmilch comes up with 1150—that emphasizes the speed of its growth. He is much more concerned with the proof that Berlin had been founded by Germans rather than by Slavs. To take both parts of the Berlin publication together: more important than the size (in numbers) and the age (in narrative) are the city's growth curve and nationality—the calculable future that the numbers show and the national identity that the histories show. The duo of number and text implied in the form of the statistical table is thus mirrored in the conjunction of statistics with history. The final referent of both parts is the territorial state, something to which Wolff points in his preface as providing legitimacy and authority to all statistical probability. Süssmilch's composite constructed from statistical numbers and historical narrative does not, however, make up a unity. Counting and narrating remain split in their evidential reference to each other. Although the territorial Prussian state is their primary source and final meaning, it remains hidden behind the difference of the two parts.

On the side of narrative statistics, well before Achenwall's Göttingen school from the mid eighteenth century, university professors (now long forgotten) had given German academic statistics a style quite different from the late humanism of Conring and Bose. Almost without exception, the new lectures on the *Notitia rerum publicarum*, the study of state affairs, were connected to the broader theme of natural law. For the diplomats of the baroque era, Conring and Bose, *prudentia civilis*, the art of governing and political decision making at court, had been the occasion for the elaboration and the application of statistics, the knowledge of singularities of a given territorial state. For the new generation, in contrast, it was no longer the courtly diplomat or princely adviser who made use of knowledge of a state's "singularities"; the goal of statistics was no longer to inform practical syllogisms in the political arena. Instead, statistics was now addressed to administrative agents occupied with providing government and care to individuals living in a state's territory. These prototypes of civil servants needed and used information about population and resources: this relationship defined the two ends of what subjectivity in the eighteenth-century nation state meant. "Once men held it to be more beneficial . . . to subject themselves to a power, so many social unions and nation states came about, that it is hardly possible to comprehend their number, and there is nearly no one alive today whom either the accident [*Zufall*, "chance"] of birth, choice, or some unforeseeable event has not numbered among the subjects of some state," the constitutional law professor Valentin Jacob Assmann writes in the academic year 1735 under the rubric *Notitia rerum publicarum*. He finds this all the more remarkable in that these countless states differ from one another so much "that I could never say of one that it is perfectly identical to any other," and that "every one of them chooses different goals for achieving the happiness that all are striving for."[64] In short, contingency in the social order is statistically constituted precisely by the "singularities" that are men and their living conditions. Naturally, Assmann assures his readers that God has ordered this contingency of nationality for everyone in a preestablished harmony. Nevertheless, an estranged view of the *sors nascendi*, the accident of birth, remains. In the first decades of the eighteenth century, students of natural law and statistics learned lessons similar to those cited by Assmann, whether those of Christian Gottfried Hoffmann, those of Eberhard Otto's *Primae lineae notitiae rerumpublicarum* (Jena, 1726), or above all those of the professor of philosophy and rhetoric Nicolaus Hieronymus Gundling, a prolific and influential disciple of the philosopher Christian Thomasius[65] and the historian and philosopher Martin Schmeitzel.[66] Lessing made this theme of the accident of birth the structural principle of his philosophical drama about *tychē* and *automaton* in the context of religion and culture, *Nathan the Wise*.[67] Fielding and Sterne and, following them, Wieland and Jean Paul [Johann Paul Friedrich Richter] all tackled the theme of the "accident of birth," which figures in *Tom Jones*, *Tristram Shandy*, and Jean Paul's *Die unsichtbare Loge* (The Invisible Lodge) and *Hesperus oder 45 Hundposttage* (Hesperus, or, Forty-five Dog-Post-Days), among other writings.

The accident of birth seems to function as the emblem of early eighteenth-century "academic" narrative statistics in the German tradition. The theme encapsulates, as it were, what one may call the dialectics of subjectivity in the Enlightenment: on the one hand, we find the foundational *subjectum* of epistemic knowledge, including statistics, and on the other, the *subjectum* of subservience to order and regulation, as, for example, the subjects enumerated and counted in public statistics. Such dialectics of statistical enlightenment is mirrored in a 1748 defense of statistics by the famous Göttingen professor of history Gottlieb Achenwall, who seeks to refute four possible objections to statistics: first, the objection that statistical knowledge is essentially incoherent; second, the argument that it is unstable, since times change; third, the assumption that everything worth knowing about the state and government is preserved in the secret archives of the courts and states; and finally, the objection that statistical knowledge can only be deduced through governmental and administrative practice.[68] Each of these four discussion points exhibits in its own way the peculiarity of an epistemic field that claims to provide stable knowledge for human subjects about the contingency of human institutions and their singularity.

But the new tendencies we observe in Süssmilch's political arithmetic and in narrative statistics still do not lead to a united field of statistics comprehending both text and numbers, narration and calculation. What is missing, as we shall see in more detail below, is the elementary unit of the fact to which counting and narration can refer in equal measure. Facts are singular occurrences in worlds framed as contingencies within a defined field of observation and notation. Narrated singularities and counted contingent events, however, do not yet offer the status of the "fact," or singularity framed as contingency. In other words, in the era when it was announced, Leibniz's "logic of probability" was still just a project. The work of the philosopher and mathematician, physicist, and aesthetician Johann Heinrich Lambert constituted a further step in elaborating and transforming Leibniz's project.

12

The Appearance of Truth

LOGIC, AESTHETICS, AND
EXPERIMENTATION—LAMBERT

Probabilities, Circumstances, and Narrative
Continuation in Goethe's *Conversations
of German Refugees*

Soon after being invited back to the University of Halle by the new Prussian king Frederick II (better known to us as Frederick the Great), Christian Wolff composed his brilliant foreword to Johann Peter Süssmilch's *Die Göttliche Ordnung in den Veränderungen des menschlichen Geschlechts* (The Divine Order in the Transformations of the Human Race), published in 1741. Though no universal theory of probability is yet available, as Wolff asserts, Süssmilch's physico-theology of the growth of world populations is the first practical attempt to come up with one.[1] In addition to honoring Süssmilch, this claim also reestablished order in Wolff's own camp by implicitly repudiating a former student of his by the name of Ludwig Martin Kahle, now a private lecturer in Halle. Kahle had taken advantage of Wolff's expulsion from Halle six years before by publishing a work titled *Elementa logicae probabilium methodo mathematica in usum scientiarum et vitae adornata* (Elements of the logica probabilium, worked out according to mathematical method for use in science and life), which purported to make good on the promise of probability logic as it had been dreamed of, putting an end to all disputes in the body politic and among the philosophers. This was sacrilege. Wolff had inherited the problem from Leibniz in its provisional, utopian state and had deliberately left things at that.[2]

Kahle's 1735 publication seizes and redefines the very status of the debate in providing a solution to the yet unfulfilled project of *logica probabilium*. He bases his claims on German Enlightenment philosophy—on Wolff himself and Wolffians such as Baumeister, and in passing also on Andreas Rüdiger—and he also demonstrates a thorough familiarity with the English experimental philosophy of John Locke and Isaac Watts. He falls back on the probability calculations of Huygens, Jakob Bernoulli, 's Gravesande, and de Moivre, and he is very cognizant of William Petty's *Political Arithmetic*. Adhering to German academic philosophy, Kahle uses the steps provided in Aristotle's *Organon* in order to analyze the Leibniz-Wolffian *logica probabilium*, the epistemic counterpart to ontological possibility: accordingly, there are probabilities first of the concept, then of the proposition, and finally of the argumentative conclusion. The certainty or probability of a concept is to be gained through the investigation of its individual features. The only thing we learn about argumentation, the proper domain of the probable in Aristotle, is the fact that the respective conclusion depends on the probability of the premises and determines the argument's certainty or probability.[3] In order to understand the logic of probability at its core, however, we must direct our attention to the chapter on the proposition, which weighs in with great bulk between the two leaner ones on concept and argumentation on either side of it. In this middle chapter, Kahle is reasoning on the mathematical level of de Moivre and Jakob Bernoulli. He introduces the quotient of the advantageous event in relation to all possible events, and discusses, like Bernoulli in the fourth book of *Ars conjectandi*, the binomial expansion for the calculation of limit values of a posteriori probability. For Kahle's logical account, whatever mathematical probability theory allows us to calculate falls into the domain of proposition.[4]

The young doctor of philosophy approaches the probability of judgment in a more abstract and, in this respect, a more successful manner than Bernoulli. He avoids reference to topical or juridical probability. Instead, he calls for identifying the elements necessary for the truth of a given proposition, and demands that its probability be calculated from their analysis. With this principle, Kahle's point of departure is due to the rival undertaking to the *Ars conjectandi*, the tradition of the Leibnizian logic of probabilities, or *logica probabilium*. By stating that everything that exists has a cause, Leibniz had substantiated his claim that probable or contingent propositions are capable of logical formulation and development in the same way as statements of truth are. Wolff then took the contention that the principle of sufficient grounds—as subsequently diagnosed in the twentieth century by Martin Heidegger[5]—framed ontology in terms of logical propositions, and carried it even further than Leibniz. At the same time, however, he turned it against his predecessor and made this logical instrument of ontology, the handling of propositions, into the very craftsmanship of philosophy. According to this Wolffian move, the reasons for a proposition to be true are basically structured in the same way as the reasons for the possibility of the asserted facts of the matter to exist.

In extending this ontological correspondence of structures from truth to probability, or as he put it, to "incomplete truth," Wolff was unwittingly preparing the way for Kahle. "When a predicate is attributed to a subject from insufficient reasons, it is termed a probable proposition," Wolff asserts. "It thus becomes clear that, in a probable proposition, a predicate is attributed to the subject because of certain reasons required for truth [*ob quaedam requisita ad veritatem*]."[6] Kahle's mathematization of the probable proposition starts out from these *requisita veritatis*. In doing so, however, he also goes beyond Wolff. In particular, he works against any residual Leibnizian intention of treating probability as a constitutive and hence a limit case, a case that remains unapproachable for analysis, rather than becoming the final province of the ontological project.

Such *requisita* are evaluated and counted out according to their dependence on or independence from one other. After this logical and analytical preparation has transferred the *requisita* into numbers, calculation then goes on and ascertains a value for the probability of a proposition, following the methods indicated by Bernoulli for compound probabilities. The most important case of such *requisita* analysis concerns a proposition whose truth-condition is given through a set of observations.[7] In this case, the proposition's validity is constituted by the claim of a universal fact or causal connection and a countable series of conditions under such claim would be true. In this case the requirements of truth, as Kahle says, are the "circumstances" (*circumstantiae*) grouped around the universal statement.

The connection of *requisita veritatis* with *circumstantiae* can be traced back to Arnauld's interpretation of the game as probability theory. Calculable probability also arises from the analysis of the circumstances under which an event occurs or does not occur in Arnauld and Nicole's 1662 Port-Royal exposition in *La logique, ou l'art de penser*. Arnauld, however, had only carried out this analysis to the point where the core of the event emerged as contingency and calculable singularity. Wolff's student goes further. Circumstances for him are variables of contingency set in a complex of interrelated occurrences in such a manner as to strictly determine the core event for which they function as "circumstances." For Wolffians, the notion of the circumstance is in fact related to moral or physical "action" (*actio*) and its "effects" (*effectus*). Such *actio* and *causa efficiens* then constitute the field in which the causality of occasion, the old topic of circumstances, can intervene. The logical operation, "because A does B, C takes place," forms the event (*eventus*, event, has two meanings in the Latin of the Wolffians: obtained effect and event). The *circumstantiae* determine the event E "because A does B, C takes place," with an iterative interlacing of $c_1, c_2, c_3 \ldots$, meaning "(always) when x, y, z is the case, E comes about." This iterative web of circumstances is called *relatio eventus* in Alexander Gottlieb Baumgarten's *Metaphysica* (1739). The term refers to an event as defined by its relation to other occurrences. Relation can mean two things here: the repeated experience of a singular event being brought about (or not) by (the failure of) its contingent circumstances; or, with *relatio* understood as a report, the protocol and narration of such repeated experiences or experiments.[8]

It is no coincidence, then, that Kahle's ontological analysis of the *circumstantiae*, or the *relatio eventus*, in Baumgarten's words, simultaneously reads like a poetological statement and a treatise on experimentation. We find ourselves in the center of the conceptual construction of early modern *historiae*, or, to cite the historian Arno Seifert, of "history as the early modern authority for empirical knowledge." In *historiae*, the mode of being and the discursive framing of all singularities in nature and culture, "history" in our modern sense as well as the "investigation of nature," are at stake.[9] To name just a few examples: in his *Della historia dieci dialoghi* (1560; Ten Dialogues on History), the Aristotelian philosopher Francesco Patrizi, a student of Francesco Robortello's, defines the smallest unit of history as simple action or elementary occurrence (*attione semplice*). If, at the end of the dialogues, he ascribes fundamental truth to this unit of action, this claim has its roots in the circumstances that guaranteed probability in the topical tradition. The familiar topoi of circumstances—perpetrator, cause, means, place, course of action, result of occurrence—give the action its definition in this order.[10] This order of the circumstances, however, applies not only to the narrative, but also simultaneously and even previously to the matter of the fact. Patrizi deliberately leaves out the conspicuously missing topos of time. If the first six circumstances result in the internal connection of circumstances (*legatura delle circonstantie*) of the simple action, then time first provides the possibility of the *legatura* of simple actions for the complex of occurrences that is history.[11] The circumstances group themselves like accidents around a substantial action. Provided with the historical ligature of singularity, this action can then be connected with circumstances and other complex occurrences. Pierre Gassendi's "scientia historica seu experimentalis" approaches such an intensified notion of *ars historica*.[12] In this context, the combination *historica seu experimentalis* is by no means as tricky as Gassendi's classification of both as *scientia*. Ascribing *scientia* to history as well as experimental knowledge presupposes a similar analysis of action and circumstances with the anti-Aristotelian scholar as Patrizi conducted it within the frame of Aristotelianism. To suggest the varieties of the early modern analysis of circumstances with a last example, we may finally turn to experimental philosophy, or more precisely, to what Steven Shapin and Simon Schaffer have called its "literary technology."[13] Right at the beginning of the *New Experiments Physico-Mechanical* (1660), which deals with the elasticity of air, Robert Boyle apologizes for a certain "prolixity" in his experiment descriptions that may stand in the way of rhetorical elegance. It is not the "plain style" recommended by Sprat and Locke that characterizes Boyle's anti-rhetorical impulse, however, but rather the process of "circumstantially relating" experiments: the report that details the circumstances and thus also necessarily entails complex elaborations. The circumstances are once again related to a central occurrence: to that which is to be observed in the experiment. They are furthermore connected such that theoretical relation (as connection) and literary relation (as report) can hardly be separated. Boyle forges various excuses for his "prolixity." His experiments, he writes, are all too new and surprising; he may want to come back at some other point to an observation that for now seems irrelevant; moreover, readers should be able to repeat the experiments for themselves after reading the report, or else be compensated by its thoroughness for the fact that they are not in a position to repeat them.[14]

In short, when Kahle's 1735 *Elementa logicae probabilium* analyzes and calculates the probability of a proposition by the product of the single circumstances and their respective probabilities, he is referring back to a formula that pervades the entire realm of early modern *historiae*. In his formula, the whole of traditional knowledge and representation of singularities is at issue: historiography, poetics, and experimentation. Kahle's treatise tackles nothing less than the great conversion of historical particularities into bundles of contingent events, events that can either happen or not. For our context we may add a further remark: the *circumstantiae*, which according to Cicero and Quintilian make up the *evidentia* of the probable narrative, also play a prominent role in the theory of the modern novel.[15]

"The number of possible variations in the *circumstantiae* of persons, things, actions, time, place, and relations is nearly unlimited." This sentence from Kahle's *Elementa logicae probabilium* could easily have appeared in any passage on the theory of the novel. "One must weigh and evaluate them according to the laws of probability logic."[16] This second sentence amounts to the proper probabilistic continuation of the first. With Kahle, the logic of probabilities comes close to the fulfillment of its original promise. In his work, it seems that logic and rhetoric, ontology and mathematics finally coincide. He upholds the interpretation according to which games of chance provide the theory of probability, and he does so without reducing this interpretation to one of its two constituents—the topical semantics of the probable or the mathematical structure of contingency. The elaboration meant to fulfill the promise, however, displays all the more clearly the ambiguity that had given Leibniz and Wolff pause about realizing its possibilities. The circumstances appear, on the one hand, like facets within the singular event that comprise its contingency in the one and only world of observation and experience. On the other hand, they allow the definition of a second world from which a set of events can be observed and, through this framing, constructed as contingent. The question remains open as to whether a more or less improbable probability or the improbability of a probability is meant. No appearance of truth has yet intervened to stabilize the relationship between probabilities and their improbability.

Lambert and Kant sought to dissolve the deceptive unity of the *logica probabilium* to the extent possible. Lambert ascertained the aesthetic and phenomenological appearance of truth in its own logical and mathematical determination as far as he could without entirely destroying the characteristic suggestion of probability logic that such determinations offer to an observer in this one world. Kant breaks down the logic of probability in mathematics and semantic discourse, but without entirely escaping the fascination of an embracing appearance of truth. Literature, too, however, has an instructive effect on the dissolution of the analysis of *circumstantiae*. By using its device of improbable probabilities, literature can make known what it means (in the one reality) to have divided the appearance of truth (between hierarchical spaces of observation).

In his 1795 *Unterhaltungen deutscher Ausgewanderten* (Conversations of German Refugees), which is at once a magisterial look back at early modernity and a turn to the near present of the French Revolution, Goethe takes up the question of the *circumstantiae*. With this framed collection of novellas and anecdotes, Goethe even makes the question the starting point for his critique of the fundamental poetology of story and history. This move is reminiscent of Johann Martin Chladenius's critique of probability. Chladenius (1710–1759) had already written of the *punctum visus*, the concept of variable perspectives, which place observers in different positions to view one and the same event.[17] For Goethe, in contrast, simply choosing this or that point of view is not the issue. *Conversations of German Refugees* emphasizes instead the very fact that observing at all means taking up a position and being bound to a point of view. At first, a character called the "old man" tells the story of the Neapolitan singer Antonelli and the mysterious cries, shots, and noises that have taken place since the death of her unhappy lover. The conversers discuss the type and possibility of the events. Brother Fritz has, as he says, a "suspicion, which, however, I don't want to tell, until I have reviewed all the circumstances again and tested my conclusions [*Kombinationen*]."[18] We can easily see where the vocabulary of circumstances and combinations comes from, a jargon that will also accompany the detective stories that develop later on in the history of literature. Instead of presenting his analysis of *circumstantiae*, however, Fritz recalls and tells a different story.[19] For its notable sparseness, a sparseness of circumstances in the first place, this is an even more mysterious story. In it, a young orphan's steps are suddenly followed by thumping and knocking beneath the floor whenever she crosses the room. This transpires in the house of a rich relative whom she attends as a servant. The head of the household orders the mysterious noises to be investigated. When the investigation remains fruitless, he threatens to whip the girl to death if such knocking is ever heard again. Immediately, the strange knocking ceases. After the abrupt end of this story, which itself had been told as part of the commentary on the first Antonelli tale, the discussion continues. Carl and the old man lead the commentary away from the combinatorics of circumstances to the equally epistemic and poetological dichotomy of wonder and probability. In doing so, they finally make a first small point of scientific theory:

"It's a shame," Carl replied, "that cases like this are not investigated thoroughly, and that to judge events that interest us so much we must always waver among different probabilities, because not all the circumstances under which such wonders occur have been recorded."

"If only it were not so very difficult to investigate," said the old man, "and to keep all the points and issues that are truly important in mind, at the moment when something like this occurs."[20]

It is not the beginning, but the continuation of the narrative process in the *Conversations* that is unfurled from this consideration. If indeed the difficulty of an analysis of circumstances and thus the difficulty of observing life as a running experiment, lies in the fact that one can never have all the circumstances at hand, then replaying the ongoing combination of circumstances in one's memory or imagination logically follows. Making use of this reservoir leads in turn to the narration of other stories, or, in an emphatic sense, to continued narration, which is why the stories in their interconnectedness are called *Unterhaltungen* here: *Unterhaltung* not only signifies the "back and forth" of conversation,[21] but also refers to upholding the continuation of communication as instituted in the framing story of the novella cycle by the baroness, the central figure of the storytellers. According to her wishes, continued communication is to be realized by narrating, and from narration to narration.[22] Maintaining and entertaining are thereby merged together.

The Appearance of Truth   313

Memory and imagination are remedies for the impossibility of keeping all circumstances present in the moment of the event, a moment that therefore always already amounts to a wonder, a mystery. They indicate a space where we can hope to access and reconstruct life as an experiment, although in reality, it can never be present as such with the requisite trials and repetitions. Certainly, the second story is no more successful than the first in revealing the proper viewpoint for studying life as an experiment. But by expressly continuing the act of narrating the first story, the second one attests to this desire and furthers the quest to realize such a life experiment. Continuing the narration appears to be in the service of transforming the simple story into an account that would enable us to look at its singular events as occurrences played out in a game and thus observable in their contingency. Narration becomes a reconstruction of the recounted occurrences insofar as the narrated story is perhaps always compromised by the blind spot from which the events of the story appear as the subject of experimentation. Every story makes us hope that, by being continued in narration, the probability and likelihood of its narration could become evident via the appearance [*Schein*] of truth.

In this line of emphasis, the discussion about changing probabilities and wonder is no longer to be conceived of under the old poetological-rhetorical concept of *circumstantiae*. The world of the topics is indeed familiar with the question of whether it is appropriate to take all *circumstantiae* into consideration, or whether it is more beneficial to introduce certain, selected circumstances into the narrative of the occurrence rather than others. But only the barrier between probable and certain knowledge defines the existence of a blind spot of observation, a viewpoint that, if made accessible, reveals singular events as occurrences in a game of chance. Old European topics had never posed the question of the nonpresence of all circumstances in one moment, because, for it, there is only one world, only one viewpoint or blind spot that all observers share. For this reason, the blind spot is not a problem of poetological-rhetorical probability; we might even say there is no blind spot for topics.

Narration, however, also always means deconstructing the place where all points and moments of experience are present, the place from which its construction could be reconstructed. Every narration, every return to memory, can for its part turn out to be a sequence of another story. In Goethe's *Conversations*, a hint at precisely this structural fact of narration and its continuation follows directly on the heels of the old man's commentary on the impossibility of having all points and moments present at once: "Scarcely had he finished speaking when a very loud crack was suddenly heard in the corner of the room. Everyone jumped, and Carl joked, "Surely we are not hearing from a dying lover [i.e., a lover of the sort in the Antonelli story—RC]?"[23]

Carl's rhetorical question plays with the possibility of the story's audience hearing a voice from a figure of the story they had been listening to. Precisely such switching back and forth between the story told and listened to into the story of the narrator and his listeners has indeed happened in this moment. The very loud bang is of the same type for the conversing emigrants as the cries and shots that seem to point to Signora Antonelli's lover. In vain do they carry out investigations about the incident from this moment up through the last novella, when the series of narratives told is taken up and mirrored by the concluding "Fairy Tale." Even if at the last minute, some explanation is provided for the chain of occurrences that leads to the splitting of Röntgen's arched desktop, which they had thought so sturdy, this whole sequence of investigation and reasoning only helps to make the framing story into one story among the others it is framing. The breaking of the desk, we may say, is the blind spot of all the blind spots in the enter-/maintaining narration, the *Unterhaltung,* of the emigrants. It is two things at once: the singular event of setting up a frame that no observer—of any order—observes any longer; and simultaneously an event within a chain of repetition penetrating from without into an observed interior space. With this cracking desk, the singular event of contingency or the contingency of the singular event manifests itself and provides a narrating analysis of the logic of probabilities.

"Probability": Examination of a Word

The main philosophical work of Johann Heinrich Lambert (1728–1777), *Neues Organon* (New Organon), has been read as a commentary on a book that, though written fifty years previously, did not exist in print at the time of Lambert's publication. In 1765, Leibniz's *Nouveaux essais sur l'entendement humain* (New Essays on Human Understanding) finally appeared in the edition of his works edited by Rudolf Erich Raspe, a year after Lambert's *Neues Organon* was published. The fact that the *Organon* offers a decisive response to the Leibnizian tradition of *logica probabilium* has, however, been largely ignored in histories of philosophy and is widely unknown in the history of probability theory.[24]

In the *New Essays*, Leibniz had continued to characterize the logic of probability, the theory of contingent truth, as a desideratum: as a yet unfulfilled undertaking with roots in the juridical forms of conditional contracts, and with an onto-theological origin in the sufficient cause of contingent occurrences. Leibniz did bring gaming theory and statistical data collection together under the roof of his great juridical-metaphysical project, but he never concentrated and merged them into one methodical process or a calculation in the manner of Bernoulli's *Ars conjectandi*. Leibniz readers like Lambert were doubtlessly familiar with such stances from many scattered passages in his published writings. The relevant passages in the *New Essays* nevertheless bring a new consideration into the mix. The individual themes of probability in Leibniz—such as the conditional contract, Huygensian gaming theory and Pascal's wager, the infinite series of numbers, and the collection of data for statistical purposes—had all pointed toward the unity of a comprehensive undertaking. They appeared as partial pieces of a still absent unity. The *New Essays*, in contrast, emphatically exhibit the heterogeneity of the individual projects, and they do so under the headings of free will and contingent truth and the logic of probability. The belated publication of the *New Essays* corrects the picture by confirming it: the heterogeneity within the project of the *logica probabilium* cannot be reduced here to the supposition that all hitherto known elaborations are only too limited or fragmentary. The heterogeneity of the parts proves to be integral to the program.

The difference between the juridical-ontological probability of Leibniz's *logica probabilium* and the technology of epistemological probability of Bernoulli's *Ars conjectandi* thus becomes fully apparent for the first time in the *New Essays*. Lambert, for his part, moves along the same trajectory as Leibniz and Bernoulli once had when they corresponded in the 1690s about the mathematics of probability. On the first pages of Lambert's 1761 *Cosmologische Briefe über die Einrichtung des Weltbaues* (Cosmological Letters on the Arrangement of the World-Edifice), probability appears, as it had once in Leibniz, as the central device to bring practical experience and ad hoc rules in astronomical observations together with the critical discussion and illumination of such practices.[25] A far-reaching difficulty clearly presents itself here to Lambert. Tinkering with the apparatus and administering the process of observation together tack a course toward a coherent theory of practice that is a theory of probability. But Lambert no longer assumes a final identity of practices and the ontological fields in which they occur. He silently rejects any such smooth fulfillment to the promise as offered by Kahle's *logica probabilium*. Precisely because Lambert is once more pursuing the project of combining Wolffian analysis with English experimentation—the project of German Enlightenment science in general—the solution of the earlier Enlightenment philosopher is clearly not even worthy of discussion for him. To calculate the probability of an event from the a priori probabilities of its *circumstantiae* does not contribute anything to the questions posed by observation and experimentation. Precisely for philosophers of science, then, a gulf opens up between empirical knowledge and logic. We may call this the phenomenological gap, to use a word (*Phänomenologie*) that Lambert coined.

Empirical knowledge and logic are now no longer merely concepts in the philosophical debate, however; rather, they refer to distinct fields of scientific work and procedure. This acknowledgment constitutes the break with a juridical-geometrical Enlightenment that Kästner, for instance, would adhere to throughout his life. The logic of the probable for Lambert is thus again a desideratum as it had been for Leibniz and a project as it was for Jacob Bernoulli: a desideratum, however, whose unrealizability is now inscribed at least phenomenologically within its very structure; and a project whose incompletion becomes a part of philosophy, the field of thought called phenomenology.

With an eye on Leibniz's juridical-theological complex of contingent truths, Lambert says, "I am not in a position to give the kind of [formalized rules of probability—RC] that might be useful in ordinary life." In relation to Bernoulli's technology of probability, he remarks that for many years he has been examining his own and others' "techniques and rules" in scientific work in order to establish a collection and classification of probabilistic problems, "which in the future I would present as notes and addenda to a theory of knowledge and discovery."[26] Nature as produced by natural science—a nature whose image will provoke such a striking reaction from Goethe in his *Theory of Colors* and elsewhere—thus for the first time distinguishes the concept of probability as its preferred object. Only as an aside does Lambert add that the new focus on probability also offers the foundation for further applications of probability, applications on various subjects of morality and everyday life.[27] With this turn in probability theory, Lambert reaches the epistemological stance that Laplace, Poisson, and Quetelet will later announce.[28]

In any case, the sober but subtle analysis undertaken in the second half of the eighteenth century by this eminent logician and philosopher of science, Lambert, results in a remarkable accomplishment. Mathematical probability around 1700 had been realized in two related, but deeply divided versions: Bernoulli's was a moral theory of legal judgment; whereas Leibniz worked out a juridical theology of free will. The difference in the two versions lies between exterior probability—characterizing the judgment on an inaccessible truth—and interior probability—free will as being marked by contingency in essence. Yet both versions still seem to presuppose an overarching, even if problematic identity of probability. Lambert's analysis can be understood as bringing out the conceptual truth of such challenged identity: a built-in difference within the identity. This difference in the identity of probability theory is the improbability of probability; and Lambert shows that the nature of the natural sciences is characterized by such paradox. Lambert traces the path of constructing probability from Arnauld to Jakob Bernoulli in reverse. Whereas that construction process consisted in interpreting the mathematical structure of chance step by step with the semantics of probability, Lambert now investigates the notion of and even the very word "probability"—*Wahrscheinlichkeit*—in order to better elucidate its meaning within mathematical theory. In the attempt to base mathematical probability on either a pragmatic or psychological structure, his new approach to the theory of science can be compared with that of Moses Mendelssohn and above all with Hume's.[29] It also resembles in this respect a peculiar attempt that emphatically reads mathematical probability back into rhetoric and the topics without being clear about the original connections. This case becomes evident in *The Philosophy of Rhetoric* published in 1776 by the theologian George Campbell, a member of the philosophical society of

Aberdeen.[30] Campbell, who had already conceived the theoretical chapter of his first book in the 1750s, drafted the foundation of a natural logic for rhetoric in sharp disagreement with Hume. The chapter of his *Philosophy of Rhetoric* in question is perhaps the most concise summary of what had been elaborated in the course of experimental philosophy under the heading of evidence. Calculating chances for Campbell forms the conclusion of evidence from experience. Observation and experiment, analogy and testimony belong to this larger notion of evidence, which is contrasted with the evidence of mathematical axioms as had been customary since 1700. The text's psychological and epistemological point consists in classifying calculation as a form of experience: the a priori calculation of chance in the toss of the die in this view appears as one mode of experiential knowledge among others.[31] Precisely or rather only through this intersection of mathematical construction and psychological validation, which represents the very problem of probability as a paradox, has the calculation of chance for once become an element of rhetoric.[32]

Lambert, for his part, does not reinsert probability into the art of rhetoric, but instead examines the word "probability" (*Wahrscheinlichkeit*) in the languages of science and of everyday life. On the one hand, "probable" functions as a metonymical expression. In "everyday life," Lambert says, we use "probable" for a whole series of various turns of phrase without bothering much about the differences between their meanings (although "probable" in a narrower sense may also occupy a distinct place among those variants):

> Thus, for example, we say "without a doubt" when no objection occurs to us. We use the expression "to all appearances" when the observation of something and its circumstances, as they seem to us, make us inclined to form a judgment. The term "believable" applies to the approval we give to a statement when we can also see the presentation of the matter. "Supposedly" clearly refers to the concept we have of a thing, particularly when it is in the future or else absent. "Probably," however, applies more to the reasons we have that a state of affairs is or will be more likely true or real than its opposite, etc.[33]

But for Lambert, since the usage of common speech does not make such precise distinctions between these various phrases, "instead of taking them all, the concept of "probability" has been singled out, and this one word's own many ambiguities have multiplied with the meanings of the others."[34] It is not difficult to recognize what it is that leads Lambert to the metonymical characterization of "probable." All the other expressions relate to the situation or performance of probable judgment in various ways ("without a doubt": we cannot think of an argument to the contrary; "by all appearances": in regarding the matter, we feel inclined to judge; etc.). In the (special) meaning of "probable," however, grounds for determining the truth value of a judgment present themselves to Lambert. This observation gives him what he needs in order to acknowledge these turns of phrase as a family of expressions at all. "These expressions collectively have something in common in indicating the type and degree of certainty with which we judge or think about a matter."[35] For Lambert, "probable" represents the expressions of uncertainty all together because he sees in it the propositional core of performative variations. The placement of "probable" in the lexicon, as Lambert describes it, is in fact the drift of all its performative variants toward the proposition. But however much "probable," according to its meaning, is a code word for a larger family in which it itself occupies the outstanding position of the epistemic notion, it is also equally a word that refers to widely various areas or cases. No semantic inquiry brings us further than that. Only "the thing itself" can help.

Probability's "thing itself," though, corresponds again to the propositional statement, not its rhetorical performances. In contrast to Hume, Lambert allows the clarification of the matter to begin with gaming theory: "The first case that suggests itself to us in this regard, and that has a very wide relevance, concerns *games of chance*, including *lots* and *lotteries*, etc., whereby rules are derived from the composition of the game in order to determine the *degree of hope* players have to win once or several times."[36]

Instances of probability in historical and juridical statements and in contexts of trust and persuasion follow as further and less exemplary cases after the game of chance as the first and widest instance. As we can easily see, these further examples increasingly reintroduce the performative circumstances of probability. The primary case of the game of chance, however, extends its assumed objectivity to the further, ever more diverging instances, thereby allowing them to become cases of comprehending probability in the first place. In short, in the first three pages of his probability chapter, Lambert repeats the construction of Bernoullian/Leibnizian probability. The (topical) meaning and the (mathematical)

320   VERISIMILITUDE SPELLED OUT

model resurface in this passage as the corresponding sides of a semantic and referential analysis of the word "probable."

For Lambert, the reference to games of chance together with the word "probable," indicating the truth value of a statement, bring into focus the changing meanings and diverging areas of probability. In the semantic usage and the corresponding subject matter, he recognizes the presupposed unity of the center: probability in the narrower sense of mathematical probability. Lambert thus converts the construction of the logic and discipline of probability (*logica probabilium* or *ars conjectandi*) into the analysis of the meaning and reference of the word "probable." But the chasm between the probability of judgment and the modeling game of chance—the chasm that in Lambert's consideration can only be recognized as the chasm between the meaning and reference of "probable"—remains the crux of the focus.

Absolutely Symbolic: Logical Investigation of Probability and Irrational Numbers

Lambert's considerations on the word "probable" can be understood as a reconstruction of the *logica probabilium*, but such an interpretation is by no means exhaustive. The reconstruction cannot entirely dispense with the problem that it makes visible in the construction.

After claiming "the first case that suggests itself [in the analysis of probability—RC]"[37] to be the theory of games of chance, Lambert deals in a masterly way with all the further themes developed out of mathematical probability since Jakob Bernoulli, de Moivre, or 's Gravesande. Those topics include: gaming theory in the narrower sense (a priori probability based on the ideality of the coin toss); the probabilistic presentation of a statistical distribution (the urn model from the fourth book of Jacob Bernoulli's *Ars conjectandi*); and finally the widely applicable theory of induction.[38] All three cases, as Lambert explains, can be reduced to counting whether and how often a certain event occurs or does not occur, and whether and to what degree a certain characteristic accords with it or not. Lambert calls this standard model "a determined type of appearance."[39]

What are we to understand by this determined type of appearance, that is, the determined lack of determinacy (of a judgment) or completeness (of a truth)? Let us assume we distinguish six possible cases a priori or a posteriori in counting the ways a die can fall, and we then assert that in one toss, a determined number has the probability of 1 in 6. In this case, says Lambert, our proposition states a probability (a determined degree of certainty), but not with a proposition that makes such a statement in a probable way (with a determined degree of certainty).

According to Lambert, the determination of probability brings about an asymmetry between stated probabilities and the probability of propositions. In the rhetorical or dialectical topos and in the poetically probable representation, there is no determinacy and therefore also no jumping around between stated probability and probable statement. In both cases, the nondetermination of what is stated is simultaneously the noncertainty of the statement; whereas in an assertive proposition, certainty is a homogeneous characteristic of what is stated and the validity of its expression. Lambert now shows that the act of counting cases of the occurrence or nonoccurrence of an event always produces the same type of validity of propositions, regardless of whether the counting takes place a priori or a posteriori, and regardless of whether it leads only to positive cases, only to negative cases, or to a proportion of positive and negative cases. The "count of cases, no matter how much one can use it to determine the degree of probability, nevertheless by itself does not provide merely probable, but rather true, certain, and determinate propositions."[40]

In more formal terms, the proposition 1/6 A is B ("1 of 6 tosses comes up as two") has no other propositional structure than A is B ("a die is a six-sided shape"). In this sentence, the determination of the degree of probability relates exclusively to the subject term, not to the construction of the judgment, which consists of subject and predicate. It also becomes clear that in view of this characteristic construction, there can be no modification of propositions of judgment at all. Judgments never have any other structure. The theory of gaming as probability theory can thus at first contribute nothing to the question of "what a probable proposition actually is, where it comes from, and how what is *probable* in it differs from what is *true* and what is *certain*."[41] Nevertheless, Lambert takes on the task of showing how judgments of probability become the source of probable propositions; and he does so without recourse to anything from rhetorical institutions or poetic affectivity. Instead, Lambert launches the apparatus of syllogisms for his undertaking:

Let there be two propositions:

¾ A is B

C is A.

The question becomes: what conclusion can we draw, since both propositions share the middle term A? We assume that both propositions are true and certain: regarding the major premise, we can be assured that neither more or less than ¾ of all of A has the predicate B; and regarding the minor premise, C is an individual we know to be A. With these assumptions, there has been no discussion so far about whether C belongs to the ¾ of A that are B or to the ¼ of A that are not B.... [W]e [can] determine the conclusion no further than the proposition that it is more presumable that B is attributable to C than not. Because among 4 A's there are always 3 that have the predicate B, and because in regards to C there is no choice, it is three times more presumable that C belongs to the A's that are B than to those that are not.[42]

According to this thought, a probable proposition for Lambert is the *conclusio* from a proposition that names the counting up of the cases (e.g., "1/6 of the throws with the die x come up as two"), and the minor premise that introduces an individual toss (e.g., "This is a toss with the die x."). The conclusion then reads, "We assume a relation of 1 : 6 for the fact that this toss of die x will come up as two." Derived from game theory, the primary reference of "probable," Lambert produces its meaning through a syllogistic process. The resulting meaning of "probable" embraces the fact of incomplete grounds for the decision of the truth or falsehood of certain propositions (Wolff).

We can now appreciate the consistency in Lambert's reasoning all the more when he stakes the logical sense of probability as a metonymy for all the other inflections, such as "supposedly," "without doubt," and "according to appearances." By this token, Lambert inverts the perspective of Aristotle's *Organon*. He does not look at probability from the point of view of pragmatic relations of knowledge (dialectic), of situations (rhetoric), or of genres of speech (poetics). Instead, he conversely attempts to represent these forms of the pragmatics of judgment purely and exclusively from the point of view of his constative logic of assertions. This turns the question "What actually is a probable proposition?" into a crux: the game of chance to be elucidated by a mathematical theory of probability is now already required in order to explain such a theory in the first place. Only the detour via the example of gaming can make possible the paradoxical form of a judgment that is probable in its structure, though judgments are defined by the fact that they are always either true or false.

We must continue our citation so as not to miss Lambert's peculiar point. In the further development of the argument, explaining the meaning of probability through syllogistic logic acquires even more momentum. The formalistic turn is not merely tacked on to the argument as its mode of presentation, but it becomes rather, in a precisely staged manner, the argument itself:

Thus if one comes to the conclusion that C is B, then this conclusion is not entirely certain, but is robbed of ¼ of its certainty: that is to say, its probability is ¾. We express this in the following manner:

C is ¾ B.

In order to avoid an ambiguity in this way of presenting probable propositions, however, we should add that the fraction placed between the little binding word [the copula: Lambert's German term here is *Bindwörtgen*—RC] *is* and the predicate B does not concern the predicate, but rather the copula.[43]

Directly following this passage, Lambert accordingly suggests another form of notation:

C ¾ is B

The illegibility of the formal notation in this proposition is, we may say, part of its meaning, a part that cannot be captured in ordinary semantics. "But if such a fraction is meant to express the degree of *probability* of a proposition, then it must be joined with the copula, whether preceding or following it." As a mere numerical formalism in the text of the judgment, the notation "3/4 is" introduces "not only the degree, but also the concept of probability, because the concepts "to be" and "not to be" do not admit any degree of intensity."[44]

According to this explanation, the concept of probability is appropriately represented when its formal notation also denotes the infraction against the concept of the copula and hence the structure of the proposition as such. The representation of the probability of a proposition must present a nonproposition. Its notation can then go on to be used for further logical operations. The formal representation in the case of the probability of a proposition is no portrayal of what it refers to. Instead, it extends the concept of the degree of probability beyond what can be said in a proposition into the exclusively formal and symbolic mode of an operator, a symbolic nonproposition.

Lambert's logical reconstruction of the construction of mathematical probability thus results in a double probability: first, in the probability that can be named and cited in a proposition that for its part is not a probable, but rather an either true or false statement; and, second, in the probability of a proposition that can be formally represented but verbally only circumscribed because the symbolic formalism presents a nonproposition. The probable proposition is no longer a single statement, but rather two "propositions" at the same time. The one proposition says what the other proposition as operator does. The one proposition lets us read that the operative proposition is now only calculation, the mathematical and logical formulation of a proposition with which further calculations can be carried out.[45]

Between the probability in the proposition and the probability of the proposition there lies a limit where the text, the alphabetical form of expression, must be given up in favor of a numerical or logical-symbolic form. The quantification of chance is not what compels this turn of the text into formalism, but rather the reconstruction of rhetorical-poetological probability. Lambert has various metaphors for the process: the chance of the counted cases "moves from the major premise to the conclusion," that is to say, from the middle term of the major and minor premises into the copula of the conclusion. Or, the graded probability "spreads out evenly through the proposition," when the copula is "afflicted" with it.[46] It is a metaphorics of influence, infringement, and even infection that leads from the probability that can be named with certainty to the impossible probability of an operative proposition and its statement. This metaphorics simultaneously destroys and protects the unity of the one probability that makes the meaning of probable argumentation and its exemplification in the game together the center of the semantic field of incomplete certainties. If Lambert's argumentation could be called a reconstruction of constructing game theory as probability, then the point that converts the alphabetic text into an operative formalism of numbers is precisely that of a provisional repeal of such reconstructed construction. Pascal's *aleae geometria*, which has been read as probability since Jacob Bernoulli and Leibniz, is not only the triumph of parallelism and mutual transparency of calculating and speaking. The successful mutual mirroring of text and numbers was indeed the great project of the seventeenth century. At the same time, however, *aleae geometria* also triggers the divergence and break between counting and saying. This moment of separation is what Lambert has to add, thus setting the trend for the eighteenth century. In the mode of infringement and expansion through the statement of probability, the pure formalism of numbers, the fraction, distances itself from the meaning of probability in order to perform it operatively. Lambert's word for this kind of operative presentation is *Zeichnung* (notation), a word that can mean "drawing" in German but also invokes the semiotic sphere of the *Zeichen,* the (linguistic) sign.[47]

According to Lambert, a mathematical concept of probability exists in the mode of poetic invention or fiction ("*erdichtungsweise*").[48] What does fiction mean in this context? Lambert derives his conclusion from the same identification between copula and the concept of existence on which the general or logical grammar of the seventeenth and eighteenth centuries had been based. For Lambert, however, such identity is only valid as an ambiguity (*Zweydeutigkeit*): being indeed claims the unity between logic and ontology, but does not guarantee it.[49] In the process of measuring probability, as we have seen, Lambert has the numbering of cases flow into a concluding proposition where measuring and gradation encroach on the "is," the copula and the concept of existence. In the metaphysical and grammatical interpretation of "to be," however, there is no gradation of the copula (and hence of the truth values of a statement) and no gradation of being.[50] This is so because, even if the name "being" ambiguously designates copula and the notion of existence at the same time, it still is one name and as such it claims to stand in for some identity of copula and existence: "Since ... *being* is an absolute unity, the fractions that can be applied to the *is* and the *is not* are only ideal and indicate the degree of probability. For in the realm of *truth* and *reality,* such fractions do not exist."[51]

Being is shot through with similarity, beginning with the foundational analogy between logic and ontology.[52] Beyond this first and irreducible degree of comparison, all ideas that are thinkable [*gedenkbar*], and all things for which the condition of their possibility is fulfilled can be compared among one another in this or that respect. The unity of the ambiguity of being is based on these two layers of analogy. Only where an idea or a thing leads to a nothing, to something in itself contradictory or impossible, do we move outside the net of the individual analogies and the meta-analogy of copula and existence. How is this emergence, this leaving the net of similarities, possible? Lambert portrays this move just as the fantasies of the poets have been described ever since Horace: thinkable concepts or real, given things are joined together in impossible ways. Lambert produces the examples of round squares and the mathematical expression $\sqrt{(1-2)}$, instances that he calls "empty fancies."[53] Nevertheless, they are not at all worthless. "We can also express them, however, with words and other signs, and in this sense they are still useful, since such inconsistencies sometimes cancel one another out among themselves."[54] The symbolic representability of the empty fancies is the opportunity, but of course no guarantee, for a possible return to the realm of what is thinkable and possible. Symbolic operations operate equally well with the thinkable and possible or the unthinkable and impossible. Alongside probability, Lambert adduces some more examples of expressions that pass, as in a loop, from the ontological through the imaginary and return to the ontological. They include expressions such as "tomorrow's yesterday," which is supposed to resolve into the meaning "today," and $\sqrt{(-(3-1)3)}$, which can be resolved in the natural number 8. The notation of the concepts and things—in Lambert's word, the *Zeichnung*—provisionally protects such "empty fancies" from falling into mere inconvertible imagination. As long as a subsequent symbolic operation can be found holding out the prospect of a return to meaning, the expression in question is not yet lost, even when at this point we are traversing a terrain of impossible combinations.[55]

Lambert goes to great lengths in order to construct the process of poetic invention in parallel ways for the alphabetic or linguistic and the numeric or mathematic form of presentation. The verbal examples operate with paradoxes (round/square; tomorrow/yesterday), whereas the mathematical examples use the square roots of negative numbers. It is easy to see, however, that the parallelism is precarious. While "round square" is and remains a logical paradox of mutually exclusive meanings, the expression "tomorrow's yesterday" uses contradictory meanings without ever risking the contradiction of the whole phrase. In this latter case, only the semantic semblance of a paradox comes about. In contrast, the mathematical example carries out an operation that leads back into the realm of real numbers through the same type of a negative square root as used in the beginning. In the imaginary realm, the parallels between speech and number are not revealed without complications. The possibility of "poetic invention" introduces a split between the semantic and operative features in the alphanumeric code, a division that often but not necessarily coincides with the border between the alphabetic and numeric.

This split is also marked in the third example Lambert cites as participating in both language and numerical representation at the same time: probability theory. Deliberately or not, Lambert gives it a very peculiar position. Although it provides an example for the imaginary dimension of symbolical operations, it does so neither as mathematical theory nor as linguistic expression. Instead, the process of mathematically formalizing the semantic content both causes the crossover into the imaginary realm and allows the return into the world of logic and ontology. Because the quantification of the middle term in the major and minor premises infringes on the copula in the conclusion and encumbers the copula with the measuring number of probability, the paradox of a graded being emerges. And because this paradox is the concept of the probable, the break in the logical notation of the concluding proposition is the only thing that can represent the concept of probability adequately.

Mathematical probability threatens the metaphysical frame in which its own program became thinkable in the first place. At certain points, at least, it departs from the parallel tracks of speaking and calculating from which this undertaking could first be formulated. Probability in Lambert's reconstructed construction extends well beyond the frame of general *mathesis* encompassing word and number, without, however, destroying it. The break between word and number is not simply to be assumed as a given, and it certainly is not final. This is the case because operationalization is not only a process that leads from words to numbers; it can also affect the very semantics of the word. In the word "probability," the purely symbolic nature of the game of chance,—which provides the word's primary reference for Lambert—can affect the semantics so much as to take on a metaphorical and operative character on its part. As we shall see, this process is strictly connected to Lambert's analysis of probability [*Wahrscheinlichkeit*] in the semblance [*Schein*] of truth [*des Wahren*].

Between "Erroneous Concept" and Error Theory

Probability occupies its special place in the *Organon* because Lambert defines its understanding and the ways to use it on the basis of what he calls its being an "erroneous concept." Erroneous, meaning misleading in an inevitable sense, is "probability" for Lambert, because it implies the meanings of blind chance and the "equal possibility of all cases."[56]

Lambert does not succumb to the temptation, from the point of view of gaming theory, to employ various games of chance literally as outright examples for probability, as Jacob Bernoulli had done. Neither does he attempt to give what Leibniz calls "equipossibility" [*Äquipossibilität*] an ontological meaning, accessible to God's *scientia visionis*, analogous to Leibniz's understanding of imaginary numbers. Instead, Lambert constructs probability as a coherent meaning from two heterogeneous strands: the mathematical theory of gaming and its juridical implications, on the one hand, and what in rhetoric and logic appears as probability and verisimilitude, on the other.[57] The "equal possibility of all cases" in the calculation of chances points to the fact that, in a game of chance, the players "according to justice" should have "equal hope of gain," or else their stake must be "proportioned" to the inequality. In view of the game, therefore, "according to justice" the conditions must "be set up in a way" for the players to have equal chances. The "equal possibility of all cases" is something that needs to be established or posited. The arrangement according to justice, the juridical institution of the game, appears as the positing or establishment [*Setzung*] of equipossibility.

Note that the word "probability" [*Wahrscheinlichkeit*] is not used here by Lambert in the context of gaming theory. To understand the game and its rules, all that is required is dogmatic juridical language, the language of the institution and its establishment or positing. The institutional world, however, is not the one that can be observed scientifically and empirically. For this reason, one should "also look to the physical circumstances that are inimical to this presupposition"—the presupposition, that is, of the establishment or positing of the game and the equal chances. Lambert thus emphasizes the rebellion of mere physics against the "erroneous concept" of chance and of contingency, and yet he continues: "Assuming this, the theory of calculating degrees of probability," above all, in Bernoulli's *Ars conjectandi*, "has already been given a solid foundation and applied in individual cases." As we can see, there is no mention of probability until the point of application in individual cases. Yet it is then immediately present, as if it had been waiting to make its entry the whole time—as if, in other words, a preexistent, general "theory of calculating degrees of probability" were now simply being applied to individual cases of everyday life.

In actual fact, probability had indeed been present all along. It had been lurking implicitly in a certain "error" in the presupposition of the game, the "error" or misleading concept of "the equal possibility of all cases." In truth, there is no equipossibility in nature: this is the fundamental "error" in all calculation of chance, and it is why games of the chance have to be "instituted" and their rules "posited" before any calculation is possible. In the rebellion of the physical circumstances against the legal dogmatism of equipossibility, the concept of an inevitable error made its hidden entry, the operable error, as it were, of mathematical probability theory.

As soon as he introduces the term "probability" and opens up Bernoulli's *Ars conjectandi*, Lambert turns to its fourth book. He discusses exclusively "to what extent the *equal possibility of all cases* assumed in games of chance, lots, etc., can take place in the real world." The model of drawing marked and unmarked slips from a barrel, where each drawn slip would be noted and reinserted, teaches the approximation of the large number: "*The longer one continues to draw out slips, the more the relation between the two kinds of drawn slips will approach the relation of those that had been inserted.* This is what Mr. Bernoulli, with the presupposition of the equal possibility of all cases, proves must happen."

How can the "mistaken presupposition" of "the equal possibility of all cases" be operable in "the real world," how can it be made reliable even outside the establishment and arrangement of fair game? Lambert's answer becomes clear neither from the pure reason of the presupposition nor from an assumed perfect arrangement of the world, but rather from the convergence of the operable error and of a clearly unavoidable imperfection of the real world. "Even in the wisest arrangement of the course of things in the world,"—Lambert repeats this anti-Leibniz Leibnizian formulation word for word in his *Anlage zur Architectonic* (Addendum Concerning the System's Architectonic)—"the equal possibility is based on the number of individual causes."

The (juridical) presupposition of equal possibility is operable in games of chance although it presupposes an erroneous concept, and in drawing large numbers from slips of paper, Bernoulli rediscovered the proportion taken over from the game with its presupposition of equipossibility. Both of these facts result from the same reason: the self-correction of the mistaken concept. In both cases, the same erroneous concept—the notion of blind chance or of equal possibility—operates, "even in the wisest arrangement of the course of things in the world," in such a way that "the number of individual causes . . . coincide so that they just as easily bring about one case as another, and compensate each other in the continuation of the game."[58] In the medium of compensation—this key concept of the eighteenth century that shows up here for the first time as an explanation of probability[59]—the judgment of probability and the frequency of events, Hacking's pair of epistemic and frequentistic probability, are connected: "Thus, however, the more probable each case is in itself, the more frequently it occurs."[60]

The equipossibility of events in the fair game constitutes an order that is to be established and presupposed against the *real world* and in the name of law and theory. The equal possibility of occurrence and nonoccurrence in the events of the real world is a second order of things formed systematically from the chaotic first order of converging bundles of causes. The fair game and the compensation of multitudinous causes, however, share one and the same structure of balancing. Justice is the semantic concept and compensation is the procedural description of the same underlying juridical-mathematical presupposition.

The blind chance of equipossibility presents itself in Lambert within a particular context. It emerges in the chapter "Das Vor seyn und das Nach seyn" (Pre-Being and Post-Being) of Lambert's *Anlage zur Architectonic*, with its discussion of sequences of events modeled after the series of numbers.[61] The context of the discussion thus already presupposes a basic structure of events and their sequence; a further local disorder or order can then be explored on the basis of that fundamental structure. What, however, is the equivalent of such a given structure of sequence and order for chance events? Lambert offers two answers to this question: the first points us to the ideal of the fair game, which can be used as a model and method for evaluating events in the real world. In a different passage, Lambert makes the reverse argument: this time, he takes his departure from compensation. The compensatory nature of things functions now as the medium of measurability for contingent events in the real world. The very fabric of nature in this case presupposes measuring the probability of events on the basis of "equal chances."

As we have seen, the theory of probability is set up on the "erroneous concept" of equiprobability. Now, on the other hand, we learn that the theory of compensation that produces a uniform medium for measuring the probability of events requires anomalies and errors to be taken into account.

Homogeneity becomes the problem to be discussed and solved between mathematics and—as Lambert would have it, with Kästner and Lichtenberg—applied mathematics. In pure mathematics, every formula and every law guarantees the homogeneity of what is derived from it. The geometrical graph follows the law of the underlying formula on each of its points. The case is different in applied mathematics, among which all operations of calculation in physics are understood. "Regarding objects of nature and artifice," two things are true according to Lambert: nature is never strictly homogeneous, but never abruptly abandons homogeneity either.[62] "Nature," in the context of applied mathematics, refers to observed nature, or even nature subject to experimentation. For this nature, the mix of what is observed and the act of observation, anomalies and errors are not exceptions to be eliminated in order to reach the truth. Instead, they constitute the rule under which "nature" only appears, which we must take into account to begin with when dealing with nature.

Compensation, then, the basic distinction for the theory of homogeneity as a presupposition of measuring events, is preeminently mixed or hybrid. According to Lambert, it all depends on "to what extent such small anomalies in part compensate for each other and in part to what extent they are either distributed homogeneously or can be seen as distributed according to a homogeneous law." In which case, "even if they are in fact not exactly distributed homogeneously," they could "nevertheless instead find a substitution so that the anomalies have no noticeable influence on the success of the calculations."[63]

Lambert adduces two examples, instances that would illustrate the primacy of anomaly in the hybrid nature throughout the nineteenth century. The first example is the analysis of the composition of a physical medium:

> Air is composed of different particles that float around in it infused, and it can easily be demonstrated that the lower atmosphere has and can have more particles than the upper. If we should wish to calculate the decreasing heaviness, density, warmth, opacity, refraction, etc., in increments in the higher altitudes, then we cannot burden ourselves with observing the situation of every single particle. Instead, we must establish something homogeneous and homogeneously decreasing for a basis. This can be done due to the multiplicity of causes and circumstances as long as we are investigating the entire sum rather than individual parts.[64]

The second example concerns those "small errors that are unavoidable in observation." The way to cope with them constitutes "a theory of its own." Such theory aims at "how to draw more reliable conclusions from experiments and observations by which something is measured: more reliable conclusions, that is, than when we apply those experiments just as carried out and with all the unavoidable attendant errors."[65]

With this theory, Lambert is finally discussing what had brought probability and statistics together since Hudde and Huygens, Leibniz and Halley. Typical of their concerns was the worry about mortality tables—for example, that if "the anomalies do not compensate" in the death registers, then the cities chosen were too small or the time frame too short. Diseases do not appear everywhere or always uniformly, and where death has carried away a large number of people, there are fewer potential fatalities in the following years. In meteorology, too, we find examples of how homogeneity must be apportioned to different periods: we must evaluate our observations about heat and cold according to the daily and yearly curves of atmospheric conditions.[66] In all of these instances, fields of continuity and homogeneity are to be determined in order to make observations, fields in which laws of steady conditions or repetition can be presupposed.

The compensation of anomalies and errors in observed and experimentalized nature forms the counterpart to mathematical probability. Probability theory relies, first, on dogmatically laying down the rules of games of chances and, second, on observation of the ensuing events. The primacy of anomaly and error in experimental nature, in contrast, presupposes a preexisting relationship between the game of chance and the semantics of probability, and hence the foundational reasons of mathematical probability. But this nature of anomalies and errors, rendered impure by observation and experimental configuration, precisely also offers a field where the unity of probability theory is no longer the problem but the solution.

Historians of mathematics such as Ivo Schneider and Oscar B. Sheynin have interpreted Lambert's error theory, and above all the theory of errors in observation and technical implementation of experiments, as the keystone in the history of classical probability theory.[67] Sheynin has shown that Lambert's remarks on the statistics of errors in the *Photometria* and the *Theorie der Zuverlässigkeit der Beobachtungen und Versuche* (Theory of the Reliability of Observations and Experiments) are the first models of what after Gauss would be called normal distribution.[68] Schneider's and Sheynin's historical reconstructions are fully justified from the perspective of mathematical probability. While Hacking and Daston pose the question of the "meaning" of probability from the point of view of the strictly nonsemantic axiomatization of modern probability theory, Schneider simply eliminates the problem of the relation between semantics and calculation from his account. Although his history of probability focuses less on the conceptions of the age in which probability theory emerged, for this very reason it is the most consistent account of its development. In error theory, the problem where Schneider finds the keystone of mathematical probability, we in fact recognize the maximal tension of the delicate construction of mathematical-semantic probability, and that means also the point of its deconstruction.

A Theory of Systems and the Aesthetics of Probability

Two separate trajectories in Lambert's mathematical and philosophical work take their departure from the issue of probability. Making note of them helps us understand how probability could become the hidden central fulcrum of his philosophy. Both projects—systems theory and the semiotic theory of aesthetic appearance—remain fragmentary or in bare outlines. They are two derivatives of probability with a promising future, even if after Lambert and with Kant they descend into latency for a long time, both in name and in fact.

The first project is called "systematology," and it exhibits the trajectory of recursive (ontological) probability or what has been termed the "improbability of probability." In his fragment on the formation and description of systems, Lambert conceptualizes the notion of a system in general. Physical, political-moral, and intellectual systems, that is, areas of continuous measurability, bodies uniformly defined by certain constitutions and homogeneous combinations of concepts—for Lambert, all these configurations count as cases of a universal theory he calls systematology. Even philosophical theories or poems can be conceived of as intellectual systems. Treaties and states appear as moral systems; and the cosmos, the solar system, machines, and buildings as physical systems.[69]

To be sure, the term "probability" does not emerge in the description of such systems and of systems as such. But in one decisive detail systematology does pick up the description of aleatory orders and the theory of anomalies and errors: Lambert speaks of a "binding force" that simultaneously propels the formation of a system and determines the configuration of elements within it. The "binding force" creates and maintains the "continuous homogeneity" of the parts of the whole system.[70] Continuous homogeneity, uniformity, or conformity characterizes the concept of the system and, in particular, that of the binding force in it. With the keywords uniformity and homogeneity, however, Lambert has stumbled upon the structure of the compensation of anomalies and errors in observed nature. On the one hand, this means that the implementation of statistically probable fields offers the typical case in which Lambert hit upon the analysis of system formation. On the other hand, the concepts of systems and their binding force as assumed in the fragment on "systematology" generalize the problem and the solution developed by Lambert in the statistical probability of anomaly and error compensation. For this reason, not every system is a case of probability theory. But every instance of "systematology" will be seen from a perspective of compensatory logic in anomalies and observational errors. Always and universally, "systematology" takes the hybridity of observed nature as its starting point; in other words, it once and for all presumes the unity of equipossibility in the fair game and of natural compensation as we find it in the basis of probability theory.[71]

The trajectory leading from the mathematical analysis of anomalies and errors to systems theory remains buried and underappreciated. In contrast, the connection of probability with appearance and its semiotics, that is to say, with Baumgartian aesthetics, is part of the broadly conspicuous development of aesthetics until Kant's *Critique of the Power of Judgment*. Lambert deals with probability in the penultimate chapter of the *Organon*'s section on "Phenomenology" [*Phänomenologie*], in the study of appearance. Previously in this context, he discusses "sensible appearance" (the appearance with which physical things confront us through our senses and perception), "psychological appearance" (the appearance with which we treat the general, abstract, and transcendental concepts in our thought and speech), and "moral appearance" (the appearance by means of which affects and passions separate us from the true conditions of the world).[72] Following the probability section comes a brief chapter, "On the *Zeichnung* [signification, but also drawing or delineation] of Appearance," Lambert's outline of an aesthetics that considers semiotics. For the first time, semiotics is here discussed under the aegis of appearance, and appearance (phenomenology) is seen for the last time under the aspect of semiotics. Lambert's phenomenology is conceived on the model of drawing in linear perspective (photometry was one of his principle scientific projects).[73] It can "be called a transcendental optics in its most universal scope . . . insofar as it generally determines appearance from the true, and conversely also the true from appearance."[74] His phenomenology construes the senses, the nerves of the brain, thought, and the psyche each as its own kind of receptor, all of whose laws of refraction and projection he investigates. Lambert's emblematic apparatus is the camera obscura.[75] Under the title "Zeichnung des Scheins" (Signification or Drawing of Appearance), he deals in his *Organon* with the material traces of such pictures on the pattern of linear perspective, using the geometric construction of the light cone as a model. Nevertheless, Lambert is not particularly concerned with pictorial representations. Instead, his favorite fields of "phenomenological" investigation include the plastic models of bodies; the picture-frame stage; the play of actors; our imagining of the thoughts and feelings of others, or memories of our own past conditions; the narratives by which we represent observed events; and finally poetry, which does not try to paint things themselves, but rather sensible and psychological appearances of things. These are all examples of the *Zeichnung des Scheins*, the process of signifying, delineating, and drawing the contours of appearance.

It is not by coincidence that Lambert places the chapter on (logical) probability at the very juncture between the appearance in perception, thought, and feeling, on the one hand, and the media and techniques of artistic representation, on the other. Logical probability—in its duplication of calculated probability and probable propositions—indeed partakes of both sides of aesthetic phenomenology. First of all, Lambert's probability clearly refers to the type and manner in which uncertain states of affairs in the world present themselves to us: they appear in the relation between desirable cases to all possible ones in a certain state of affairs. To this extent, the calculation of probability is the formalized version of a schema of perception and thought; it functions in the manner of the physical receptors of humans. Secondly, the probable proposition for Lambert is also the symbol per se of a pure operation of thought, an operation without reference in the world: this is the case of the copula of a probable statement, a statement whose truth value is afflicted with a fraction instead of being either One or Zero. The probable proposition ("A x/y is B") shares this significant feature with an expression such as $\sqrt{1(1-2)}$. They are both symbolic per se, that is, merely operative through *Zeichnung*, signification and drawing, which happens by interrupting the connection between the semantically linguistic and the numerical type of writing. Accordingly, Lambert calls the concept of probability a fiction, a poetical invention. The symbolic *Zeichnung* per se is the epitome of fiction and invention.

In man's physical and psychological receptors, the semiotic nature ascribed to appearance is only a way of understanding its functioning.

Because there are no material traces that can be invoked, the "language of signs"—as Lambert calls it—remains a modeling device here, whose meaning is not specifically in question. What we perceive or think is articulated in an idiom different from that of true worldly relations. But he never disputes that it is the true things and relations in the world that are represented even in the figurative idiom of our senses. It is the truth that appears in the idiom of the senses. Only with the artificial and artistic media that leave material and tangible traces does the *Zeichnung* specifically come into play, a moment of operation of its own relative to its function of serving the purpose of appearance. That can mean that the *Zeichnung* displays appearance for its own sake—like poetry, which represents how things and relations appear in our thoughts and feelings. In this case, it comes down to a kind of duplication, and we might ask ourselves whether the *Zeichnung* is adequate to the appearance. Or it can mean that the signs lead an unruly life of their own, and we must still learn to see them as the media for the appearance of truth—this case is exemplified by perspectival arrangement of the stage. The significance or drawing of *Zeichnung* comes forward in its own right in this case, and it has to be read and decoded in order to grasp the appearance of truth through it.

Logical probability for Lambert is located at the juncture between our physical receptors and the technical media of art, and it participates in both. It can even be said to form the general expression of its relation—its unity and its division. Probability is *Wahr-Scheinlichkeit* (true-appearance or veri-similitude), the metaphorical name of phenomenology. Phenomenology, or "transcendent optics," deals with nothing other than the relation between truth and appearance, or, as Lambert also likes to say, the translation from the language of appearance into that of truth. The metaphor of the appearance of truth makes explicit the double nature of phenomenology: the unity of and the division between epistemology and media theory, between a subject-centered theory of receptors and a theory of artistic and technological media.[76]

13

"Probable" or "Plausible"

MATHEMATICAL FORMULA VERSUS
PHILOSOPHICAL DISCOURSE—KANT

PHAENOMENOLOGIA GENERALIS

In February 1765, after a half year's work as a civil servant in the Prussian Bureau of Surveying and Building and a month as a member of the Berlin Academy of Arts and Sciences, Johann Heinrich Lambert finally gave "free rein" to his "long-harbored desire" to exchange thoughts and ideas with the Königsberg philosopher Immanuel Kant. His letter to Kant politely addresses issues of intellectual priority and enthusiastically invites cooperation in the grand project for the "improvement of metaphysics."[1] At the beginning of the missive, Lambert offers his *Neues Organon* as a mirror of identification; Kant "would find his own likeness in most of my book."[2] At the end, he suggests "dividing up between ourselves the composition of individual parts of a plan we seem to share."[3] Kant's initial response is one of equally courteous evasion. Five years later, however, he sent Lambert a copy of his inaugural dissertation, *De mundi sensibilis atque intelligibilis forma et principiis* (The Form and Principles of the Sensible and Intelligible Worlds). In distinguishing between the world of the senses and of the intellect, Kant presents a position "that . . . I shall never have to change."[4] Every question of cooperation or competition is laid to rest. From now on, Kant and Lambert are separated by precisely what might have unified them: the appearance of truth. In the accompanying letter from September 1770, Kant proposes the outlines of what would later be called the transcendental aesthetic in the *Critique of Pure Reason*. Here, as a challenge to Lambert, it is termed "general phenomenology [*phaenomenologia generalis*]," "a quite special, though purely negative science." "[S]pace and time, and the axioms for considering all things under those conditions, . . . are actually the *conditions* of all appearances and of all empirical judgments."[5]

Lambert takes up the challenge and develops his own view of the appearance of truth. Nowhere else is his method explicated so clearly as in this instance where Lambert defends himself against Kant's qualification of *phaenomenologia generalis* as a "purely negative science." What for Kant, in his specific understanding, is a matter of "critique," is for Lambert the issue of a philosophical procedure. With the twentieth-century philosopher Hans Blumenberg, we might call this procedure a "metaphorology."[6]

In formulating his own position, the Kantian *conditio subjectiva* serves as Lambert's point of departure: "Since I cannot *deny reality to the changes* [i.e., the 'changes' that phenomena undergo due to space and time—RC], until someone teaches me otherwise, I also cannot say that time (and this is true of space as well) is only a helpful device for human representations."[7]

Lambert follows the Königsberg philosopher in differentiating between the thinking I and—as Kant would have it—the "way of representing myself as object."[8] Accordingly, space is "a simulacrum" of the "world of thought." But for Lambert, what is said of space also applies to the relation between things in physical reality; so much so, in fact, that the two sides bear "perhaps . . . a still closer resemblance . . . than merely a metaphoric one."[9] This "peculiar reality" emerges since space is something "that we cannot define by means of words used for other things, at least not without danger of being misleading."[10] No meta-language is prepared to raise the "simulacrum" of space, space in "metaphorical" usage, out of the relations in the world of real things.[11]

The metaphysician can and should apply the argument of the appearance of truth for critical purposes. Separating the figurative from the literal meaning in the name of appearance serves as the criterion of success for such an application: "If he is successful, he shall have few contradictions arising from the principles and win much overall favor."[12] The critical separation between appearance and reality, it seems, can never be deduced through mere theory. The operational use of terms, however—the kind of use we encounter in Lambert's theory of probability and its status—is precisely the model for how to proceed critically with the appearance of truth. Thus, we can reflect on the consideration that, between the metaphorical and literal use of terms, there is a third kind of use that is purely operational. This latter mode means to release the purely technical rhetoricity of metaphors, and it constitutes a use of terms that is always possible and even unavoidable. In connection with the distinction between "the simulacrum of space and time [*simulachrum spatii et temporis*] in the intellectual world" and "actual space and actual time," he thus asserts: "Our symbolic knowledge is a thing halfway between sensing and actual pure thinking. If we proceed correctly in the delineation of the simples and in the manner of our synthesizing, we thereby get reliable rules for constructing signs of things that are so highly synthesized that we need not review them again and can nevertheless be sure that the sign represents the truth."[13]

Lambert demonstrates this moment between "sensing" and "pure thinking"—the moment of a rhetorical or technical procedure that might be called metaphorological[14]—in a similar manner as in the case of the probable in the *Organon*. In both cases, imaginary numbers and their operative character help him to make the point:

No one has yet formed himself a clear representation of all the members of an infinite series, and no one is going to do so in the future. But we are able to do arithmetic with such series, to give their sum, and so on, by virtue of the laws of *symbolic* knowledge. We thus extend ourselves far beyond the borders of our actual thinking. The sign $\sqrt{-1}$ represents an unthinkable nonthing. And yet it can be used very well in finding theorems.[15]

The common purpose that Lambert offers to Kant and that Kant decides to evade stands in contrast to the issue of probability. The theory of probability plays an important role in the "improvement of metaphysics" for both philosophers. In both cases, the theory of probability has to do with metaphysics' task of separating appearance from truth. The old project of the *logica probabilium* is shot through with the paradox of the *improbability of probability*: calculable probability in an observed world is revealed to depend on the incalculable probability of the world of observation. Lambert and Kant no longer hold this paradox open in hope of future solutions, as Leibniz or even Wolff did; nor do they attempt to ignore it. Both men instead see this paradox as a significant case of *phaenomenologia generalis*. For Lambert, probability is even exemplary for the analysis of the appearance of truth. He takes up the paradox one last time within the confines of *logica probabilium* when he accounts for the paradox of the improbability of probability with the theory of operational terms. The distinction between truth and appearance can be just as easily supported as suspended once we accept that they may be carried out with operative terms. Operative terms, however, follow the technical rhetoricity of metaphorology. This case even exhibits the primary, full meaning of the appearance of truth in probability (*Wahr-Scheinlichkeit*: appearance of truth). Lambert thus once again upholds the unity of the *logica probabilium*, and at its core he finds an appearance of truth that is pure operation, metaphorological rhetoric or technique. Kant, in contrast, does not merely separate probability and its mathematical operations from the real world in which any kind of calculation takes place, which is the position Lambert ascribes to him. Instead, Kant shows that there is not just one type of probability that comprises both the mathematical calculation and our everyday observations and their verbal expression. His radical critique breaks off the very project of a *logica probabilium*; and consequently probability once and for all falls out of the central position it had occupied in earlier Enlightenment philosophy. The moment when Kant rejects his predecessors' claims that mathematical and discursive probability are ultimately one and the same is deeply connected to the beginnings of the new, critical analysis of the appearance of truth.

The cracks of this shattering were first heard in the lectures on logic that Kant taught from the early 1760s on, which, just as Lambert held fast to Baumgarten's *Metaphysica,* were based on the Wolffian Georg Friedrich Meier's handbook *Auszug aus der Vernunftlehre* (Excerpt from the Doctrine of Reason).[16] In it, Meier reiterated Wolff's definition of *cognito probabilis, verosimilis*: "If we assume an uncertain cognition . . . , then we recognize more and stronger reasons to accept it than to reject it, and then *our cognition is probable.*"[17]

Kant finds this observation ambiguous, just as Lambert had before him. Wolff, Meier, and Baumgarten had conceived of the *requisita veritatis* as the logical elements of a proposition, as well as the ontological conditions for a certain state of affairs to exist. Lambert then placed his own distinction between the certain knowledge of probability and the paradox of probable knowledge, which thinking makes use of only in terms of a technical operation, precisely in the problem hidden in this conception. In doing so, he gave up on the parallels between the possible and the probable. The probability that we calculate from a certain point of observation had a different status for Lambert than its transformation into a general conclusion, which pulls this fraction to the copula and applies it as part of a statement about the world. Nevertheless, Lambert understood the transition from the major to minor premise as a regulated operation, an operation that coincides with the nature of the appearance of truth (or its "metaphorological" structure). In order to construct his notion of probability (*Wahr-Scheinlichkeit* or appearance of truth), Lambert thus holds onto the frame of a world that is firmly installed, even if contingent, a first world that comprises all possible worlds within it without realizing everything ontologically possible. He keeps to the assumption of such a first world as an operational device, and yet he abandons the claim that it possesses a substantial entity.

Kant picks up from this development in Lambert when he uses the two available Latin words, *verisimilis* and *probabilis*, for making a distinction between what is *apparent* or *plausible* (*verisimilis*) and what is *probable* (*probabilis*). Through its play on the word *Schein*, this argument clearly attests to a Lambertian origin,[18] but it also represents a decisive turn away from Lambert. A cognition is "plausible" for Kant as long as we regard only a number of grounds for or against its truth. The cognition only becomes "probable" when we additionally know how the insufficient grounds for or against it relate to the total or sufficient ground, that is, the sum of all possible grounds.

This distinction can be found in all the sources for Kant's lectures on logic: from the annotations in the marginalia of Kant's own copy of Meier's *Auszug aus der Vernunftlehre* and the various transcriptions of the lectures to Jäsche's 1800 textbook compilation of Kant's *Logik*.[19] It does not really matter, though, to whom we attribute the many minor variations we find. They may be due to the different degrees in understanding and capacity for expression of the note-taking students, or they may reflect certain doctrinal changes on the part of Kant. In either case, they show

Kant taking up Wolff's concept of a homology of the possible and the probable in order to shatter its theoretical foundation.

Sometimes the "plausible" judgment seems like a mere preliminary stage of the "probable." It then only involves considering the grounds for something and comparing them with opposing grounds, without, however, taking pains to develop a measure to compare the two sets of grounds. The mental impact of the positive grounds outweighing the negative ones is the tipping point. Only by comparing the impression these grounds make with the sum of all grounds can an immutable and certain judgment of the "probable" be reached.[20] "If the elements of probability are homogeneous, they are numbered (in mathematics); if they are non-homogeneous, as in philosophy, they are weighed, that is, evaluated by their effects—this, however, happens only after the obstacles in the mind have been overcome," Kant writes. "The latter, however, do not provide a relation to certainty, but only between one plausibility and another."[21] In the few cases where Kant connects "plausibility" with a legal consideration of evidence, and from there goes on to consider mathematical procedures even for those "plausible" examples, he refers to Leibniz. For Leibniz, doing so had not been a problem, since he still assumed the final unity of *logica probabilium*, that is, in Kantian terms, the unity of "plausibility" and "probability."[22]

Requiring a standard for measuring and numbering also points to Leibniz. According to Kant, insufficient grounds can only be counted and hence expressed by fractions of probability where certainty is to be understood as one, and the sufficient reason supporting it as the sum of all insufficient grounds. For the individual insufficient reasons to be countable (and hence homogeneous), we must presuppose the cognition of the sufficient reason; and in order to form the cognition of sufficient reason, we in turn have to determine the sum of all individual reasons (and, with it, the homogeneity of all those reasons). Certainty, or sufficient reason, as one, is the standard of probability.[23] This means, however, that the old project of probability is splitting up. The split occurs between mathematics (that is, the construction of objects through the understanding and intuition) and discursive philosophy (that is, the givenness of objects for experience): "The mathematician can determine the relation to sufficient reason; only the philosopher has *verisimilitudo*. Subjective, practically adequate."²⁴ With this, Kant touches upon the deepest layer of Wolff's and Leibniz's theory of the probable: the principle of sufficient reason can only have a logical-ontological meaning when, in principle, one does not distinguish between observation with a world presupposed in general and observation of particular conditions in the world. This is again possible if, with Wolff, we imagine the world as a particular form; a world into which nevertheless all the special worlds we construe can and must be inserted. For this operation, Leibniz had forged the formula of the best of all possible worlds. Lambert then turned this cosmological model into a metaphorological one. Kant finally allows the constructivism of mathematics and the experience of philosophy to diverge radically.²⁵

"People have made a fuss about a logic of probability [*logica probabilium*]. But this is impossible."²⁶ With this dictum, every connection between the appearance of truth—the *verisimile*, which Kant's theoretical philosophy knows as plausibility and dialectic illusion—and probability—the *probable*, which becomes an exclusive matter of mathematics—is suspended and cut off. The linguistic consonance in German of *scheinbar* (plausible, apparent) and *wahrscheinlich* (probable) no longer indicates a theoretically relevant circumstance. Nevertheless, the continuity between the plausible and the probable remains fascinating even for Kant. In his copy of Meier's *Auszug aus der Vernunftlehre*, he notes: "difference of probable cognition from the cognition of probability. widows' funds. Whether a comet will hit the earth. Thunderstorms. Insurance" (no. 2616); "Extracting the truth from a multitude of observations. How the inclination to a certain vice relates to the two sexes. Confessional. How many lots are in a lottery box" (no. 2617); "The art of decoding" (no. 1618) "Guessing, from the number of real cases (of white and black balls), the number of both kinds" (no. 2619).²⁷ It is impossible to discern how exactly these notes and examples should relate to the distinction between discursive *plausibility* and mathematical *probability*. In any case, they compromise the distinction between letters and numbers that Kant's critique of reason will presuppose from this point on.

The fascination compromising the distinction reason had just introduced evolves in two directions of thought: first, a peculiar reflection in Kant's notes relates to the motive of the improbability of probability. "No (universal) rules of probability can be offered except that the error of several will not all fall on one single side, but must instead be a reason for agreement in the object."²⁸ With this, Kant connects the average of a large number of cases of error—Lambert's principal theme in the case of technical observations—with probability, but he does so only in an intricate manner.²⁹ We have clearly hit upon a critical point: the coincidence of probability and the theory of errors emerges as possible or even ominous. By following this path, Kant would have returned to Lambert.

A second direction of compromising fascination points toward cosmology and the philosophy of history: "In the whole of the infinite series of causes every action is determined, although, if it is to take place, it is free. For it is indeed not determined from among the grounds that humans can cognize" (no. 2610). "Whatever is coincidental in each single case must be necessary in the whole; thus what a human does is naturally coincidental in each single case; but in the whole, he does what God wills" (no. 2611).³⁰ These notes in Kant's copy of Meier's *Auszug aus der Vernunftlehre*, which sound like excerpts from Süssmilch, will not reappear in either the lecture transcripts or in Jäsche's textbook. If the statistics of large numbers seems to cancel itself out even more fundamentally than the theory of errors, then this is due to the fact that behind the large numbers, the well-known philosophical questions of the norm and the law became visible. Kant's commentary on this matter is preserved in the lecture transcriptions. There he states that, for him, probability is eventually not a matter of theoretical or practical philosophy, but of the faculty of judgment: "Passing judgment on probability is no work of the understanding, but rather of the power of judgment."³¹

Philosophy's fascination with the probability of the average and of large numbers is the same fascination that the power of judgment and its verbal discourse have with the language of numbers barred from it. Thus citations of statistical probability turn up in Kant at two prominent points having to do with the power of judgment: once in terms of aesthetic judgment; and once in terms of judgment in politics and the philosophy of history. This does not mean, however, that we are dealing with aesthetic and teleological amendments to the theory of probability, as is the case with Lambert. In both instances, Kant's citation of statistics is a way of offering help where discursive philosophy reaches its limits and of forging a hypothesis despite the lack of given principles. The mathematical construction of probability appears in the discourse of judgment as an image. This image displays evidence, the kind of evidence, namely, that is characteristic of mathematics in contrast to philosophy. As such, it functions as figural evidence of theoretical evidence. To discuss Kant's theory of probability, we must therefore take a different turn than we did in the commentary on Lambert. In order to describe the figural evidence of theoretical evidence, we must turn not only to the argument but also to the textual quality of philosophical critique. The number in the text on the faculty of judgment functions as an image, and it does so, as we shall see, as an irreplaceable image.

The Place of Statistics in Aesthetics: The Normal Idea

Statistics enters Kant's *Critique of the Power of Judgment* as the "aesthetic normal idea." This is chronologically the later of two appearances of statistical numbers in Kant's major works. In the paragraph "On the ideal of beauty," Kant proceeds to explain the ideality of form from an aesthetic point of view. He does so under the heavy burden of the first sentence in the paragraph, which proclaims the principle of the critique of judgment, including the judgment on probability: "There can be no objective rule of taste that would determine what is beautiful through concepts."[32]

The "standard" for aesthetic judgment of form is to be arranged with this requirement, a "model image" of the imagination, which would exist "fully *in concreto*," but only "in the idea of the one who does the judging."[33] Kant introduces the "standard," which he calls the "normal idea," with the following directorial note: "In order to make it somewhat comprehensible how this happens (for who can entirely unlock its secret from nature?), we shall attempt a psychological explanation."[34] What follows is no mere example, but rather an explanation; yet no explanation entirely adequate in the matter, but rather one that "makes it somewhat comprehensible how this happens." A rhetoric of provisional modeling introduces the normal idea. Thus is the way prepared for the entrance of an appearance of truth into the text of philosophy that, at first sight, does not seem very different from Lambert's.

"In a way that is entirely incomprehensible to us, the imagination . . . knows how to reproduce the image and shape of an object out of an immense number of objects." In doing so, it "as it were superimpose[s] one image on another," and hence knows how "to arrive at a mean that can serve them all as a common measure." Additional modifications of this "as it were" follow:

> Someone has seen a thousand grown men. Now if he would judge what should be estimated as their comparatively normal size, then (in my opinion) the imagination allows a great number of images (perhaps all thousand) to be superimposed on one another, and, if I may here apply the analogy of optical presentation, in the space where the greatest number of them coincide and within the outline of the place that is illuminated by the most concentrated colors, there the average size becomes recognizable, which is in both height and breadth equidistant from the most extreme boundaries of the largest and smallest statures; and this is the stature for a beautiful man.[35]

The example requires three steps that remind us of Lambert's investigation of the word "probable": first, an ascertainment in the mode of psychological conjecture ("in my opinion": the experiment of psychotechnics); second, an analogy to geometrical optical construction ("if I may": the analysis of perception); and third, the conflation of both models of explanation under the aegis of statistics ("perhaps all thousand": the probability of large numbers and averages). "One could get the same result ... if one measured all thousand men, added up their heights, widths (and girths) and then divided the sum by a thousand."

When Kant in conclusion calls "this shape [of] the normal idea of the beautiful man" the "average man,"[36] we are no doubt reminded of Quetelet's statistical investigation thirty to forty years later, as well as of what Cesare Lombroso would ascertain statistically, and finally of what would be implemented with the help of Galton's quincunx in the random generator. Seen as a precursor of later statistical techniques, the citation of statistics in the context of the *Critique of the Power of Judgment* can be called "a lonely idea."[37] The statistical and technical implementation indeed begins precisely with Kant's rhetorical implementation of the figure of probability, which is the operative image of the normal idea. A rhetorical figure, the image of the average man indicates the future emergence of technology at precisely this juncture.[38] The formulas of hypothetical argumentation make it clear, however, that Kant is only gesturing toward such image of the statistical average. Though banned from the discourse of theoretical and practical reason, statistical probability returns as an excluded element in the discourse of judgment, that is, as the figuration of an image.

The normal idea is certainly only one moment in the structure of the aesthetic ideal, which is defined as the "*purposiveness* of an object, insofar as it is perceived in it *without representation of an end*."[39] The entire structure of the undertaking develops from the double meaning that can be read out of Kant's classical "aesthetic ideal": in it, a true appearance can be distinguished from the appearance of truth. This means that the ideal realization of the aesthetic process—the outer perfection of form according to unknown rules—is to be distinguished from the task of providing ideals with intuitions—the attempt to substitute forms for theoretical concepts that, in themselves, lack intuitions. The normal idea stands for the former. Statistical averaging generates form without substance or reference. "Yet there is still a distinction between the *normal idea* of the beautiful and its *ideal*."[40] This distinction is introduced by a physiognomic view that invests the plain form of the average man with interiority and exteriority, and consequently entails considering the body as the field of expression for moral and intellectual interiority.[41] Categorically, however, the relation of the inner to the outer in the reflecting activity of the understanding consists precisely in the fact that the inner has no relation whatsoever to what is on the outside.[42] Precisely and only the form of the normal idea—the form without substance or reference—steps in to offer an instance of the ideal of beauty. To present the ideal in full, form would have to function as the exterior expression of an interior. Such an operation, however, contradicts the nature of interiority. A form has to be generated before and without any reference to interiority, a form that, however, as if after the fact, can offer itself for the meaningful form, the form that is the exterior of an interior. The normal idea of beauty can therefore be called the form of form. "Normal idea" and "aesthetic idea," as we see, are not coequal sides of the aesthetic ideal: The ideal of beauty, Kant says at one point, requires "two elements: . . . the [aesthetic] normal idea" and the "idea of reason,"[43] which he later in turn calls the "ideal" of the beautiful.[44] The entirety of the aesthetic idea thus consists of the normal idea and itself as ideal, that is, of its mere presentation and itself as something presented in the presentation.

The turn that comes about as a result of the asymmetry between the normal idea and the aesthetic ideal directs our attention to the fact that we have arrived at a decisive point in the *Critique of the Power of Judgment* as a whole. The paragraph "On the ideal of beauty" concludes the analysis of beauty under the category of understanding called "relation," the category that in the case of aesthetics more precisely means purposiveness. Purposiveness, however, which here is discussed as one of the categories of beauty, that is, as purposiveness in perception, is simultaneously the fundamental characteristic of teleological judgment and hence of the overall title of the *Critique of the Power of Judgment*. With the unity or disunity of the (aesthetic) normal idea and the (teleologically purposive) ideal of beauty, we might argue, the unity or disunity of the aesthetic and teleological judgment are thus also at stake.[45]

The normal idea offers a structure for teleological perception and presentation. Cognizing shape as an expression of reason and morality requires first of all the cognition of the perception and presentation of shape in its own teleological structure. In this context, we should consider that, for Kant, aesthetics and the teleology of nature and history by no means present symmetrical parts for discussion. The *Critique of the Power of Judgment* does not consist of two parts of equal importance and status. The theme of the teleological judgment is the more comprehensive one de jure. It shows how, in the order of knowledge and of the world, reason would act if it could only have the necessary data from nature and history at its disposal in a legitimate manner. Teleology is the completion of philosophy, however, without valid legal basis, constantly stimulated and activated by various series of observations and collections of data. Conversely, aesthetic judgment is de jure a mere annex to the teleology of judgment. It lacks the occasions to be activated in nature and history. Shapes of artistic quality do not simply arise in the natural world; whereas the beauty of nature, as impressive as it may be, would never constitute the need and the occasion for establishing aesthetic judgment. Aesthetics does, however, have its own legal realm: the compass of the judgment of taste. Even if a person never knows whether the object she judges is art, she can always know that what she is uttering, attacking, or defending is an aesthetic judgment. Although subordinate in function and without a clearly defined subject area to refer to, it is aesthetic judgment that first creates its own procedural realm. The teleological judgment crystallizes as an activity of its own only in the special case of the judgment of taste, and only the judgment of taste can achieve valid acts according to its own rules and in its own name. Only because teleological judgment comprehends this particular part, the aesthetic judgment, does it become the third critique, the *Critique of the Power of Judgment*, thereby attaining the status of a comprehensive bracket for the Kantian system as a whole. Judgment of taste is the first to furnish the concepts of purposes and teleology with an operation and a procedure of its own. Aesthetics, the area of the imagination and of presentation, makes up the part of the teleology-and-aesthetics whole that is oriented toward operation and procedure; and the entire complex of the three critiques relies in turn on aesthetics for such procedures. In other words, the way the normal idea of beauty relates to the ideal of beauty is a miniaturized mirror image of how aesthetic judgment is related to the teleological judgment in toto. As much as the normal idea offers a practical way to achieve the form of form, aesthetics proper lends the model of a procedural realm to the power of teleological judgment understood broadly.[46]

We started from the question of how the normal idea may be able to function as a presentation of the aesthetic ideal. If this is in fact the question of the critique of judgment in theory, then we can now say that in the text of the *Critique of the Power of Judgment*, this query is transformed into the corresponding one: how does the citation of statistics present the concept of a normal idea? Outside of the kind of figuration provided by the numerical presentation of probability and in the alphabetical code of the critical discourse, the normal idea is written only as an effect of an image that is barred from intuition. The normal idea must make it possible for us to imagine "the greatest purposiveness in the construction of the figure [*Gestalt*]." Hence it must present us with a purposiveness that can only appear in the one judging, and there only as a "technique of nature." The normal is thus an image that corresponds to an idea of reason that could not exist without the model of this very image.[47]

The irreplaceable figurative evidence of theoretical evidence, the normal idea per se, is the graphism of the table, the *méthode visuelle* of statistical diagrammatics or thematic cartography. Kant's word for "presentation" [*Darstellung*], *hypotyposis*,[48] is nothing other than the very term from Greek rhetoric for which the Latin *evidentia* had served as a translation in the history of rhetoric. The term refers to the "placing-before-the-eyes" (*ante oculos ponere*) whose rhetorical and mathematical history is related to the construction of geometrical figures in a way with which Kant was as familiar as he was with the specifically modern connection to the tables of statistics.[49]

The Citation of Statistics in the Philosophy of History: Population Lists

The second conspicuous citation of statistics in Kant can be found in the context of the critique of teleological judgment. It does not occur, however, in the second half of the *Critique of the Power of Judgment*, as we might expect, but rather in one of the essays on the philosophy of history that Kant published in the 1780s. While the statistical model of the norm of beauty has often been understood as an anticipation of coming experimentalization and probabilization in human sciences, Ian Hacking and François Ewald take Kant's indication of population lists at the beginning of his *Idea for a Universal History with a Cosmopolitan Aim* as a recourse to actual contemporary practices.[50] What happens in Kant with both instances, however, is the citation of a statistical probability, and in both cases the citation concerns the mathematical theory and practice of probability in the context of philosophical discourse, where it appears as plausibility [Kant's German term is *Scheinbarkeit*].

In 1784, the year when the *Idea for a Universal History with a Cosmopolitan Aim* appeared, the argument for statistics was a widespread phenomenon in Prussia under Frederick the Great and the politics of how to present public affairs. Three months later, for instance, the senior minister Ewald Friedrich von Hertzberg would dedicate his speech on the occasion of the king's birthday to population and economical statistics. In Hertzberg's *Laudatio*, the praise for what the statistics indicate, the economic aid programs after the Seven Years' War, seems to rival the king's concern for the institutional management of such statistics.[51] The citation of statistics in Kant's essay on the philosophy of history certainly occurs in a very different genre and style, but it coincides structurally quite well with the politics of representation in the panegyric to Frederick's government.

Kant writes: If one considers "the play of the freedom of the human will *in the large*," it is to be hoped that in "what meets the eye in individual subjects as confused and irregular, . . . a steadily progressing though slow development" becomes apparent.

Thus marriages, the births that come from them and deaths . . . seem to be subject to no rule in accordance with which their number could be determined in advance through calculation; and yet the annual tables of them in large countries prove that they happen just as much in accordance with constant laws of nature, as weather conditions which are so inconstant, whose individual occurrence one cannot previously determine, but which on the whole do not fail to sustain the growth of plants, the course of streams, and other natural arrangements in a uniform uninterrupted course.[52]

The example of mortality tables and population lists follows directly from the introductory sentence of this treatise on the philosophy of history. It claims that "human actions," the appearances of free will, "are determined just as much as every other natural occurrence in accordance with universal laws of nature."[53] Human actions first become visible as natural occurrences when they are compared with events in nature, for instance, with meteorological tables. Teleological interpretation concerns human actions insofar as they can be presented through a law of nature. Kant's citation of statistics is located in a line that extends directly from William Derham's Boyle lectures from 1711 and 1712 (and the German readers of Derham, Süssmilch and Kästner) to Laplace and Quetelet. Statistical probability, which had once taken its departure from its applications on the *vita civilis*, now makes its return, via the detour of natural laws—particularly of astronomy[54]—to what will be called *faits sociaux*, the facts of society.[55]

The data of human history, which accrue like those of astronomical observations, can in Kant's case no longer be represented in the tight correspondence between statistical diagram and historical narration of Süssmilch's *Der Königlichen Residenz Berlin schneller Wachstum und Erbauung* (The Rapid Growth and Construction of the Royal Residence of Berlin). Kant's demand for such data in such quantity—which it is, moreover, impossible completely to exhaust—signals a new stage in the argument for statistics on historical processes. In the image of their probabilistic calculation, Kant's statistics shows what no text of history can narrate. With this probabilistic statistics, Kant refers to the constitutive gesture for philosophical history as such. Voltaire had already, in fact, introduced moral statistics and mortality tables into the discourse of history in a brief text from 1744. He later connected the piece with other similar texts and in 1756 used the compilation as a kind of foreword to *Anecdotes sur le czar Pierre-le-Grand* and *Histoire de Charles XII*, which contrast the accounts obtainable from traditional archival sources with statistical information from the most recently published periodical mortality registers. Voltaire sees archives of treaties, expert legal opinions, and historical reports as insufficient for "rigorous" history. In them, he finds only scattered "incidents," not the underlying, more durable kind of knowledge supplied by economical and population statistics. "Was Spain richer before its conquest of the New World than today?" he asks. "How much greater was the population at the time of Charles V than under Philip IV? Why did Amsterdam have hardly twenty thousand souls two hundred years ago? Why does it have two hundred and forty thousand inhabitants today? And how can this be known with certainty?"[56]

History reevaluated from a statistical point of view, for Voltaire, is philosophical in a methodological and a political sense.[57] The numbers of statistics provide the basis from which he can articulate his methodological mistrust of all histoires that exist only as narratives. The stories of ancient history accordingly offer an example that borders on mythology for Voltaire. The new history that will one day provide us with "rigorous" instruction, is political. Its epistemic and institutional potential emerges precisely at the point where Voltaire writes:

In a few years, we will know how Europe is actually populated . . . : for at the end of each year in nearly all the large cities, the number of births are published, and according to an exact and certain rule, provided by a Dutchman as talented as he was untiring [Voltaire is referring to Willem Kerssebom or Nicolaas Struyck],[58] we know the number of the inhabitants by the births. This is an object of curiosity for anyone who would like to read history as a citizen and as a philosopher.

With his citation of statistics, Kant does not merely refer to Voltaire's constitution of the discourse of philosophical history in general.[59] More precisely, he even cites the peculiar duplication of methodological critique and current politics. Kant in fact reassessed universal history in its entirety by making its current and fundamentally pending realization the necessary and necessarily lacking political part of itself. Politics thus becomes philosophical history's inherent supplement.

Large-scale academic projects embracing natural law, history, and statistics corresponded in the Germany of the 1760s through the 1790s to Voltaire's ingenious but relatively obscure launching of a discourse of philosophical history. The protagonists of this work, whom Kant clearly has in mind in his essay, were the Göttingen professors August Ludwig Schlözer and Johann Christoph Gatterer. By looking somewhat more closely at their research projects, we can easily see that Kant's citation of statistics picks up, not merely on Voltaire's twinning of philosophical and political critique, but also its German counterparts. The formula of their work is to redefine statistics from the standpoint of the sovereign state by strictly constructing the status of the data. In Gatterer's 1773 *Ideal einer allgemeinen Weltstatistik* (Ideal of a Universal World Statistics),[60] sovereign statehood is declared the condition for the possibility of society becoming an object of history and statistics. Savage peoples have neither history nor statistics; only travelogues and geographical descriptions bear witness to them. Neither do "subject" or "tributary" peoples have history or statistics, because they "do not constitute any particular body politic themselves." "At the most, their own numbers can appear in world histories or world statistics, which follow a merely geographic plan."[61] It is no wonder, then, that of the nearly thirty historical-statistical peoples Gatterer counts, over twenty are in Europe. Asia has five, Africa one (Morocco), and America is a single white spot, "divided between Europeans and savages."[62] These determinations of the proper object of study return again in the topical frames in which entries can become data. "Land and people" comes at the beginning, followed by "forms of state and government," then "resources" and "industry."[63] In Gatterer's view, "land and people" relates to the following topoi asymmetrically, since neither land nor people are actually frames of historical or statistical "data." They only become such when the people have taken possession of the land and corrected its natural deficiencies, and insofar as this possession and cultivation have taken place under public governmental auspices.[64] Statistical description implies the historical time of previous settlement and cultivation processes, just as the narration of history presupposes the statistically descriptive unities of sovereign states. Statistics thus lies at the foundation of history, because statistical "data" only exist in the space of history.

Schlözer, for his part, refers to legal issues and the technicalities of public census to constitute what he for the first time calls "data" in a terminological sense. Throughout all his obvious competition with Gatterer and the historiographical approach,[65] Schlözer offers the corresponding pragmatic theory of methodologically constituting data. As he remarks in 1804, with regard to the Société statistique in Paris and the Statistical Bureau in Berlin, statistics "acquaints scholarship and government with each other without arrogance or shyness."[66] Schlözer's method of constructing data requires three steps: the selection of "national curiosities" from the large pool of "regional curiosities"; the distinction and preparation of the chosen "curiosities" to be further developed; and, finally, the ordering of the "national curiosities" according to the topoi of the state and its purposes.[67] The first and the third step offer nothing new. Ever since the German baroque statisticians Conring and Bose, *notitia rerum publicarum* had related particulars to the topics of the state as a whole by means of what was ultimately a poetical device. The seemingly superfluous intermediate step in Schlözer is derived, however, from British political arithmetic. In it, Schlözer dissolves the poetological structure of singularity turned generality and transforms it methodologically into the operative unity of the data. Functioning according to the principle *relata refero*, statistical data are precisely not bundles of signs with referential semantics. With his intermediate step of constructing data, Schlözer breaks off the poetological relation of singularities to the general found in baroque statistics.[68] These are two exemplary points of departure for Schlözer's constructivist move: for one thing, there are subject areas in which "curiosities" exist from the beginning only in the form of relatio, that is, in the form of a report (e.g., the statistics of corsets, which no searching gaze is allowed to observe, or of fashion, which cannot be identified by simple observation); and there are other areas that require a public institution even to create the possibility of a field of observation (deaf-mutes can only be counted insofar as there are institutes for the deaf and dumb). In any event, a sure piece of data is obtained only when, "as is demanded by most, it is expressed in numbers."[69] In this point, Schlözer finally meets his opponent Gatterer: data can only be reliably ascertained by government. Private persons can only collect it. Collecting, however, presupposes "a readily available reserve of statistical materials, and social institutions that are no longer entirely undeveloped."[70] The section on the "collector," the person putting observations together, is thus preceded by a section on the "creator," which is a state or society. With the fiction of a state that colonizes and cultivates a barbaric land, Schlözer finally develops the narrative of the "creation" of data. In it, he mentions swarming statistical emissaries, provincial prefects who carry out continuous surveys of their districts, and periodic revisions of the comprehensive statistical works.[71] In the statistical "datum," the civilized state establishes and publicizes itself.

We should be aware of all this with respect to Kant's application of population statistics to universal history in his 1784 essay in a gesture reminiscent of Voltaire. The construction of the datum is the first and foremost element in this discussion, a construction carried out on the basis of the emergence of society or statehood. Indeed, for Kant, achieving the "perfect state constitution" is the final "purpose of the human race," toward which such philosophy of history steers. Orientated toward the theory of the state, though not sociology,[72] Kant prefaces his *Idea for a Universal History with a Cosmopolitan Aim* by citing statistics in evidence, but never returns to it again. This early introduction and subsequent disappearance of statistics is reminiscent, again, of Voltaire's *Essai sur les mœurs* (1769), to which *La philosophie de l'histoire* (1765), with its insistence on statistics, was in the end a foreword.[73] In an important contrast to Voltaire, however, Kant's statistics does not lead us to the historical topos of the character of peoples and manners. For Kant, statistics and history are rather connected in the human subject: "Individual human beings and even whole nations think little about the fact, since while each pursues its own aim in its own way and often contrary to one another, they are proceeding unnoticed, as by a guiding thread, according to an aim of nature, which is unknown to them."[74]

The formulation, "one often contrary to another," appears directly following the citation of statistics, and initially it seems to serve only to supplement the example of the population lists. From the essay's third proposition on, however—by becoming "antagonism" and "incompatability"—the phrase "one often contrary to another" turns into a notion and an argumentative and narrative model.[75] As parenthetically as this narrative model, the statistics citation also introduces the methodological *mot d'ordre* into Kant's essay when the text adds "as a guiding thread." "As a guiding thread" can be read as encapsulating a mere comparison, but also as a conceptual determination.

This double character continues to shape the notion of the "guiding thread" and finally makes it itself something like a guiding thread for the *Idea for a Universal History with a Cosmopolitan Aim*: factually, it can mean the "guiding thread of nature," but it can also point methodologically to the attempt to "find a guideline for such a history." Kant sums up the twofold use of the "guiding thread" when he finally speaks of the quasi-transcendental nature of "this idea of a world history, which in a certain way has a guiding thread *a priori*."[76] In short, "antagonism" and "guiding thread" are the code words and notions under which Kant transfers and develops the citation of statistics into the further discourse of philosophy, and the philosophy of history in particular.

The statistics of large numbers, the argument of mathematical probability that in the text of philosophy immediately turns into mere plausibility, is indeed virtually a transcendental or what has been also called a quasi-transcendental.[77] Possibly, it is only an example in the preamble, a figure of speech that temporarily and illustratively stands for what in the later nine propositions will be called "transcendental." In that case, it stands for the antagonism as a guiding thread, for the methodological word valid for the factual circumstances themselves, as they emerge from the example of statistics in the preamble. Or perhaps the guiding thread of the antagonism is a concept that relies on statistics and that makes it possible to cite mathematical probability in a philosophical context. To put it more succinctly, either statistical probability stands as if, as a *quasi-subtitution*, in place of a transcendental concept of history, or this concept is itself the quasi-transcendental of statistical probability in history.

The Quasi(-)transcendental and the Double Play of Probability

Two readings of Kant's use of the example of statistics in the philosophy of history are thus possible; they are, moreover, both unavoidable.

The preamble to *Idea for a Universal History with a Cosmopolitan Aim* begins with the critique of practical reason. The relation to law is here inherent in every single action; a particular action is good or bad. Whoever views the individual actions alongside one another as they occur sees only irregular appearances that are uncoordinated without any discernible ordered relation. Only when considered "in the large" can regularity be cognized in "what meets the eye in individual subjects as confused and irregular." The fact that the individual action is singular in its moral core and has nothing to do with the rule of statistical averages is a positive finding for the moral philosopher. It confirms that every single action is a free act, as it should be. The regularity "in the large" must remain hidden in each respective action. This discovery is the result of the critique of practical philosophy. After Kant has provided the citation of statistics and added the commentary voicing the actual treatise, however, the perspective on the actions and their possible regularity or irregularity changes abruptly. He writes, namely, that it seems as if "individual human beings and even whole nations . . . each pursues its own aim in its own way, and one often contrary to another." From now on we are dealing with intersubjective action, action that can be told in stories and of which history is made: "One cannot resist feeling a certain indignation when one sees their doings and refrainings on the great stage of the world and finds that despite the wisdom appearing now and then in individual cases, everything in the large is woven together out of folly, childish vanity, often also out of childish malice and the rage to destruction." There is no other choice for the philosopher, "who, regarding human beings and their play in the large, cannot at all presuppose any rational aim of theirs—than to try whether he can discover an aim of nature in this nonsensical course of things human."[78]

In place of the irregularity of individual actions offered and required by morality, motivations can now be discerned all too easily by the historian. In the game that is a play on the great world stage, one finds reason even more readily in small things than "in the large." To insist on a plan for this play "in the large" would amount to a subreption of providence. This argument encapsulates the teleological-aesthetic critique of history and its narration.

Two different constellations of rule and nonrule can be discerned here: (1) From the beginning, statistics sees the game that plays out "in the large" as actions that repeat themselves, cultural repertoires (deaths, marriages, births). From the point of view of practical reason, Kant reminds us here of the singularity of each respective moral action that lies unattributably hidden in the repertoires of actions that can be counted for statistical purposes. (2) The play "in the large" that plays on stage means precisely individual actions. They are, of course, not considered in their moral character of freedom here, but rather as actions of which stories

"*Probable*" or "*Plausible*" 359

and history consist. An intention that could be implied would in this case have to be derived from the context offered by history and its narration. A universal and philosophical history can only exist when both constellations are treated as a single field. This is precisely the program undertaken by Kant's *Idea for a Universal History*.[79]

Statistical behavior in the sense of a regular distribution of chance in weddings, births, and deaths is, we might argue, quasi-transcendental. It is a poetical image that mathematics donates to philosophical discourse. In contrast, the fact that in acting "one often contrary to another," humans are following an intention of nature that can be comprehended statistically and only statistically, can be called quasi-transcendental. This state of affairs pertains to a regularity that could be used as a model for narrating histories of providence if such regularity could in fact be taken as a rule of experience. The identity of the quasi-transcendental statistical-pictorial reading with the quasi-transcendental historical-providential reading, if there is any, lies in a statistics of historical actions. One would have to be able to see through the countability of repertoires of contingent actions and relate them to the moral causes of each singular action accordingly. And one would also have to be able to count out and average the conflicting actions on stage the way statisticians do with marriages and reproductive behavior in a given population. The statistics of historical action, the contingency of singularities, is quasi(-)transcendental. In other words, the instance of the statistical average in Kant's philosophy of history suggests an ideal (nonempirical) point of reference of historical development; and in that point, the probabilistic account of human actions without regard to their moral intention (social statistics proper) would coincide with that of contingent causalities precisely of the intention within each singular human action.[80]

This argument can be made even clearer with a view to the image or concept of play and game (the German *Spiel* has both meanings) in Kant's *Idea for a Universal History*. The word *Spiel* especially pervades the preamble and the beginning of the main part. In the opening lines of the preamble, Kant mentions the "*Spiel* of the freedom of the human will in the large," in which a rule of nature should be discovered. Toward the end, he writes that "regarding human beings and their *Spiel* in the large," no "rational aim of theirs" can be presupposed.[81] The formulations appear to be identical.

Nevertheless, a different meaning corresponds to the different way in which they introduce the regularity of nature and ignore the purpose and the rules of humans. The "great stage of the world" has been mentioned recently before the second instance.[82] Thus also with "human beings and their *Spiel* in the large," we should understand a theatrical play. The spectacle of history on the world stage, particularly when seen "in the large," pushes the confusion and disorder before our eyes to an extreme. In this case, only a glance behind the scenes, a view able to detect the hidden plan of nature, could conclude that there is any order and regularity in the stage events. The "*Spiel* of the freedom of the human will in the large" from the beginning of the text, which is not determined by context, reminds us in contrast of games of chance. In both cases in the game or play (*Spiel*), freedom makes its entrance in the appearance, and thus, in Kant's terms, in the realm of the contingent, the irregular and the lawless. But the one *Spiel* "in the large"—the game of chance—introduces the law of chance precisely in playing and through playing; while the other *Spiel* "in the large"—the stage play—only emphasizes the confusion as long as we do not decide to read the play as a theatrical production. The statistical *Spiel* "in the large" becomes sort of, in the modus of the *quasi,* a category of a law of freedom: the *Spiel* in which freedom comes into appearance allows a law to be calculated and to turn into an image for the law of freedom. The antagonistic theatrical *Spiel* "in the large," in contrast, shows that the category of a law of freedom bears the *quasi* as an inauthenticity in its inner construction: the increasing confusion in the actions played out on stage is only connected to a possible plan, which could direct the stage play, by the authority of an invisible dramaturge.

Wherever the word *Spiel* appears elsewhere in the *Idea for a Universal History*, it remains undecided between the two extremes of the stochastic game and the theatrical play. It can thus emphasize the notion of freedom from purpose and of irregularity in a purely negative way, without indicating either the metaphorical path of probability or the metonymical path of theatricality and of a play within the play.[83] The expanse of *Spiel*'s modeling power from gambling to histrionics corresponds to how games and plays had been subject to consideration in theology, law, and public order since early modern times. This mix of meanings never caused a conceptual problem until Pascal or Leibniz. As long as the contingency of the world owes its existence to an establishment of justice and order just the same way as the dramaturgy of world historical events does, there is no fundamental difference between the game of contingency and the play of singular actions on the stage. Kant's critique of the *logica probabilium*, however, dissolves this idea of a legal establishment as the presupposition for mathematical probability. Mathematical probability and philosophical plausibility go their separate ways. For this reason, the play and the game of probability in the *Idea for a Universal History* can no longer simply be of one and the same nature, although the composition of the text demands precisely that once more. Without the possibility of crossing over from the game to the play and vice versa, there can be no philosophy of history, and no critique of the philosophy of history either.

Evidence Modified

By alluding to births, marriages, deaths, and their *Spiel* "in the large," Kant is also taking up, along with Voltaire and the German academic statisticians, the very means to present the evidence of such regularity, the practice of drawing statistical tables. The statistical work we might think of in this context as a model for Kant introduces the mathematical operation of the geometric series into population statistics. Once the mathematical procedure has entered the construction of a statistical table, the evidence of such a table can no longer be simply translated back into a descriptive or narrative text. This, we may argue, is exactly what happens between Kant and a specific chapter in Süssmilch's *Göttliche Ordnung*.

"On the Rate of Propagation and the Time Required to Double the Size of the Population," chapter 8 in book 1 of Süssmilch's *Göttliche Ordnung*, which expounds on the theory of population statistics in general, provides the reference point.[84] Together with chapter 9, which deals with the "greater and powerful hindrances of the multiplication of the human race," including natural catastrophes, epidemics and wars, chapter 8 concludes the analytic-statistical part of book 1. Five more chapters follow, which draw conclusions from the latter for what Foucault called modern biopolitics. In chapter 8, however, not only book 1, but arguably the entire work reaches a theoretical climax. Hitherto Süssmilch has presented population figures and discussed lists, principally from Prussia and its provinces, resolving them with a view toward magnitudes such as the ratio of the sexes, the mortality rate, the rate of marriages and births, and the ratio of births to deaths. The numerical and textual devices in these chapters remain in the precisely defined framework of pre-state statistics, statistics as they were available to the Protestant pastor. Statistical representation and its mathematical manipulation operate in the frame of tabular evidence. This is the familiar tableau of number columns as we know them from the early modern mortality lists down to the Prussian population tables. The calculations undertaken with the numbers belong to basic operations of arithmetic such as converting equations, averaging, and proportions.

The consonance of representation and knowledge is revealed not only in the form of the operations, but also in their content. On the one hand, magnitudes such as the ratio of sexes, spousal fertility, and the ratio of births to deaths are the statistician's target or output values. From them he draws conclusions for cosmological laws that God has decreed once and for all, and for his recommendations in health, social, and economical policies that the king should set in motion here and now. At the same time, however, the ratios of the dead to the living, of births to the total sum of the population, of spousal fertility to the number of marriages are also operative intermediate values. In the context of pre-state statistics and their limited resources for collecting data, the function of these numbers and operations is to process the available data in such a way as to ascertain the population of a given nation at a certain point in time. Ever since Graunt and Petty, the statistical ratios have served for the extrapolation of data sets that are spatially and chronologically discontinuous onto the entire population. In the magnitude of the whole population, datum and calculation, historical ascertainment and political end, input and output variables coincide. Converting the output value and input value is the conceptual principle of evidence in Süssmilch's statistics. Every proposition of his statistics can simultaneously be read as an entry in the semantic topics of state and cosmology, and as an operative step in the construction of these topics. Thus, again and again in Süssmilch's text, we find the old trope of "placing-before-the-eyes," the indication of a figure that had been so energetically applied by Leibniz to the rhetoric and to the graphic design of his so-called state tables: "The following table will place before our eyes the time frame required for the population to double with the current mortality rate."[85]

"*Probable*" or "*Plausible*" 363

As we shall see, however, the statistical presentation in chapter 8 involves a further mathematical processing of the numbers. By that token, the twofold, numerical as well as discursive evidence intensifies and transcends the order of mere statistical representation to the evidence of universal history. Such evidence will then no longer offer itself to the reading gaze. It rather coincides with the technicality of the mathematical process.

The "time required for the population to double" constituted an important theme in political arithmetic at the time. In its context, it served as a measure for evaluating birthrates, fertility, number of marriages, average life expectancy and other factors of population statistics for administrative and political consideration. The time required for a population to double in size was in this sense regarded as a political time, the time of national progress. These issues—which Süssmilch treats so impressively in his publication on the growth of Berlin—dominate the first half of chapter 8.[86] In the second half, the older early modern sense of universal history comes into the foreground.[87] Universal history at that time had been an ancillary science in biblical hermeneutics. The measurement of a population's multiplication referred to the thorny problems of biblical chronology, the measurement of the times from the Fall to the Flood, and from the Flood to the history of Judea.[88] Süssmilch, too, still calculates in this manner with the fertility of paradise, the two- and three-hundred-year life expectancies of the patriarchs, and the manifold multiplications required from the time of the survivors of Noah's ark until the nation of Judea. Universal history, as always in the eighteenth century, thus extends both progressively (politically) into the future and regressively (historically-apologetically or cosmologically in natural history) into the past.

Süssmilch goes one important step beyond such progressive and regressive arithmetic of statistical numbers in chapter 8 of book 1 of *Göttliche Ordnung*, however, leading into the analysis of geometric series. In the foreword to the work, Süssmilch has already mentioned his "valued friend and colleague, Professor Euler," who has provided him with significant "support . . . in the calculation of the redoubling."[89] And it is not simply a matter of doing justice to the true author or of lending authority to his own undertaking when Süssmilch explicitly attributes the construction of three tables of population growth to the "most worthy director of the mathematical class" in the body of his text.[90] For Süssmilch, Leonhard Euler's name signifies the step beyond the literal-arithmetical space of evidence into purely mathematical analysis.

That Euler's co-authorship extends throughout the entire first part of chapter 8 is shown by a comparison with his *Recherches générales sur la mortalité et la multiplication du genre humain* (General Investigations into the Mortality and Multiplication of the Human Race), presented to the Royale Academy of Sciences in Berlin in 1760 and published in its *Mémoires* in 1767. The model for the calculation of population growth, by now grown classical, appears in Euler's *Recherches*, just as it is found paraphrased by Süssmilch in chapter 8. In order to obtain a general formula for the development of the population, two respectively independent magnitudes must be brought into relation. The first is the mortality rate; the second, the rate of fertility or reproduction. With the algebraic representation of these two magnitudes, Euler created the prerequisite for the general analysis of population statistics regarding any given determination of place or time on the basis of a sufficiently large amount of data. Euler noted the mortality rate as a descending series with the terms R = {N–(1)N, (1)N–(2)N, (3)N–4(N) . . . (n–1) N–(n)N} (where N equals the number of children born in year 0, and {(1), (2), (3) . . . (n)} represent the series of decreasing fractions for (n) < 1, which correspond to the subsets of those still living in the years {1, 2, 3 . . . n}). Independent of this is the constant function of the fertility or reproduction: F = α M (where M stands for the number of persons living in a given time span, for example, a year, F represents the respective number of children born, and the coefficient α thus equals the fertility factor assumed to be constant). If one now considers both the function of mortality and that of fertility for an assumed span to be constant, and does not take further external factors into account, it then follows that the number of births stands in a constant relation to the total sum of those living.[91] For Euler, solving demographic problems thus requires knowing the independent functions of the mortality and fertility rate and representing the respective values in a geometric series.

This mathematical analysis of population growth is addressed, or better, presupposed in Süssmilch's chapter 8. The text of the chapter lets the "valued friend and academic colleague, Professor Euler," make his entrance only at the point where the tables are displayed in which a second-order evidence, based on Euler's series analysis, are placed before readers' eyes. These tables show functions of functions. The first table demonstrates how the time spans necessary for the doubling of the population relate to the assumption of mortality and fertility (in this case indicated by the ratio of deaths to births). The following graph curve, which represents the function for the proportion between births and deaths for any given year, departs from and, in fact, transcends the usual space of presentation in *Göttliche Ordnung*, a space that is constructed symmetrically from text and number. The large table on which *Euler* represents for *Süssmilch* the values of a geometric series—which models the construction of a population where certain suppositions are taken for granted about the number of children (six per couple), age of marriage (20), bearing age (two twin births at 22, 24, and 26 years of age), and age of death (40)—then concludes the withdrawal from the space of traditional tabular evidence. Süssmilch shows Euler's table after having displayed a table of his own, which, in the spirit of the old universal histories, begins with a single couple and represents the population growth with ad hoc assumed rates of multiplication. "Professor Euler," he then writes, has now completed the table of the function "on my request according to the planned method." "It is too beautiful for me to leave out."[92] The series construction of the phases of multiplication and the model construction of population growth from an original couple clearly correspond to the political-progressive and the cosmological-reconstructive parts in Süssmilch's chapter 8, making it the methodological seed of a universal history, a philosophy of history in the alphanumeric texture of political arithmetic.[93]

This brief sketch of Süssmilch's famous Euler chapter in *Göttliche Ordnung* outlines how far population statistics had developed at the point when Kant provides his *Idea of a Universal History* with a citation of population statistics. However Kant's essay from the mid-1780s relates to Süssmilch's and Euler's works between 1740 and 1760 in detail, Euler's purely mathematical and graphical evidence of statistical probability, which was institutionalized in life insurance, provides the basis for the philosophy of history. In the text of Kant's treatise, the allusion to Euler's graph functions as an image of evidence. Kant's pivotal text on the history of philosophy and the argument of antagonism would have been impossible without it.[94]

While from an epistemological point of view, Kant's citations of statistics in his aesthetics and the philosophy of history are indicative of the period after Buffon and Euler, with regards to material presentation, they belong precisely in the initial years of its actual, modern history. Even if with this last remark we step outside the citational context proper as discussed in this chapter so far, in order to understand the new evidence beyond the classical Süssmilch tableau, it is important to turn at least briefly to the various techniques of material presentation. The so-called *méthode graphique*, which got its start around 1800 and whose roots lie in the 1780s, can be characterized most readily with a negative definition. The resulting tableaus are representations that have left behind the realm of the two traditional forms of evidence: no longer are they classical tables or geometric diagrams assigned to a mathematical formula as a figure of intuition. Previously, tables and geometric figures constituted the two elementary forms of connection between numbers and visible representations, between the discrete and the continuous. A multitude of new methods of presentation—difficult to classify generically—developed, however, and they replaced the representational achievement of the first two simple kinds of evidence with analogous elements of different sorts. They introduced new emblematic and cartographic features of representation as well as textual and figurative ones. August Wilhelm Crome offers the best example for the entry of statistical magnitudes into cartography. The synopsis of his efforts, which were published in 1818 while Crome taught *Cameralwissenschaften* (economics and administration) at the University of Giessen, bears the (Schlözerian) title *Allgemeine Übersicht der Staatskräfte von den sämtlichen europäischen Reichen und Ländern, mit einer Verhältniss-Charte von Europa, zur Übersicht und Vergleichung des Flächen-Raums, der Bevölkerung, der Staats-Einkünfte und der bewaffneten Macht* (General Overview of the Strength of States in All European Kingdoms and Countries, with a Relational Chart of Europe, for the Survey and Comparison of the Area, the Population, the State Revenues and Their Armed Power). In the foreword, Crome celebrates his newly enriched art of cartography in terms of the old evidence, yet in a markedly modified manner. He speaks of "placing-before-the-eyes" and extols how "what words and numbers express with dead letters is represented *pictorially* and as if *alive* by drawing on such maps."[95] Crome moreover differentiates in a new way between symbol and icon: "The study of geography and statistics is greatly eased when the form of the *representation* is not merely *symbolic* (in the narrowest sense of the word), but is rather also made intuitive. For sensualization through pictorial representation . . . is a very powerful means to facilitate the study of *geography* and *statistics* as well as of *natural history*."[96] With this, Crome summarizes his works from the preceding two decades: his 1782 "product map of Europe," *Europens Produkte*, the 1785 *Größen-Karte* (Map of Quantities), and the 1792 *Verhältniß-Karte* (Map of Relations).[97]

William Playfair's comparative symbolic presentation of statistical data, a work published shortly after 1800, has become more influential and better known. Playfair used the basic elements of circles and bars in his diagrams. They are meant to indicate the territorial area, population, and state revenues of the European nations. The circles are all arranged side by side, with a vertical line bisecting each to show its respective diameter, and they represent with their areas the very embodiment of area—the territory of sovereign states. The contents of the circles thus represent in their proportions to one another the proportion of their square mileage, which is additionally inscribed in them. The different hues dividing the circles into segments or concentric circles indicate, in the case of Russia and Turkey, the distribution of territory on different continents. The heights of two bars erected tangentially to the left and right of each circle represent the two basic assets of the state: the red bar on the left shows the number of inhabitants; the yellow bar on the right, the state revenues. A matrix of vertical lines running parallel to the diameter line and thus intersecting the bars perpendicularly, indicates the relation of magnitudes that are extracted from the bars' heights. Finally, a line connecting the ends of the two vertical bars on each circle symbolizes the idea of a desired balance between population and state revenues.[98] Playfair, too, developed his "entirely new" method of diagrammatic representation from a combination between cartography and statistical evidence. He did not, however, enter the statistical magnitudes in the maps through symbols, but instead represented cartographic magnitudes by figures, and statistical magnitudes by lines. Playfair had published several cartographic works in the 1780s. In 1800, he translated a tabular, comparative representation of European states, which—in the spirit of Gatterer's synoptic method—a certain Jakob Gottlieb Isaak Bötticher had released in 1790; Playfair characterized his own, diagrammatic statistics as an innovative further development from these tables.[99] Bötticher, the author, had come up with these statistical surveys, these modes of evidence before the transition to the post-evidentiary *méthode graphique*, in the 1780s for instruction in the private school he founded in Königsberg in 1778. His instructional duties there had brought him into a lifelong close acquaintance with a philosopher at the local university, Immanuel Kant.[100]

14

Kleist's "Improbable Veracities," or, A Romantic Ending

Studies of Law and Cameralistics

Heinrich von Kleist, who was to be a famous German Romantic writer and dramatist, studied law and cameralistics—public administration—at Viadrina University in Frankfurt on the Oder in the years 1799–1800, while also attending lecture courses in history and receiving private tutoring in ancient languages. Some sixty years earlier, a Viadrina law and mathematics professor, Johann Friedrich Polack, had invented the *mathesis forensis* on a Leibnizian model, a discipline combining law and various forms of calculation. Since that time, of course, new fields had emerged, such as the German model of political science, *Staatswissenschaft*, as well as statistics, applied mathematics, and *mathématique sociale*, all of which had taken the place of the science of forensic calculation in their own ways. Other places, meanwhile, such as Göttingen, Halle, Berlin with its academy, and especially Paris, had become leaders in those new disciplines. In Frankfurt on the Oder, however, even during the brief span of Kleist's academic career,[1] the university still adhered to the prerequisites of the Enlightenment political mathematics that had flourished here. In the terms of the seventeenth century, these prerequisites had been constituted by the discourse of natural law and the mathematical discipline of geometry. The university's perceived outmoded policies meant that the Prussian king would have no qualms about closing the school in Frankfurt a decade later in order to found the Friedrich Wilhelm University in Berlin with new political fanfare. None of this undeniable traditionalism, however, could prevent a certain headstrong student from drawing his own pointed—even extravagant—conclusions from the well-preserved core constellation of academic subjects.

The precarious relation between law and calculation, each of which presupposes and fulfills the other, would remain influential in Kleist's work up through the innermost construction of *Penthesilea*, his most radical work for the theater, which reimagines the episode in the Trojan War pitting Greeks against Amazons. Kleist's Amazon queen Penthesilea unhesitatingly embraces the ruthless governmental logic of Kleist's own time, which in the play is mythically shrouded in the "holy law" of Amazon society, with population statistics at its very core ("Whenever, after yearly calculations, / The Queen wants to recover for the state / What death has taken from her . . . ").[2] It is of note here that the Prussian Statistical Bureau, which had just been founded according to the plans of the liberal economist Leopold Krug, brought out its first representation of a state fiscal balance sheet for the fiscal year 1804–1805. It was, however, also the last one before the bureau perished, along with the fiscally balanced Prussian state, in the military catastrophes at Jena and Auerstedt. Because of his 1805–1806 assignment at the Königsberg *Domänenkammer* (provincial council), Kleist was intimately familiar with this avant-garde practice of state mathematics. In 1807–1808, as Kleist was working on *Penthesilea*, Baron vom Stein finally managed to reopen the *Statistische Bureau*, which was completed in 1810. This time, it would operate according to the civil servant Johann Gottfried Hoffmann's more conservative "statistical" conception,[3] which entailed the traditional design of embedding the numbers within a framework of narrative topoi.[4] At the point in the play when the above quotation occurs, after having explained the workings of the Amazon state, Penthesilea has to repeat her narration of statistical numbers word for word. Achilles, smiling distractedly, had not listened to her (verses 2033–35). Just as she herself, smiling in equal distraction, had not listened when Odysseus attempted to acquaint her with basic legal principle of war, the law of friendship and enmity. According to Odysseus's maxim, since she was presently attacking the Trojans, she would have to form an alliance with their enemies, the Greeks.[5] In these two tableaus from the play, we witness how erotic bedazzlement both divides and unites the two early modern conceptions of the state. The legal idea of the state defined above local sovereignty and the relationship between sovereign states (the play codes this concept in Greek and masculine terms) is juxtaposed with the conception of governmentality, the form of political organization that turns on itself and focuses upon care and discipline and, with them, upon the relation of the rulers and the ruled (the play expresses this concept in the Amazonian myth, and hence in feminine terms).

Eighteenth-century universal *mathesis,* the construction of discursive mathematics, remains self-evident in Kleist, as it both precedes and forms the basis for any possible "Kant crisis." An epigram published in the daily newspaper edited by Kleist, the *Berliner Abendblätter,* reads: "Humans can be divided into two groups: those who understand a metaphor; and (2) those who understand a formula. Those who understand both are too few to make up their own group."[6]

A sentence from Kleist's famous 1805 "friendship letter" to Ernst von Pfuel has been placed alongside this fragment: "I can find a differential and make a verse; are those not the two goals of human ability?"[7] Such statements presuppose an early Enlightenment, Wolffian writing space, a space composed of numbers and letters alike. They lead at least back to the 1760s, when the Berlin academy published its prize essay question on evidence in mathematics and metaphysics. The sentences of the fragment and from the letter, however, are not meant simply to be taken literally. In the letter, Kleist is describing his aptitude for a cameralistic appointment, which should provide a basis for living together with Pfuel. It sounds like a self-imposed demand to someone who wants to be able to be a good husband to his friend. The unity of poetry and calculation set up as an ideal in the fragment also has a false ring to it. After all, in the preceding first fragment, Kleist speaks of a mistake committed by Tycho Brahe out of ingenious fantasizing: a mistake, according to Kleist, that cost "more effort of mind" than the mathematical truth of a Kepler.[8]

Dissolving Poetological Probability

Kleist's "Unwahrscheinliche Wahrhaftigkeiten" (Improbable Veracities) plays out a kind of étude for the alphanumeric claviature of the counting as well as narrating evidence of probability. The brief text tells three stories. The narrator announces them as examples demonstrating that "probability . . . is not always on the side of truth." After the third story is told, one of the listeners identifies its source as the "appendix to Schiller's *History of the Revolt of the United Netherlands.*" The listener also mentions that in the appendix it was accompanied by a commentary: "the author notes expressly that a poet should not make use of this fact; the history writer [*Geschichtsschreiber*], however, because of the irreproachable nature [*Unverwerflichkeit*] of the sources and the agreement of the witnesses is compelled to take it up."[9]

Before Christian Friedrich von Blanckenburg—a Prussian officer, like Kleist and the narrator of the three "Unwahrscheinliche Wahrhaftigkeiten"—published his *Versuch über den Roman* (Essay on the Novel) in 1774, what in hindsight came to be called the "theory of the novel" had not enjoyed a status as its own form of writing and thinking. The main venue that afforded any space or opportunity for a discussion of the novel and its theory was the novel itself. We find reflections on the genre mostly in the guise of forewords to novels, digressions, or chapter prefaces. Friedrich Schlegel's *Brief über den Roman* (Letter on the Novel) was arguably first to give the theory of the novel a determinate place and a defined meaning. This becomes evident especially after Schlegel made it part of his larger account of the study of literature and criticism, *Gespräch über die Poesie* (Dialogue on Poetry), the defining 1800 publication that shaped our very notion of literary theory. In contrast, Blanckenburg's book seems like a vast essay that expands a special and coincidentally neglected topic of literary criticism into book length. "Unwahrscheinliche Wahrhaftigkeiten," with which Kleist filled an entire issue of the *Berliner Abendblätter* on January 10, 1811, offers still another instance of narrative theory, a theory that leads back to the times before Schlegel's philosophical and historical theory of the novel. The insights of Kleist's text do not, however, appear in an argumentative text before, within or about a narrative work; instead, they consist in a narration of narrative theory put to the test again by narration.

In the title, at the beginning, and at the end of the story of the three stories, various speakers offer three different versions of the theory of narration. Once again it is the differentiation between history, philosophical knowledge, and poetry from the ninth chapter of Aristotle's *Poetics* that stands in for a missing poetics of the novel. And once again the rhetorical sophism of improbable probability is superimposed on this differentiation. The title "Unwahrscheinliche Wahrhaftigkeiten"—a piece of novel theory formulated by the narrator even prior to the beginning of his text—clearly alludes, in its paradoxical form, to the rhetorical sophism of improbable probability: the German word for "veracity," *Wahrhaftigkeit,* is linked to the word for "probability," *Wahrscheinlichkeit.* On the other hand, by writing "veracity" instead of "probability," Kleist also reveals the connection to the passage in Aristotle's *Poetics* where history is distinguished from both poetry and philosophy. "Improbable veracity" characterizes the historical report vis-à-vis the two other positions in the triad.

At the beginning, the narrator—the narrator narrated in the text—insists that truth often conflicts with what is probable. His claim obviously applies the sophistical paradox to the distinction from Aristotle's *Poetics,* insofar as it refers to the relation between philosophical—and thus universal—truth and the probability of poetry. One of the listeners at the end of the stories offers a (fictive) citation from the work of the poet Schiller and the historian Curth on the Dutch struggle against Spain. According to this invoked authority, an event that must be accepted as valid by the standard of historical criticism can nevertheless be improbable. This insight applies the sophistical paradox to the same distinction in Aristotle's *Poetics* insofar as it refers to the relation between the probable—poetry—and the *kath'hekaston* (each individual event) of history. The probability of poetics (defined by the distinction between history, knowledge, and poetry) thus itself becomes the material for a story whose form, according to its title, is determined by the improbable probability of the rhetorical sophism. We might call this procedure a rhetorical analysis—a taking apart, isolating and testing—of rhetorical and poetological terms. With the second version of the paradox (according to which "probability is not always on the side of truth"), Kleist inserts a self-citation from his most famous novella, *Michael Kohlhaas,* into "Improbable Veracities."[10] While the novella, which Kleist had written about six months previously, mostly follows the style of a historical chronicle from Luther's times as it recounts a case of extreme intricacies in law and justice between the individual and state power, the episode containing the formula of improbable probabilities is an entirely fictitious addendum, and openly so. In this episode,

Kohlhaas, the hero erratically fighting with the authorities of his time over what he deems to be his rights, meets a gypsy who hands him a fateful prophecy about one of his mighty adversaries, the Saxon elector. The novella does not simply wrap up with this Romantic passage—which has always been suspected of being a superfluous deus ex machina addition to it.[11] Rather, this episode alone makes the whole narrative more than a mere retelling "from an old chronicle":[12] it enables the "chronicle" of Kohlhaas's legal case to be concluded and thus narrated as a novella or short novel in the first place. The paradoxical poetical formula signals the ending of *Kohlhaas* therefore as the moment when Aristotle's distinction between poetry, philosophy, and history—and hence the text's fictional status—is at stake. This circumstance is all the more ironic since the gypsy's mysterious note, which Kohlhaas swallows on the gallows before the eyes of the prince elector, contains a historical prophecy for the Saxon sovereign. The slip of paper clearly foretells the loss of the electorship and part of his lands after the defeat of the Protestants he leads in the so-called Schmalkaldic War against the emperor. Fiction is thus distinguished from history in the Aristotelian sense of particular events by a constitutive element of prophecy; and the prophetic power of determining "an ending" in turn provides the first basis for stories and histories to be told in a narrative sequence of events. Poetry's difference from history, the poetical formula which for its part is pervaded by the paradox of improbable probability, proves to be an elementary form of the theory of the novel. The theory of the novel will no longer be able to separate literature and history, despite their primordial difference. It will rather contain them together with their difference in itself.

Critics have at least twice noted that Kleist's short piece "Unwahrscheinliche Wahrhaftigkeiten" plays through a basic constellation of modern narration and, surprisingly, even narration as it is characteristic for the modern novel. Because of the unusual stakes and qualities of these studies and their authors, it may be in order to briefly address the two readings and their circumstances. In his 1936 Kleist study, Clemens Lugowski, an eminent if problematic German critic of the 1920s and 1930s, insists on the short text's importance. Recalling his reading has its own implications, since the study on Kleist is deeply marked by what might be called intellectual fascism. Lugowski's earlier book *Die Form der Individualität im*

Roman (Form, Individuality and the Novel) had been nothing short of a pioneering work. Rather than approaching the novel from a historical and philosophical point of view—the unavoidable perspective from Schlegel through Lukács—Lugowski instead offered a theory of the novel based on its narrative technique. This had been an entirely fresh and inspiring undertaking, a way of approaching the question of the novel that the narratologists of structuralism would later reinvent. The Kleist book, however, with its anti-Goethean title *Wirklichkeit und Dichtung* (Reality and Fiction) and its rather dubiously glorified style, goes on to make Kleist the antithesis to everything Lugowski's earlier narrative theory of the modern novel had championed. Lugowski's important finding in his first book had been to analyze all modes of creating novelistic context based on motivation—whether it be providence or chance, psychology or social milieu—into a network of narrative figures: rhetorical devices that motivate the novel's plot, secure the identity of its characters and establish the coherence of their actions. Long before Barthes, the characteristic "reality" of the novel thus appears to be achieved through narrative means. If that is so, novelistic reality can be seen as an artistic and even artificial construction. In his second book, Lugowski resumes this analysis and gives it a polemical spin: in the Kleist book, he invents a literary tradition of unmediated reality that he now plays out against the literary construction of reality, particularly in the French novel: a construction that relies on the rhetoricality of narrative motivation and its rationalistic and, as Lugowski now implies, alienating effect.[13] The title "Unwahrscheinliche Wahrhaftigkeiten" and its two variations at the beginning and end of the text clearly cast a powerful spell on Lugowski and strengthen his idea of the unmediatedness of reality. He is not interested in the fact that the title, opening, and conclusion of the text in effect strip apart and then reassemble the traditional poetological and rhetorical terminology in the course—or rather as the course—of narrating the three stories. The readings of Carol Jacobs and Cynthia Chase from the 1970s and 1980s, who also attribute a unique importance to "Unwahrscheinliche Wahrhaftigkeiten," are as distant as possible from Lugowski's Germanic act of critical violence. Quite in the sense of Paul de Man, these readers undertake an analysis of a rhetorical construction of reality, in this case the reality of consciousness through the "rhetoric of Romanticism." In their studies, Lugowski's theme of novel and anti-novel and his questions about narrative construction and historical event do in fact reappear. But the engaging commentaries of Carol Jacobs and Cynthia Chase investigate the paradoxical nonfunctioning of "Unwahrscheinliche Wahrhaftigkeiten" and its peculiar reality effect in place of, and in sharp contradistinction to, Lugowski's idea of an unmediated reality. Jacobs reveals this paradox in the doubling of *Geschichte* as "story" and "history" and juxtaposes the telling of stories (fabulation) with the discipline of writing history. The historical discipline distinguishes itself from storytelling, according to Jacobs, through its alleged control and mastery of the narrative process, while relapsing into fabulation at the same moment.[14] Chase for her part traces this condition again to a more fundamental condition of the underlying paradox. She is interested in the divergence of cognitive probability (tropology) and the fashioning of an utterance as the appearance of truth (performance).[15] The two readings analyze and reflect upon the abrupt opposition that intercedes between the construction of the novel and the singularity of the event, the very opposition that we recognize between Lugowski's constructivist first and ideological second essays on the novel and its theory. Like Lugowski, however, Jacobs and Chase both reveal themselves to be spellbound by the title, "Unwahrscheinliche Wahrhaftigkeiten." No one mentions the fact that the title and its variations in the text are not merely carrying out rhetorical chess moves. Instead, rhetoric and meta-rhetoric—rhetorical technique and the citation of traditional rhetoric—intersect in every one of these moves. It is not until the cited rhetoric and the citing rhetoric are seen in their interplay with each other that we may recognize the space that is simultaneously historical and constructs history—the space of "Romanticism," of the philosophy of history and the theory of the novel. This is the space where "story" and "history" substitute for each other in turn and where cognitive tropology and performance coincide.

"Unwahrscheinliche Wahrhaftigkeiten" recalls the probabilistic theory of the novel as it emerged in the eighteenth century, and it does so as if in an abbreviated formula with its very title. The question of how first to distinguish and then again bring together the event and the law of causation, or, in logical terms, the singular and the universal had been thematized and mastered by the figure of the improbability of the probable ever since Fielding and Wieland, if not already since Bodmer and

Gottsched. These writers effectually interrogate the framed immanence of probability for an observer who himself has a place outside of the game of probability and thus at the same time, possibly, within the game of a different probability. They thus redefine poetical probability in terms of the novel's "improbable probability" and, at the same time, render that paradox manageable through the differentiation of two levels (inside versus outside the game). The situation of such "deparadoxification," as we might call it, is indicated in Kleist's title and, at the end of the narrative, in the quotation from Aristotle's *Poetics*. On the other hand, "Unwahrscheinliche Wahrhaftigkeiten" also exposes the "deparadoxification" of improbable probability to a test that can succeed or fail—the test constituted by the narrative itself and its plot. Both narrator and listener in "Unwahrscheinliche Wahrhaftigkeiten" act out, by narrating and listening, poetic probability, and they do so in the paradox of improbable probability set up with the title by the authorial narrator. One question thus remains open about this story composed of three attempts to tell stories: namely, whether and in what manner the authorial narrator's control over narrating and listening endures in the story of narration, and hence what makes it the theory of the novel in a nutshell? At stake, therefore, are the status of the event and above all the issue of how the categories of the singular and the contingent should be related to one another for an appropriate understanding of the event. The *singularia* of history and the *kath'hekaston* of Aristotle's *Poetics* can only be characterized as events in the strict sense of book 4 of the Port-Royal *La logique*, when they are observed under the category of contingency. According to this logical interpretation, events can either take place or not take place, be or not be. In the context of Aristotle's *Poetics*, however, this is not at issue. Singularity is here opposed to that which exists unchanged and is known as substantial form or regularity. With his interpretation of game theory as probability, Arnauld only defines and brings into play singularity as a contingent event. Kleist's "Unwahrscheinliche Wahrhaftigkeiten" exposes the epistemological status of the event to the literary play of probability: "improbable probability" is one formula for the status of the event in literature, the figure of framing contingent events according to Arnauld; the distinction between history, knowledge, and poetry is the other, classical one.[16]

The Three Improbable Veracities and
Their Epistemological and Historical Profiles

The officer narrates three stories:[17] The first takes place in the Rhine campaign in 1792. A bullet apparently hits a Prussian soldier in the breast. To the amazement of all the officers, he continues marching forward. In the evening, the examining doctor offers an astonishing diagnosis: the bullet must have ricocheted off the soldier's breastbone and then slid along a kind of semi-circular groove that it furrowed beneath the skin halfway around his body until it ricocheted again off the backbone, after which it exited perpendicularly from the soldier's back. The narrating officer himself was a witness, though admittedly not of the event but rather of its effects. It becomes increasingly unclear, and thus less significant, whether he is claiming to have witnessed what happened directly or not. First, he notices the soldier, who has an entry wound in his chest yet continues to march forward; he then seems only to note how other officers observe the wounded soldier and order him to be treated. Finally, he reports the doctor's evening examination like an authorial narrator without any indication of how he found out about it. The *res gestae* are narrated in a story that is more and more indifferent to the testimony of the events on the part of the narrator.

The second story plays out in 1803 along the bluffs of the Elbe. A huge stone block is in the process of being felled by a slowly working mechanical contrivance. The narrating officer waits along with other curious onlookers for several days to observe the remarkable event. The block finally falls, however, while the sightseers are eating lunch at an inn. Not until evening are they able to view the site and reconstruct the circumstances of the fall. The event he concentrates on is the "result of this fall": a barge floating along the Elbe has been thrown out of the river onto the opposite bank. The officer presents himself emphatically as an eyewitness, but as an eyewitness only to the barge's final location on the far bank. To explain the connection between cause and effect, between the fall of the block and the displacement of the barge, the officer cites several relevant numbers. He mentions the width of the strip of land beneath the cliff that the block must have struck, the length and breadth of the barge, and the weight of its load. The numerical data create the atmosphere of a very particular kind of explanation of the connection between cause and effect, its modeling in an—albeit dubious—experiment: "just as a piece of wood remains on the edge of a flat vessel when the water on which it drifts is shaken." The officer also indicates what kind of effect is at hand here for a discussion in terms of experimental physics ("experimental physics" being the name of a lecture course Kleist took with a Professor Wünsch in Frankfurt on the Oder).[18] The basis for the officer's explanation of the barge's lifting is "the pressure of the air," thus evoking the family of—admittedly very different—experiments that Otto von Guericke and Robert Boyle had made famous during the previous century, which still had a prominent place in the textbook on experimental physics by J. C. P. Erxleben with which Kleist was familiar.[19] In the narrative, an experimental contrivance consisting of the shallow vessel filled with water and the piece of wood provides the model for the event whose effect is so carefully observed and whose circumstances are at least rudimentarily measured. In the experimental model, this setup should make the connection between cause and effect "probable." The central story proves thus to be *historia* in the old sense of the word, the sense of natural history. It shows paradigmatically what observation of nature means in modernity. The event to be wondered at is "itself" missed in this case. Instead, the observation turns into measuring something which is declared to be the effect of the event and then contriving a model for the sake of such measurement.[20]

The third story takes place in the seventeenth century during the Dutch war for independence from Spain. The duke of Parma, who is besieging Antwerp, has blocked the river Schelde with a bridge of ships. The citizens of Antwerp set loose several fireboats to float down to the bridge and destroy it. At the moment of the explosion, a young cadet at the side of the duke of Parma is thrown from the left bank of the Schelde to the right without suffering any injuries. Precisely here, where no personal testimony of the narrator is possible, the central event moves to rhetorical actuality. The officer narrates the explosion of the bridge and the catapulting of the cadet in the historical present. An investigation and explanation of the event is neither part of the story, as in the first case, nor added by the narrator as an experimental analogy, as in the second one. Instead, the actuality of the historical in the historical present is itself an explosion of narration. The narrator takes his leave of the company after ushering in this effect—the narrated effect and the effect of his narration—and he leaves it to one of the listeners to add a claim of historical probability. The historian, according to the testimony of the listener, had recorded this story in his history book because the number and quality of the witnesses—the classical warranty of probability—proved the value of the probable.

Most interpretations of this text only pay attention to the relation of the first to the third story, of the "history" to the "story," as Carol Jacobs puts it. They can thus remain in a space of purely discursive, humanistic probability. Such a reading gives short shrift, however, to the conflict between the singularity of event and the universality of knowledge, even if we take only these two stories into consideration. A reading that remains in the compass of story and history, in other words, fails to comprehend the conflict between the measure of "complete trust" that the officer imputes to the anecdotes and the disbelief of the listeners. It is certainly easy to overlook the fact that with the probability of testimony at the end of the third story, we are dealing with a question of measurement. In the second story, however, this connection becomes conspicuous as soon as we notice its scientific theme and tone. Once again with this second, central story, Kleist completes the ensemble of early modern *historiae* that take as their subject singularities of all kinds—not only from history proper but also, especially, from natural history.[21] Once again here "Unwahrscheinliche Wahrhaftigkeiten" follows Goethe's *Conversations of German Refugees*, to which Kleist perhaps even alludes with the dating of his first story in the Rhine campaign of 1792. Only from the point of view of the experiment are the categories of cause and effect in their specific modern meaning inscribed in the relation between the particular event and the regular causality as represented in the fictional story. And only in the repetitive structure of the experiment does the singular simultaneously become contingent, something that can be thus or different, can take place or not. Only through the central position of the observation of nature and its experimental simulation, that is, can the paradox of improbable probability even aspire to govern the difference between the individual, the universal, and the probable.

Thus, taking the second story into account, one can only concur with Carol Jacobs and Cynthia Chase as to the dysfunctionality of the classical discourse of the probable in "Unwahrscheinliche Wahrhaftigkeiten"; at the core of Kleist's narrative triptych, the computation of chance becomes probability theory.[22]

The Romantic Ending of the History of the Three Improbable Veracities

Modern probability theory starts by isolating an event, then construes it as a center around which factors and circumstances are organized that make the event more or less probable. In this core, narrative and experimental, topical and numerical probability communicate with each other. In this disposition—which Arnauld was the first to conceive of rhetorically and mathematically—the event and its probability are bound up with an underlying improbability. It is the type of improbability that also characterizes the constitution of the game of chance, the artificial arrangement in which contingency can first be observed. In Kleist, the event makes its appearance in this same manner of improbable probability, and it does so in all three instances with the sudden eruption and violence of its occurrence: as a gunshot; the toppling of a stone block from a cliff; an explosion. In none of the three stories, which all circle around the probability of an event, is the event actually present. In the first story, though the narrator himself "notices" that the soldier is wounded, he never sees him shot. The same seems to be true of the officers who "couldn't believe their eyes at this strange sight." In the second story, the circumstances surrounding the narrator's missing his chance to witness the event are carefully orchestrated: he hikes to the bank of the Elbe in order to see the block quarried out of the side of the cliff, but when it finally falls, he is sitting at lunch in an inn, and is clearly in no great hurry to reach the site. He does not even mention other witnesses. As if to substitute for what he fails to see, the officer emphasizes all the more, with deictic reference to himself, his autopsy of the effects left behind by the event. "These eyes saw [the barge] in the sand—what am I saying? . . . they even saw the workers who were laboring with levers and cylinders to get it afloat once again." In the third story, which derives from historical tradition, there can be no direct testimony of the narrator himself to begin with. After the departure of the narrating officer, when a listener pipes up in defense of "the irreproachable nature [*Unverwerflichkeit*] of the sources and the agreement of witnesses," he is clearly referring only to the formal quality of the tradition that allows it to be taken up in a work of history. What began in the second story is thus continued here: the blatantly lacking testimony is worked out as if in an act of transference in the relation between narrator, narration, and audience. The necessarily missing presence of the historical event is replaced by the historical present. The rhetorical device in this instance even stages the event itself as an event in the performance of the speech: "At that moment, gentlemen, in which the vessels float down the Schelde to the bridge, there stands, observe well, a cadet officer on the left bank . . . : now, you understand, now the explosion takes place, and the cadet . . . stands on the right bank."

The word "now" refers to the point in time of the historical present of the story and in the actual presence of the narration of the story. In this "event itself," it seems, narration and what is narrated, the performance of the narrating speech and its cognitive content, coincide. But precisely this coinciding is revealed as the deception of hermeneutic belatedness: "Did you understand?" the officer repeats. The listeners have indeed committed an act of—hermeneutical—understanding: they give voice to their disbelief, just as the officer had foretold to them at the outset. Thus, though the stories may not seem successful, narration by no means fails at this point. On the contrary, the contract between narrator and audience precisely included the improbability of veracity as its premise. The improbability of the perception or comprehension of the event "itself" is inconclusive; and the acknowledgment of such inconclusiveness is the very presupposition of narrating and listening.[23] Nothing makes that scenario clearer than the acting out of the missed event "itself" in the story, in the narration, that is, of the history that has become the story.[24] In contrast to what readers such as Lugowski persuaded themselves to see, Kleist pointedly does not insist on an unmediated reality—either in this passage or anywhere else.

In "Unwahrscheinliche Wahrhaftigkeiten," the probabilistic event is both a dramatic singular event and a contingent one marked in a series of repetitions. The phantasmatic present, in which the explosion occurs and the body of the cadet is thrown from one bank to the other in the third story, clearly combines the explosiveness and military procedure from the first story with the spectacle of nature to be pursued experimentally from the second story. In the second story, the invisibility of the event "as such" creates as much of a spectacle as do the seemingly coincidental causes that bring it about. The officer makes the trek to the riverbank day after day in hopes of watching the stone block fall; then at the decisive moment, he happens to be at lunch. This tale amounts to a science story par excellence: it is a story from the laboratory of the experimental physicist. What by chance remains unseen—the fall of the boulder to the strip of land beside the river and the barge's propulsion to the far bank of the Elbe—is mirrored in the series of observed events in an experiment with a shallow bowl of water in which a piece of wood is floating. This series of observations, however, has been subject to calculation at least since Thomas Simpson and Johann Heinrich Lambert in the eighteenth century, and is for its part constituted by a succession of contingent events.

The series of observations is offered as a substitute for the coincidental chance of the missed original event. And in this series and its specific order, there is no longer any place for an original event that could possibly be missed. The first story displays an inconspicuous style in narrating the improbability of the probable event. The narrator is present and sees the soldier who has been shot in the chest; other officers cannot "believe their eyes at this strange sight." The text does not suggest that the narrator or the other officers observed the moment of the shot's impact; yet neither does it exclude the possibility. The question simply does not matter in how the story is offered to us. Already in this first story we are confronted with an observing gaze that operates after the fact. Certainly, the doctor's close focus does not, as in the second story, isolate cause and effect in order, then—experimentally—to relate them. Instead, he reconstructs a coherent chain of events—that is, he tries to narrate the story of the bullet's path through the soldier's body. The doctor's examination takes place here and now, at the present time. He proffers his tracing of the bullet's path as a judgment that others may consider false or nonsensical and that he himself can take back, but it is and remains unique in the act of his utterance and, as the case may be, of his written record. No blatant switch takes place here from the missed event to a constructed series of events. Instead, there are two stories, and the latter, the doctor's narrative reconstruction of the bullet's path, is meant to explain the former, the officer's account of the curious wound. What is clearly and superficially true of the explanatory story counts equally, though less obviously, for the former one, too. The story defines the scheme for the isolation of the event, and, with it, for the analysis of its probability: it defines an explanation that is reconstructive from the beginning. For this reason it is impossible to perceive by sight whether one sees the event or not. The isolation of the event as an event is—perhaps already in viewing it—a matter of the narration of the story and of its structural belatedness.

What happens, we may ask further, when Kleist connects the first two stories as examples of the same thing and brings them to a point of culmination in the third, the historical story? Contingent improbability of the probable event radicalizes and limits the narrative improbability of the probable event. It radicalizes it in its core structure, because contingency presupposes seriality and repeatability. Contingency, however, can only be attributed to events for which the possibility of repetition can be asserted. Only under this assumption can we say that such events can occur or not occur. This characteristic splits the identity of the singular event. In contrast to the Aristotelian *kath'hekaston* of history that fiction transforms into the poetics of probability, the singular event that is defined as contingent is no longer simple and one in itself. On the other hand, defining events as contingent requires the implementation of measures that introduce and control the possibility of repetition. A field must be delimited in which events can systematically occur or not occur, and positions must be established from which to observe such events in a competent manner. Frames of overall systems must develop in which such positions and the probability values of the occurrence or nonoccurrence of events can be stabilized. Achieving this goal of systemic observation and the specific hermeneutics of the observer presupposes the definition of a boundary between the observer and the field of observation. The analysis cannot stop here, however. Once the Aristotelian configuration of singularity and law of causation is transformed into the fields of contingency and observation, the (in each case) singular event from Aristotle's *Poetics* returns, as it were, and it makes a subversive comeback. Each field of contingent events and their observation again requires the narrative of its own development and functioning. Narrative theory's improbability of the probable event, however, precisely disallows any hermeneutic subordinations, systemic hierarchies, or systemically defined positions of observation. It rather brings into play in each case a new figuration of the event, one that, for its part, precedes any computation and measurement. The narrative figuration of the event remains, far beyond the question of the "blind spot," a presupposition for and an undermining element in the construction of the contingent event and its frames of observation.

What we see is an intertwining and conflict of narrative and contingent (im)probabilities of events. More precisely, we could describe this process as an interlocking harmony and antipathy between the narration of the story, which takes place in each case here and now, and the narrated stories, which only the orders of possible worlds can allow to be thought. The orderly sequence of the three stories, the three forms of *historia* as story, experiment, and history, pervades and traverses the tale that is told about what happens between the narrator and the listeners. It is this latter story that we actually read on the surface level of the text. This story of the three stories told develops from the first-person narrative at the beginning, which strictly separates author and audience, continues via the partial contact between the narrator and the listeners in the experiment story, and culminates in the phantasmatic acting out in the case of the history story and its narrative present.

To follow what happens between the narrator and the audience, we must take notice of a kind of reversal from the position of narrator to that of listener in "Unwahrscheinliche Wahrhaftigkeiten." From the onset, the narrator acts like the reader and hermeneutical expert of his own stories, while the listeners seem mentally to assume the position of the narrator at the genesis of narration. The narration of the three stories takes place, as far as the narrator and listeners are concerned, with the following peculiar premises: "The officer . . . explained . . . beforehand that in this particular case he made no claim for their belief [that of the listening company]. / The company in return promised him their belief in advance."[25]

Thus we see an exchange of roles "in this particular case" in terms of settlements made "beforehand": the disposition generally assumed proper to the addressee of a persuasive speech, disbelief or ignorance, is taken up here by the speaker. The listeners in contrast take the speaker's initial intention to persuade or convince his addressees as the premise of their listening. The two sides promise or pretend that they want to avoid the conflict characteristic of their rhetorical situation, but they both effectually precipitate it in a radicalized form.

Neither in the form of knowledge shared by the two parties nor in the form of a contract—an act of establishment to which both sides commit—is there any kind of agreement above or beyond this disunity from which both sides proceed. The company "exhorted" the officer "to speak, and listened." The apparent final failure of the narrative after the third story, with the disbelief of the listeners and the provocative departure of the narrator, in fact ensures the prophesied disunity. Discord is the point of departure for both narrator and audience, but no sublation of the conflict by any knowledge or agreement is allowed for. The inconclusiveness of the truth of the event is not (or not only) the problem; this very indeterminacy also and simultaneously constitutes the structure of narration.

To take the readings of Carol Jacobs and Cynthia Chase even further, we might say that the situation of presupposed disunity is indeed characteristic of the hermeneutical structure of the whole text. The narrator resolves to keep the stories free from any persuasive semblance of truth in order to make them even more unassailable as the pure cognitive appearance of truth—a truth that, in this case, is identical with the contingency of events. The audience, however, has already conceded in advance what no narration can admit: the cognitive probability of what is narrated, with its observer-specific hermeneutics, cannot help but be fulfilled through narrative performance—in particular, in the vein of a persuasion that is inherent in all "lifeworld" communication. There is no truth in the language of history—not even in the theory of probability—that has not already lodged a claim to help truth come into appearance. Indeed, in order for any account to be accepted as true, it requires a splendid and effective appearance. There can be no historical or scientific truth untainted with the persuasive tactics of rhetoric. At the same time, however, mathematical probability and the rhetoric of probable narration do not share any common measure. Verisimilitude, or probable narration, continuously brings into play new figures that then make a measure of probability possible: a measure, moreover, that the theory of mathematical probability already presupposes as a given. The probability of and in the narration of "Unwahrscheinliche Wahrhaftigkeiten" continually produces its own contradiction. In citing mathematical probability in philosophical discourse, and thus implying the split between the probability of mathematics and the plausibility of philosophy to be complete once and for all, Kleist takes Kant's quasi-transcendental "game" a step further here. But more importantly, he also moves further into the direction of Lambert's "false concept." Lambert had legitimized the purely operational nature of probability as an intermediate phase: probability would only be provisional, a remainder of incomprehensibility in the metaphor, and an operative formality of the irrational number; always, however, in the end, it would eventually be restored to the identity of notation, *numeratio*, and full meaning. The disunity, shared but not agreed upon, in the hermeneutical functioning of Kleist's narrative triptych is a way to integrate and further develop both Lambert's and Kant's responses to the challenge of probability.

For the sake of clarification, it is worthwhile to return once again to Kleist's citation of his own major novella *Michael Kohlhaas* in "Unwahrscheinliche Wahrhaftigkeiten." The Saxon prince elector in the novella has just discovered the existence of a mysterious prophecy written on a piece of paper that Kohlhaas possesses. In this note, a gypsy in the Brandenburg city of Jüterbog has written down her prediction of Kohlhaas's tragic future after his negotiations with the Brandenburg elector. The chamberlain receives an order to procure this note for the elector. To this end, the chamberlain searches around for an old gypsy who resembles the description of the Jüterbog seer. He hopes that Kohlhaas, who has only seen the gypsy momentarily when she unexpectedly hands him the note, will be deceived by the similarity. Introduced as the prophetess of Jüterbog, the second gypsy has orders to demand the slip of paper from Kohlhaas. The novella's conclusion is "Romantic" also in the traditional sense—common up through Kant—of romance, the epic genre of strife and love preceding the modern novel. In "Unwahrscheinliche Wahrhaftigkeiten," it depends on the final twist that the gypsy whom the chamberlain uses in order to deceive Kohlhaas is in fact the selfsame gypsy who originally gave him the note. At this point, the citation of the improbability of the actual event resurfaces, and the narrator of *Kohlhaas* positions himself in the same way that the narrating officer in "Unwahrscheinliche Wahrhaftigkeiten" does vis-à-vis his audience: "and just as probability is not always on the side of truth, something had happened here that we are obliged to report, though we must allow anyone who wishes to doubt it the freedom to do so."[26]

The reader of *Kohlhaas* can no more know what to do with this freedom than the listeners in "Unwahrscheinliche Wahrhaftigkeiten" do. It is quite unclear how far this freedom should extend. If the reader does not believe that the chamberlain's gypsy is identical to the Jüterbog soothsayer, then he will have a hard time believing in the conversation between her and Kohlhaas and all the further details of the concluding episode. The narration reports how the chamberlain informs the supposedly false gypsy about all the circumstances of the prophecy in Jüterbog, so that she will be in a position, though an interloper, to parry every turn of conversation with Kohlhaas. She could also, as indeed happens, become a double agent, switch sides, and warn Kohlhaas instead of defrauding him at the duke's behest. But the supposition that she might in fact not be the real gypsy introduces a new and even greater inexplicability and unaccountability for the reader. Nothing differentiates this conclusion of the story from the previous ones, which report an occurrence in the *intentio recta* of narration. If the reader accepts the proffered freedom not to believe the improbable identity of the final gypsy as the Jüterbog fortune-teller, then he not only places a great burden of inexplicability on the psychological and motivational ending of the story but also undermines his trust in the *intentio recta* of the narrative in general.

What is the improbable truth that is at stake? It is improbably true that, out of all the old gypsies, the chamberlain chooses the actual prophetess from Jüterbog. This coincidence, however, does not only show the value of an improbable probability in the world of the ascertainment of facts. The episode is already composed of appearance and truth in its construction: Kohlhaas is meant to be deceived by the semblance of standing before the actual Jüterbog soothsayer, who—improbably—really is the soothsayer. What the chamberlain plans tactically and persuasively as the mere appearance of truth is actually the truth, but in such a way that it can never appear to him as true. This extreme instance of a paradox is possible only because of the politics of similarity that is at play here: the harder the chamberlain tries to find a gypsy similar to the real one, the less will he be able to recognize this real one when she is standing before his eyes. The appearance of truth with which he is operating makes him incapable of seeing when truth appears in and as this appearance. The role of the gypsy as a double agent brings this intersection of all differences to a climax. Exemplifying the open-ended logic of the probable in Romantic narration, the novella's conclusion—which is precisely what makes it Romantic—is thus staged by the intertwining of harmony and conflict in the appearance of truth.

Conclusion

Lambert's logical and metaphorological investigation into the term "probable" (*wahr-scheinlich*) and Kant's quasi-transcendental theory of play are complementary. They exclude each other but can be understood reciprocally as theorizing what the other one does, or as completing in fact what the other sets up in theory. They thus offer a final commentary on Leibniz's repeatedly renewed announcement of *logica probabilium*, the theoretical enterprise that culminates in the concept of equipossibility. Jacob Bernoulli had already used it methodologically, but without, however, discussing its philosophical meaning: according to Bernoulli's mathematical insight, only what can be traced back to the situation of the equal possibilities of the occurrence or nonoccurrence of an event can be calculated in probabilistic terms. Leibniz, on the contrary, had wanted to protect equipossibility, as a cosmological concept of free will, precisely from its merely statistical implementation: in the Leibnizian concept, logic and ontology are so tightly intertwined that the idea of a causality out of free will corresponds to the principle of logical equipossibility. Lambert's appearance of truth and Kant's distinction between the probable and plausible, each in its own way, remove the substantial content of this ontological construction without erasing its former place and function in philosophy. Lambert formulates a theory of mere—operational—appearance in probability. Kant, meanwhile, in practice applies his statistical examples in the context of aesthetics and the philosophy of history in this manner of mere appearance; but in theory, he shuns once and for all the unity of probability. For Kant, the probability that can be claimed discursively and the probability that can be calculated mathematically can by no means be one.

What then, we may ask, is the status of a probability theory that produces results by its mathematical operations but no longer knows the meaning and theoretical status of this selfsame probability? The problem has lingered implicitly ever since Jacob Bernoulli and came into the open with Lambert and Kant. The so-called epistemological problems of probability, which philosophical mathematicians since Condorcet have regarded as a problem in need of a solution, have become a privileged way to address the puzzle. These thinkers asked whether the cognition of the probable is a priori or a posteriori; in other words, whether it is a matter of the frequency of events or the structure of judgment. Such questions result from the process of the interpretation, after the fact, of gaming theory as probability. While probability in the logical and rhetorical sense is clearly either subjective or objective, a characteristic of chance or a state of mind of whoever is judging, the same alternative does not apply to the game and the chances in gaming. When Ian Hacking convincingly assumes that the two interpretations, subjective and objective, must both be always possible for probability to emerge, he indirectly draws the consequences from the primordial ambivalence that resides in the interpretation of gaming theory as probability.

Hence the debate about whether the meaning of probability is "subjective" rather than "objective" is the very symptom of a probability theory that develops out of its context in philosophy and endeavors to acquire its own, mathematical, status. Such discussions are signposts for the only path that leads out of the thorny question of how to understand the *logica probabilium* and its entanglement with philosophy and mathematics, argumentation and calculation. The only remaining way to proceed is simply by letting the question go and ceasing to feel obliged to give an answer to it.

A good case in point for this kind of forgetting the question one began with is how Siméon-Denis Poisson is supposed to have finally distinguished "objective" from "subjective" probability. The differentiation he proposes clearly has to do with the insight into the difference between the example of the game and the representation of probability, rediscovered as it were one hundred fifty years after Jacob Bernoulli. Poisson defines probability first subjectively, that is, topically: "The probability of an event is the measure of reasonable grounds [*la raison*] we have to believe that it takes place."[1] This includes the possibility that different people, according to their individual status of knowledge, may have variously strong grounds for or against believing in the occurrence of an event. With this backdrop, Poisson hits upon his famous differentiation between *chance* and *probabilité*, which he admits are "almost synonymous . . . in common speech." However, he employs *chance* to refer to the "events themselves . . . independently of the knowledge we have of them" and *probabilité* to refer to judgments based on what people claim to know. Within only two further paragraphs, Poisson is making the newly distinguished *chance* into the presupposition of chance calculation necessary for measuring probability theory's *probabilité*: "The measure of the probability [*probabilité*] of an event is the ratio between the number of cases favorable to this event to the total number of cases that are favorable or the contrary, and all equally possible, or that all have the same chance [*et tous également possibles, ou qui ont tous une même chance*]."[2]

"Subjective" *probabilité* and "objective" *chance* are to be juxtaposed only as long as we do not actually engage in measuring probability in games of chance. This measurement, then, is nothing other than the establishment of a game on the basis of justice and the equipossibility of the indicative event for winning or losing. In this establishment—or this introduction of an "objective" sphere through and for "subjects"—the "objective" *chance* provides the unity of a standard for the "subjective" *probabilité*. When through these argumentative moves Poisson first disassembles and then reassembles the interpretation of gaming theory as probability, the mathematical theory becomes self-reflective. In Poisson's very definitions, probability simply but perfectly redescribes the process of its own constitution.

This self-dissolution and recomposition of probability made its first glorious entrance onto the public stage of learning and science in the lecture on probability that Laplace gave in Paris in 1795, the inaugural year of the *École normale supérieure*.[3] This is the moment when probability theory, with all its great uses for human knowledge, society, and the nation state, first makes its claim to be an autonomous mathematical theory; it was also the point when, with the start of the *École normale supérieur*'s first cycle of lectures, the French Revolution showed that it was consolidating. Laplace later expanded the text of his lecture in the 1812 *Théorie analytique des probabilités* (Analytic Theory of Probability), and again in the 1814 *Essai philosophique sur les probabilités* (Philosophical Essay on Probabilities). In the 1814 version the emblematic sentence reads: "The theory of chance consists in reducing all the events of the same kind to a certain number of cases equally possible, that is to say, to such as we may be equally undecided about in regard to their existence, and in determining the number of cases favorable to the event whose probability is sought."[4]

This sentence—presented to listeners later on, in the course of the second lecture, as the first principle of probability[5]—equates the "objective" game with "subjective" probability by highlighting the precise point of methodical equipossibility. The careful wording avoids a formulation that would allow the possible and the probable to meet directly. Instead, they are joined under the aegis of the pragmatic, the probabilistic action: equal possibility appears as the result of a methodical act, the act of reducing occurrences in the world to equally possible cases; and the probability of belief manifests itself in the indecision between choices of what one should do or believe. The "that is to say" that divides and binds possible and subjective probability articulates how to procure strict hesitation or undecidedness for methodological reasons.[6] Such an articulation, for the first time, enables possibility and probability, the old Leibnizian equation of *possibilitas* and *probabilitas*, to connect with each other. While for Leibniz, this connection only had its place in brief allusions to the juridical practice of conditional contracts and to the great unattainable truth of creation, Laplace includes it right in the initial definition of probability. The definition's wording thus repeats its own history and establishment.[7] This shift from foundational history to resurfacing explication is related to a different, previous move that had been fundamentally cosmological. Laplace is known to have worked throughout his career with a strong version of determinism for both the physical and moral (or, in today's terms, psychological) worlds, which were thought to be derived from a single formula, just as Leibniz, the author of the *Théodicée*, had once imagined it. In the *Essai philosophique*, Laplace actually identifies Leibniz by name and quotes the "axiom known by the name of the principle of sufficient reason." But the celebrated indebtedness to Leibniz reveals only half of the truth. The "axiom" of sufficient grounds for Leibniz had been the formula of free and personal creation, a personal freedom that as such, however, left no other choice than the creation of a particular world, namely, the best of all possible ones. For this reason, already in this world, only a classicist design can be read in Leibniz as the individual style of its creator, and even this design is nothing other than what should in any case be derived from the formulas of the geometer God. But Laplace eradicates even this last trace of the world's legibility.[8] Characteristically, in the version of the *Théorie analytique des probabilités* in which Laplace is no longer addressing either the audience at the *École normale supérieur* or the larger public at which the *Essai philosophique* was aimed, he omits Leibniz's name and the principle of sufficient grounds. In its place, he introduces the famous total intelligence later known as "Laplace's Demon." This comprehensive intelligence embraces two and only two things: a complete set of rules for the world system's construction; and all relevant data from a single moment of that system. From these two points, the formula of the entire world system, including all its possible changes and states can be derived, and Laplace's superior intelligence is this world's ideal reader.[9] The total intelligence, though later dubbed a demon, is no longer characterized in theological and metaphysical terms, and it is no longer marked by perfect insight into truth. The total intelligence is characterized, rather, by perfect engineering and an unlimited capacity for the collection and preparation of data. Consequently, this superior intelligence is now no longer simply contrasted with what actual empirical people know, whose incomplete insight into things then can be measured in degrees of probability as had been the case during the era of Leibniz and Jacob Bernoulli. "Probability is relative, in part to this ignorance, in part to our knowledge."[10] Clearly Laplace is no longer thinking only of individuals, whose limited insight is insufficient to comprehend the cosmos and whose ignorance can by probability theory still be brought to a compensatory epistemological triumph. Instead, he is thinking of data-collecting entities and institutions. Of course, individuals, being all too human, may not be in a position to construe world formulas. But what fundamentally and forever differentiates them as mere individuals from the total intelligence is the circumstance that they could never collect all the data necessary and convert them into complete data records. Statistical bureaus, such as the institutions founded in Paris and Berlin, can offer a modicum of assistance with this problem in the fields of politics and society.

However, a concluding section in the *Essai philosophique* that provides a transition to the *Théorie analytique* follows immediately after the passage on human individuals as deficient agents of data collection. It constitutes a final word of sorts: first, as concluding *captatio benevolentiae* [appeal for approval] addressed to the audience of the *École normale supérieur*; and second, as an argument conclusively distinguishing an "analytical theory" from a "philosophical essay." In fact, in the 1814 publication of the *Essai philosophique*, the latter becomes a mere introduction to the former. It is, as we shall see, equally a last word on rhetoric, spoken before or even precisely in the turn with which probability diverges into an autonomously mathematical theory.

The relation between ignorance and knowledge that characterizes probability in general, according to Laplace, also structures the situation of a public presentation before a large number of listeners in a particular way. The speaker achieves different degrees of credibility according to how well the listeners are informed about the matter at hand and to what extent they hold the expressed representation of the matter to be trustworthy and competent. The measure of their belief in the speaker can be calculated according to the ratio of their ignorance to the knowledge they impute to the speaker. With regard to this law of probability for Laplace, there is no way of overcoming an implied lack of knowledge and no need to do so, as is the case with the "self-incurred immaturity" that in Kant's famous definition will finally be mastered by enlightened insight. For Laplace, the ignorance and superstition of times past were just as much based in the rhetorical misuse of probability as, currently, the present general enlightenment stems from using probability well: through advancing science in general and probability theory in particular. His preamble to probability at the *École normale supérieur* proclaims it to be the truth of rhetoric. Rhetoric, we are to conclude from Laplace's words, is valid only as long as we believe in probability instead of developing the means of measuring it: but the days of such a lack of computing power, the days of an old regime of knowledge, are about to come to a quick end just when Laplace is lecturing to the first class of the *École normale*. To this end, the definitions with which the second lecture begins will be of considerable assistance. The first and most fundamental of these definitions, in fact, has already been heard, though without having been recognized as a principle. With this move and transition, rhetoric persuades itself against itself and into becoming mathematical probability. The mere decision to believe this or that, based on an assumed deficit of one's own information as compared to the speaker, is to be converted into the permanent reduction of situations, things, and matters to the core of contingency and indecision, to the situation of undecidability between yes and no, occurrence and nonoccurrence. Already within rhetoric, probability has thus begun to set itself up in rhetoric's place and to become the new rhetoric or communication strategy of a data-collecting society. This does not mean merely that this passage in the *Essai philosophique* is a holdover from the old *logica probabilium*, which the *Théorie analytique* will demonstrate to be eradicated. It also means that the old question of how the contingency of chance calculation is connected with the meaning of the probable continues to pose itself further in a self-reflexive theory of probability, and finally even in a purely axiomatic version of mathematical probability. Like the history of the topics and of literature, the history of science distinguishes between a founding phase of probability since Pascal, classical probability since Jacob Bernoulli, autonomous statistical probability since Laplace, and the axiomatization of probability around the beginning of the twentieth century.[11] And yet the internal, constitutional history, the act of determining the relation of contingency and sense, structure and meaning, is capable of no such story with a beginning, middle, and end.

Notes

Unless otherwise indicated in the Bibliography, translations are by the author and the translator.

INTRODUCTION

1. See Cassin, *L'effet sophistique.*
2. In his *Probability and Opinion*, E. F. Byrne provides the authoritative discussion on probability qua opinion in the ancient—Aristotelian—tradition and modern (mathematical) probability theory, but he makes no comment on or defense of the way he connects the analytical distinction between ancient verisimilitude and modern probability with his concern to resurrect aspects of ancient—rhetorical and dialectical—verisimilitude in modern probability theory.
3. See Krüger, *Probabilistic Revolution.* However, Krüger and his collaborators understand the "probabilistic revolution" to include the entire development of probability in modernity, whereas this study addresses only its first emergence in the late seventeenth and full eighteenth centuries.
4. For the scientific history of probability, see Ivo Schneider, "Entwicklung des Wahrscheinlichkeitsbegriffs"; for the literary history of probability in the eighteenth century, see Patey, *Probability and Literary Form*; for the philosophical history of the concept of probability, see Cataldi Madonna, *La filosofia della probabilità nel pensiero moderno.*
5. Blumenberg, *Paradigms for a Metaphorology*, chap. 8, "Terminologization of a Metaphor: From 'Verisimilitude' to 'Probability.'"
6. These remarks only touch the surface of Blumenberg's argumentation, however. While at first sight Blumenberg in fact seems to think of metaphor and concept as opposed terms, a more precise reading of his work shows that he himself had clearly indicated a more nuanced view. According to the more refined and adequate version of his theory, metaphor and concept are rather moments in a continuous process of terminologization in the history of science; for such a reading of Blumenberg see Campe, "From the Theory of Technology to the Technique of Metaphor."

7. Blumenberg, "Concept of Reality and the Possibility of the Novel."

8. Barbara Shapiro offers the best overview on the themes of probability outside of mathematics in her *Probability and Certainty*; for the thematic fields around mathematical probability, see Daston, *Classical Probability in the Enlightenment*; for the genesis of a theory of the novel in the modern sense, see Kimpel and Wiedemann, eds., *Theorie und Technik*.

9. Foucault, *Order of Things*. The terminological history of knowledge undertaken in this book, however, is different from Foucault's analysis of the formation of discourses. This study does not intend to examine discourses in their entirety, but rather individual terms that presuppose the cohesion of a discourse.

10. One might argue that reality as represented in the novel is its first and foundational "mythical analogue"—the distinct coherent substratum, on which the narrative forms of the novel rely, and from which they are generated. For the notion of the "mythical analogue," see Lugowski, *Form, Individuality, and the Novel*.

11. Jacques Derrida resumes the phenomenological distinction between historiographic history and the historicity of origin and constitution in his *Edmund Husserl's "Origin of Geometry,"* 132. While stating the mutual exclusivity of history and historicity and, in particular, the independence of historicity from history, Derrida surprisingly opens the window, as it were, on history: "This does not mean . . . that it is impossible or useless to try an extrinsic and 'parallel' historical approach to this subject." What then is the possibility and use of such extrinsic history, history that is "conducted with the certainty that everything is spoken of then except the reduction itself"? This book tries to make methodological use of Derrida's improvised insight, and to present and think through the relation of unrelatedness between constitutive history (historicity) and historical history.

12. Remarks on probability and the history of its meaning are scattered throughout the work of Niklas Luhmann; for social history, see Luhmann, *Risk*; on distinguishing art from science, id., *Art as a Social System*, chap. 7.2.

13. Krüger, *Probabilistic Revolution*, convincingly demonstrates the rich variety of uses (see esp. vol. 2, *Ideas in Science*).

14. Ewald, *L'État providence*, 143.

15. Link, *Versuch über den Normalismus*, 2ff., 132–141, articulates this opposition twice. First, he contrasts the norm of the law with the normalism of social and political order. Whereas the law is expressed in imperative declarations, normalism is based on probabilistic and statistical practices. Then, within the field of normalism, Link distinguishes "protonormalism," which uses arguments of usual behavior for formulating binding norms, from "flexible normalism," which allows for a free play between standard values and deviation.

CHAPTER I

1. Bernoulli, *Ars conjectandi*, 109. Mises, "Grundlagen."
2. See Schneider, "Entwicklung des Wahrscheinlichkeitsbegriffs"; Hacking, *Emergence of Probability*; Daston, "Mathematics and the Moral Science" and *Classical Probability*.
3. See Shapiro, *Probability and Certainty*, and for literary history, Patey, *Probability and Literary Form*.
4. See Schneider, ed., *Entwicklung der Wahrscheinlichkeitstheorie*.
5. See Hacking, *Emergence of Probability*, esp. chap. 2.
6. Blumenberg, *Paradigms of Metaphorology*, chaps. 1 and 8.
7. For a contrasting view, see Byrne, *Probability and Opinion*.
8. Daston, "Rational Individuals Versus Laws of Society."
9. Link, *Versuch über den Normalismus*.
10. See "Digesta," bk. 11, pt. 5, "De aleatoribus" in *Corpus Iuris Civilis*, ed. Behrends et al., 3: 22–24.
11. Pseudo-Cyprian, *Liber de aleatoribus*, in *Patrologia Latina*, 2: 907.
12. Augustine, *City of God*, bk. 5, 1–11, 187–206.
13. Games (of chance) and plays (or spectacles) are often discussed together in moral and theological literature. The German word *Spiel* is even a homonym capturing both meanings. Despite their obvious difference, games and plays, the nature of contingence and the practice of performing, seem to belong together. In part II, we shall come back to this linked aspect of plays and games in the chapter on Kant.
14. See the remarks on Kant's use of the word *Spiel* in pt. II, chap. 14 below.
15. Nietzsche, *On the Genealogy of Morals*, 1.15, 48–52.
16. See, e.g., Tertullian "De spectaculis"; Saint John Chrystostom, *De inani gloria et educandis libris* (Paderborn: F. Schöningh, 1968), 6 and 30–31.
17. Roccha, *Trattato . . . : Contra i giuocchi della carte e dadi*, 54. See also Pseudo-Cyprian, *Liber de aleatoribus*; Isidore of Seville, *Etymologicarum* 18.42–69; Aquinas, *Summa theologiae*, bk. 2, pt. 2, qu. 168, art. 2.
18. Daneau, *Brièe remonstrance*; Gataker, *Of the Nature and Use of Lots*.
19. Calvin, *Institutes*, trans. Allen, 1: 226 (1.16.6).
20. This reading follows Jacobs, *Prädestination und Verantwortlichkeit bei Calvin*. In Calvin's first edition of 1539, predestination and providence (in this order) make up the dual theme of one and the same chapter of the Institutes (initially chap. 8, later 1.16). Calvin changed this fundamentally in the 1559 edition, where he treats the two themes in separate chapters: first providence (in relation to the doctrine of God and Creator); and then predestination (in relation to the doctrine of salvation and Christology). According to Jacobs, the themes of predestination and providence involve each other; predestination can be called a special

instance of providence, *providentia specialis*, but providence is part of what predestination's purpose is in God's plan. Jacobs, *Prädestination*, 69.

21. Calvin, *Institutes*, 1: 229 (1.16.9).

22. Daneau's *Briève remonstrance* and Gataker's *Of the Nature and Use of Lots* remained authoritative well into the eighteenth century.

23. Frain Du Tremblay, *Conversations morales;* Wittich, *Exercitationes*.

24. A similar argument can be found in Christian Wolff's *Vernünftige Gedancken von dem gesellschaftlichen Leben der Menschen und insonderheit dem gemeinen Wesen* in id., *Gesammelte Werke*, 1st ser., vol 5.

25. In addition to Thiers, *Traité*, and Wittich, *Exercitationes*, see La Placette, *Des jeux de hazard*, in id., *Divers traités*. Within this discussion of physical motion, calculable action, freedom of action, and divine providence, Wittich leaves open the question of whether games of chance belong to particular providence or not. Since we cannot take anything more for granted than ordinary or general providence, we are not allowed to use games and lots as means of interrogating God's will (Wittich, *Exercitationes*, 78).

26. La Placette, *Des jeux de hazard*, in id., *Divers traités*, 218.

27. Ibid., chap. 6.

28. Five years after the second edition of Barbeyrac's *Traité du jeu*, a German translation also appeared: *Herrn Joh. Barbeyracs Tractat vom Spiel*, trans. J. W. Lustig (Bremen, 1740).

29. See Othmer, *Berlin und die Verbreitung des Naturrechts*, 60–93. Barbeyrac's opponent Jean Frain Du Tremblay, who had been a member of an appeals court before he published on juridical and literary matters, was also knowledgeable about game theory.

30. Pufendorf, *De jure naturae et gentium*, translated here from Barbeyrac's French translation, *Le droit de la nature et des gens*, 1: 109.

31. See Barbeyrac's note ibid., 1: iii.

32. Ibid., 1: 108. See Barbeyrac's translation of the chapter title: "Des contracts ou il entre du hazard."

33. Ibid., 1: 111, Barbeyrac's translation: "dans toute Guerre publique & reglee."

34. Ibid., 2: 108.

35. Ibid., 2: 111.

36. See Mesnard, *Pascal et les Roannez*.

CHAPTER 2

1. Hacking, *Emergence of Probability*, 63ff. This assessment draws on Hacking's premise according to which the emphasis on decision emerges from the wager's literary framework, the defense of faith by means of reason. For Hacking,

a universal art of conjecture or theory of probability is already laid out in theological apologetics.

2. Pascal's general method, in contrast to Fermat's combinatorics and especially Huygens's arithmetical formula, represents the older method as found in theories of balance and game of chance as early as Luca Pacioli, Geronimo Cardano, and Nicolò Tartaglia. See the selections and commentary in Schneider, ed., *Entwicklung der Wahrscheinlichkeitstheorie*, 1–40.

3. Henri Gouhier explicates various relations of the wager with the apologetics of faith, which—as the contemporary historian Nicolas Filleau de la Chaise reports—Pascal set up as the final goal of the texts collected in the *liasses* (Gouhier, *Blaise Pascal*, 297–306). Gouhier argues convincingly that the argument of the wager should be read as an epistemological abridgment of the entire project of the *Pensées*. This even leads him to think of the wager as an introduction to a completed version of the Apology for the Christian Religion, the work of which the *Pensées* fragments were meant to be a part.

4. Pascal, *Œuvres completes*, 102b.

5. Ibid.

6. Ibid.

7. Gouhier, *Blaise Pascal*, chap. 5; Mesnard, *Les "Pensées"*; Topliss, *Rhetoric*. From among the extensive literature, I have also looked especially to the following for guidance: Brunet, *Pari de Pascal*; Orcibal, "Fragment *Infini–rien*"; Lönning, *Cet effrayant pari*.

8. Hacking, "Logic of Pascal's Wager"; see also Thirouin, *Hazard*.

9. The critical assumptions implied in this brief synopsis cannot be laid out fully here. With the *communis opinio* of researchers, I take for granted that the treatises *De l'esprit géométrique* and *De l'art persuader* (*Of the Geometrical Spirit* and *The Art of Persuasion*) were conceived as parts of a single text. For this argument, see Schobinger, *Blaise Pascals Reflexionen* and de Man, "Pascal's Allegory of Persuasion."

10. French citations are from Lafuma's edition of Pascal's *Œuvres complètes*. The cited *Pensées* will be followed first by their Lafuma number and then the Brunschvicg number; the page numbers will be given in parentheses: first from the Lafuma edition, and then from the English translation by Krailsheimer. Here, e.g.: L418; 233 (550b; 150).

11. *Pensées*, L418; 233 (551a; 152).

12. In his *Arithmetices*, Pierre de La Ramée (Petrus Ramus) writes for example: "With number, we have to look at notation (*notatio*) first and at counting second." Regarding *notatio*, he continues: "For the numbers to be notated and written on the counting frame (abacus), there are the ten markers 1, 2, 3, 4, 5, 6, 7, 8, 9, 0. The first one means One, the second Two, the third Three. . . . The circle, the last marker, does not have any meaning of its own. If put down at the right side,

it has the power to enlarge the other markers" (Ramus, *Arithmetices libri duo*, 1, 3). See Krämer, *Berechenbare Vernunft*.

13. See Marin, *La critique du discours*, 265ff.

14. "Notre âme est jetée dans le corps, où elle trouve nombre, temps, dimensions, elle raisonne là-dessus et appelle cela nature, nécessité, et ne peut croire autre chose" (*Pensées*, L418; 233 (550a; 149).

15. This takes up a modified version of Gouhier's interpretation, which differentiates between demonstrative proof and reasoned argumentation in the wager text (Gohier, *Blaise Pascal*, 279–287), but especially the reading of Marin, who emphasizes the cleft between persuasive rhetoric and cognitive demonstration in the *Pensées* in contrast to the integral logic of Arnauld and Nicole (Marin, *Critique du discours*; see also id., "Die Fragmente Pascals"); Sarah Melzer develops Marin's lines of interpretation even further in her *Discourses of the Fall*.

16. Pascal, *Pensées*, L418; 233 (550a; 149).

17. See the authoritative work in this respect, Schobinger, *Pascals Reflexionen*, 11ff., 309–317.

18. "That is . . . whatever movement, whatever number, whatever space, whatever time there may be, there is always a greater and a less than these" (Pascal, *Minor Works*, 42).

19. "Such is the admirable relation that nature has established between these things, and the two marvellous infinities which she has proposed" (ibid.).

20. Pascal, *Pensées*, L199; 72 (526b; 91). See also Marin, *Critique du discours*, 105–111.

21. Pascal, *Minor Works*, 47.

22. In his commentary on this passage, Marin notes that the recourse of thinking in the sciences of measurable magnitudes to their foundations, a recourse that occurs in wonder, also means ultimately deflecting it from scientific knowledge proper. But for Pascal, Marin adds, wonder does not destroy science. Indeed, after his statement about the character of the two infinities as image, Pascal continues in fragment L199; 72 to claim that an infinite number of scientific problems, for example, geometrical ones, can be solved by reflecting on infinity (Marin, *Critique du discours*, 110; L199; 72, 526b).

23. Pascal, *Minor Works*, 47. The formulation of *juste prix*, the "just value" of men in the sight of the world and of the creator can be found in *Pensées* L199; 72. The formulation develops out of the pure specification of relation *au prix de* ("measured by"): "Let man, returning to himself, consider what he is in comparison with [*au prix de*] what exists . . . let him learn to take the earth, its realms, its cities, its houses and himself at their proper value [*juste prix*]. What is a man in the infinite?" (526a; 89) The specification *au prix de*, applied to men and leading to their *juste prix*, goes back to the preceding formulation that characterizes the relation between the imagination and things imagined: "it is no good inflating

our conceptions beyond imaginable space, we only bring forth atoms compared to [*au prix de*] the reality of things" (526a; 89).

24. "There is . . . a relationship between man and all he knows. He needs space to contain him, time to exist in, motion to be alive" (Pascal, *Pensées*, L199; 72 [527b; 93]).

25. Ibid., L418; 233 (503a; 149).

26. With a view to the *Réflexions sur la géométrie en general* (Reflections on Geometry in General), Schobinger draws similar conclusions about the ambiguity of the infinite. In particular, he shows that Pascal uses this ambiguity for epistemological and anthropological purposes (*Pascals Reflexionen*, 11ff., 309–317). One must add, however, that the argumentation of the *Réflexions sur la géométrie en général* is even more pointed in this respect. When it comes to the discussion about zero and one (unity), hence on the side of the minima, and only here, Pascal hits upon the difference between the two meanings of the infinite. The argument runs as follows: one (unity) has determinate peculiarities in respect to the series of natural numbers greater than one; for this reason it has been excluded from the series of numbers since Euclid. But one (unity) is not "heterogeneous" with the series of numbers in the same way as zero (for which there is no corresponding sign in Latin numerics). One (unity) is the final element in the series of whole numbers in division; zero, however, is the value of the unattainable limit of infinite convergence of division in the series of fractions less than zero. Zero is, as Pascal concludes, radically "heterogeneous" in respect to the measurement and the magnitude of numbers from whose operative possibility the convergence on zero emerges (see Schobinger, *Pascals Reflexionen*, 70–73).

27. See, e.g., Aristotle, *Topica* 3.6, 120a33ff.; 4.2, 123a12ff.; or 6.5, 142b7ff.

28. The entire passage, noteworthy in this context, reads as follows: "We call highest and maximal propositions those propositions that are universal and known and manifest to such an extent that they need no proof but rather themselves provide proof for things that are in doubt, for those propositions that are undoubted are generally the principles of demonstration for those propositions that are uncertain. Propositions of this sort are 'Every number is either even or odd' and 'If equals are subtracted from equals, equals remain,' and others whose truth is known and unquestioned" (Boethius, *In Ciceronis Topica*, 33).

29. "We speak of a universal axiom when what follows is always true not only with respect to any antecedent and throughout, but when the relation can also be reversed; as is the case with 'man is a rational being,' 'a number is either odd or even,' 'the wolf is born to howl'" (Pierre de La Ramée [Petrus Ramus], *Dialecticae libri duo* 63, cited from Schmidt-Biggemann, *Topica universalis*, 48).

30. Pascal, *De l'esprit géométrique*, § I, 349.

31. Typically, Aristotle uses the topos of "even-or-odd" to show that something does not have the characteristic of a number. See the definition of the soul

as a self-moving number in *De anima* 404b29, and the commentary in *Topica* 6.3, 140b1ff.

32. Pascal, *Pensées*, L418; 233 (550a; 149).

33. Fragment FB 233 displays a structure in detail that Marin (in reference to this and other passages) claims for Pascal in general. According to Marin, geometry and science more often than not relate to other, nonscientific, discourses in Pascal in more than metaphorical ways. The relationship is instead structural, based on the fact that geometry and science are characterized by their constitutive boundaries and, hence, by the possibility of transgression. Marin, *Critique du discours*, 268.

34. "But by faith (*foi*) we know his existence (*existence*), through glory (*gloire*) we shall know his nature" (Pascal, *Pensées*, L418; 235 (550b; 149).

35. Ibid.

36. See McKenna, "L'argument 'Infini–rien.'"

37. Pascal, *Pensées*, L418; 233 (550b; 149).

38. The findings on this particular fragment are close to the general remarks and theses of Marin on the problem of the 'je' in the *Pensées*, which he formulates in the context of the theological and moral annihilation of the "I" at Port-Royal and in Pascal (Marin, *Critique du discours*, chapter 9, "Du discourse: sujet et manières," esp. 347ff.).

39. Pascal, *Pensées*, L418; 233 (550b; 150).

40. This seems to be implied by most readers of this passage: see, e.g., Kirsch, *Blaise Pascals "Pensées,"* 93–112.

41. Pascal, *Pensées*, L418; 223 (550b; 150).

42. Gouhier stresses the point that, in contrast to the first part of the fragment, calculation in the subsequent section creates its own terrain, the terrain of what is to be measured by it. This interpretation is close to the reading proposed here. In this instance, however, the artificial character of the calculation of chance, which does not simply require rationality, but instead creates it, coincides in a unique way with a discursive situation that allows taking recourse to calculation in the first place (Gouhier, *Blaise Pascal*, 287–297).

43. Pascal, *Pensées*, L418; 223 (550b; 150).

44. See Pascal, "Lettres à Fermat sur la régle des partis," *Œuvres complétes*, 43–49, esp. the letter of August 24, 1654, 46–49, and "Divers usages du triangle arithmétique dont le générateur est l'unité," ibid., 54–63, esp. § II, 55–57.

45. The concept of the rhetorical here follows Hans Blumenburg's formula of an "anthropological approach to rhetoric," or the related notion of "speaking-of-finitude" in Pascal in the account of Behrens, "Zur anthropologischen Fundierung."

46. Pascal, *Pensées*, L418; 233 (550b; 150).

47. See Daston, *Classical Probability*, 1–19; and Kirsch, *Blaise Pascals "Pensées,"* 93–122 (Kirsch's reconstruction is employed in the following with some modification).

48. Pascal, *Pensées*, L418; 233 (551a; 151).

49. Gouhier, *Blaise Pascal*, 290–296.

50. This admittedly somewhat awkward interpretation seems in hindsight to be supported by Pascal's comment on the third model: "That leaves no choice; wherever there is infinity, and where there are not infinite chances of losing against that of winning" (Cela ôte tout parti partout où il n'y a pas infinité de hasards de perte contre celui de gain). *Pensées*, L418; 233 (551; 151). The case of an infinite gain with an infinitely great chance of losing, however, only turns up in "interpretation b" of the second model.

51. Pascal, *Pensées*, L418; 233 (551a; 151). "No choice," as Krailsheimer translates the phrase 'Cela ôte tout parti' into English, means, more specifically, "no choice based on what one has to expect, that is, what follows from calculation in this case," with *parti* meaning "one's appropriate share or portion in a given matter."

52. Pascal, *Pensées*, L418; 233 (551a; 151).

53. The three cited passages are located in *Pensées*, L418; 233 (551a; 151–152).

54. Pascal, *Pensées*, L418; 233 (551a; 152).

55. For the question of the unity of the text, see Gouhier, *Blaise Pascal*, 247–279.

56. Pascal, *Pensées*, L418; 233 (551a–b; 152).

57. Pascal, "Triangle arithmétique," *Œuvres complétes*, 50–54; and "Traité des ordres numériques," 63–69.

58. Pascal, "Triangle arithmétique," §§ II and III, 55–62.

59. In his final objection, the atheist, who is trying to weasel out of the game, begins by arguing, "that it is uncertain whether you will win, that it is certain you are taking a risk." The reply begins by stating that nothing other than the actual definition of the game is paraphrased in such notions. According to this argument, the objection amounts to a merely tautological denial of the game: "Every gambler takes a certain risk for an uncertain gain." Pascal, *Pensées*, L418; 233 (551a; 151).

60. See "Leibniz on Estimating the Uncertain," trans. Melo.

61. Bernoulli, "Meditationes-Artikel," 131b, in id., *Werke*, 3: 68ff.

CHAPTER 3

1. This formulation comes from Dieudonné, *Abrégé*, 2: 281.

2. After the first draft of *Van rekeningh in spelen van geluck* had been submitted to the publisher van Schooten, but before the final editing of the text, Huygens explored the positions of Pascal and Fermat through correspondence with an acquaintance from his first sojourn in Paris, Claude Mylon, and then with Pierre de Carcavy. Through these middlemen, Huygens presented Pascal and Fermat with a gambling problem, and thus induced the—otherwise unapproachable—authorities on gaming calculation to reveal their positions in detail. See the letter

of May 20, 1656, in which de Carcavy announces Huygens's contacts with Pascal (Huygens, *Œuvres complètes*, 1: 418ff.); Huygens's letter to de Carcavy of June 6, in which he presents Pascal and Fermat with his questions (442–446); de Carcavy's letters of June 22 and September 28, in which he communicates Fermat's and Pascal's answers to Huygens (433–434 and 492–494); and finally Huygens's satisfied response on October 12, 1656 (505–507).

3. Struik, *Land of Stevin and Huygens*, 89. Huygens's life and work are treated in detail in chapter 8.

4. Huygens, *Fragment*, 116–121.

5. In fact, however, the reverse division also exists: in his later works, Stevin composed problems of logic and of theoretical mathematics (algebra and arithmetic) in Dutch and wrote on themes of technical, applied mathematics in contrast in French and Latin. In doing so he translated his works partly himself, and partly let others undertake the translations.

6. Huygens, *Fragment*, 42–46.

7. Stevin formulates this based on the prevalent humanistic theory of stem and root words: Flemish, he claims, has 1,428 different roots; Greek, 220, and Latin, 150; Flemish, moreover, is supposed to exhibit 742 monosyllabic root words, while Latin has 5 and Greek none at all. See Dijksterhuis, *Simon Stevin*, 126–129.

8. Having introduced Stevin's "Uytspraeck" as a model, Huygens turns to Hugo Grotius, whom he urges to translate legal studies into the language of the people, as Stevin did for arithmetic. Somewhat paradoxically, the conclusion of the passage implies that in law, as in arithmetic, the domestic equivalents of barbaric foreign technical terms are already at hand in the usage natural to the lower classes, women, and children. Huygens, *Fragment*, 43, 45.

9. For the conversion to decimal notation and calculation in the theoretical mathematics of the fourteenth and fifteenth centuries, see Krämer, "Kalküle als Repräsentation," 112ff. Krämer shows that in decimal notation, the structural order of the *notatio* coincides with the operative requirements of the *numeratio*. In this respect, Stevin's proposed decimal notation corresponds to Pascal's wager.

10. See the commentary of Gericke and Vogl in Stevin's *De Thiende*, 40–56; as well as Dijksterhuis, *Simon Stevin*, 17–19.

11. Husserl, *Crisis*, 43ff.

12. For Frans van Schooten (the son), see Struik, *Land of Stevin and Huygens*, 84–86.

13. The Dutch text, *Van rekeningh in spelen van geluck*, can be found in Huygens, *Œuvres complètes*, vol. 14, here 61. Huygens's translation fragments are enclosed in his letter of May 6, 1656, to van Schooten (*Œuvres*, 1: 414). Van Schooten's translation (F. van Schooten, *Exercitationum mathematicarum libri quinque* [Leiden, 1657]) is quoted here according to the text of bk. 1 of *Ars conjectandi* by Jacob Bernoulli, "De ratiociniis in ludo aleae," in id., *Werke*, 3: 110.

14. For the translation of *kans* as *expectatio*, see Freudenthal, "Huygens' Foundations of Probability."

15. For the Latin version, see Huygens, "De ratiociniis," in Bernoulli, *Werke* 3, 109; for the Dutch translation of the foreword, see Huygens, *Van rekeningh in spelen van geluck*, in id., *Œuvres complètes*, 14: 57). Huygens enclosed the Latin and the Dutch versions in a letter of September 28, 1657, to van Schooten, ibid., 2: 59ff.

16. For the poetical works of Constantijn Huygens, see Meijer, *Literature*, 141–148; concerning his early works and intellectual personality, see Colie, *Some Thankfulnesse to Constantine*. See also Alpers, *Art of Describing*, 10–25.

17. In a letter of May 5, 1656, to van Schooten, Huygens speaks about using the vernacular because certain words and phrases had not "been available" to him before the completion of the work. Now, after its completion, however, he delivers test translations of the beginning and of a few further passages.

18. Disclosed in the "Avertissement" (foreword) to Huygens, *Œuvres complètes*, 14: 5n13.

19. Huygens to van Schooten, May 6, 1656, *Œuvres complètes*, 1: 413.

20. Huygens again addresses the frivolous character of *Van rekeningh in spelen van geluck* by writing that if van Schooten has no time to translate his *ludicra*, he can undertake the translation himself. Besides, he continues, it is of little importance whether the treatise appears now, later, or even not at all (letter of July 25, 1656, in *Œuvres complètes*, 1: 460). This clearly contradicts his appraisal of the work to other correspondents (as in a letter of September 7, 1657, to R. F. Walter de Sluse, ibid., 2: 55).

21. See, e.g., Huygens to de Sluse, July 27, 1657, in id., *Œuvres complètes*, 2: 42.

22. The translation follows Freudenthal, "Huygens' Foundations of Probability," 114. *Rechtmatigh spel* and *ludus iustus*—translated by Freudenthal as "equitable game"—mean the emblematic "fair game." As appropriate as this translation or Freundenthal's choice would be, I have decided to render these expression most of the time "just game." The argument in the following relies on the notion of justice as the source of the meaning and the semantic content of our "fair play."

23. Huygens, *Van rekeningh in spelen van geluck*, in id., *Œuvres complètes*, 14: 61. Cf. the partial German translation in Schneider, ed., *Entwicklung der Wahrscheinlichkeitstheorie*, 41–43, and the French translation in Huygens, *Œuvres complètes*, vol. 14.

24. Daston, *Classical Probability*, 24.

25. Ibid.

26. Huygens, *Van rekeningh in spelen van geluck*, in id., *Œuvres complètes*, 14: 63.

27. Aristotle, *Nicomachean Ethics*, 86 (5.1132a).

28. Aristotle's notion of equality (*isos*) does not belong to a determined field of science; measure, law, justice, guilt, and innocence bear, however, a conceptual

family resemblance, according to Leyden, *Aristotle on Equality and Justice,* 32–40; see on this, e.g., Aristotle, *Metaphysics* 5.15.1021a.

29. Riffaterre, *Semiotics of Poetry.*

30. If we compare this *norma studiorum et vitae reliquae* with the course of education described by the father in his own autobiography, a greater emphasis on legal studies becomes noticeable in the plan for the sons. In both generations, however, the seriousness and professionalism with which the fathers prepare and carry out the private education of their sons is striking.

31. Constantijn Huygens, *père, á ses fils,* to Constantijn and Christiaan Huygens, May 9, 1645. Norma studiorum et vitae reliquae praescripta," in Christiaan Huygens, *Œuvres complètes,* 1: 4ff.

32. Vinnius, *Tractatus quatuor,* 3rd ed. Vinnius's area of specialty was civil law, or *jurisdictio* (a modern term, uncommon in Roman law) in the narrow sense.

33. In this regard, see also Dießelhorst, *Lehre des Hugo Grotius.*

34. "Etsi conventiones nudae, quantumvis voluntate sufficienter significatae, legum Romanarum vinculis adstrictae non sunt, ad earum tamen observationum naturae & gentium jure adstringimur: Nihil enim aequitati naturali magis consentaneum est, quam stare conventis, placita, fidemque servare" (Vinnius, *Tractatus quatuor,* 3rd ed., here the tractate *De pactis,* 150).

35. On justice and the method of probability, see Vries, *De calculus der billijkheid.*

36. See Leyden, *Aristotle on Equality and Justice.*

37. Huygens, *Van rekeningh in spelen van geluck,* in id., *Œuvres complètes,* 14: 69.

38. The fact that, in Huygens, the third proposition or the formula of chances, is only arrived at by way of the detour of the first proposition, the proposition of justice, renders his "calculations rather ponderous." See Dieudonné, *Abrégé,* 2: 280.

39. Huygens's statement of the problem and Fermat's solution as reported by Carcavy immediately precede this discussion. Huygens is obviously attempting to pin down the Parisian mathematicians. According to Huygens, Fermat could only have come up with his solution, which is in accord with his own, by consciously or unconsciously employing the same theorem; see the letter to Carcavy, July 6, 1656, in Huygens, *Œuvres complètes,* 1: 442. Huygens argued similarly when Carcavy communicated Pascal's answers to him. But indeed Pascal had explicitly agreed with Huygens's solution of the chance theorem; see ibid., 492–494, 505–507.

40. Huygens, *Van rekeningh in spelen van geluck,* in id., *Œuvres complètes,* 14: 1–3, "Voorstel" (proposal).

41. In the Dutch original, Huygens always uses the formulation of the *rechtmatigh spel* (lawful or rightful game) for the numerical formula of his theory. In his model translations, he is even more exact and careful, however. In the

definition of chance, he translates *rechtmatigh spel* as *aequa conditione certans*. In a preceding passage about justice in gaming, where the Dutch has no clear indication of the justice or the fairness of the "equal," the Latin version supplies *aequum*. And at the end of the introduction, where Huygens gives a preliminary suggestion of the just game, he writes *in lusu qui alteri damnosus non sit* [who, in gaming, is not inflicting harm to the other player] (Huygens, *Œuvres complètes*, 1: 414).

The three Latin formulations carefully carry out the chiastic construction of game-justice (*aequa conditione certans*) starting from the fairness of the "same conditions" in the game (*aequum est*), on the one hand, and the just game (*lusus qui alteri damnosus non sit*), on the other. Frans von Schooten follows Huygens's own translation by making these three passages from the introduction all turn entirely on the term of *aequus* (fair, just, equal). He translates: *aequum est—aequa conditione certans—aequus ludus*. See Bernoulli, *Werke,* 3: 110.

42. Spinoza, *Letters*, letter no. 38, 213–214.

43. Ibid., 213.

44. Spinoza, "Berechnung des Wahrscheinlichen," in id., *Algebraische Berechnung des Regenbogens*. Hans Christian Lucas and Michael John Petry argue for Spinoza's authorship, which had long been disputed, discussing the text together with the letter to van der Meer quoted above, see their introduction, ibid., *xviii*ff.

45. For the reference to [Spinoza,] *De nagelate schriften van B.d.S.* (1677) in this context, see Lucas and Petry's introduction, ibid., *xxxiv*ff.

46. Hacking, *Emergence of Probability*, 98.

47. See Spinoza, *Theological-Political Treatise*, 195–207.

CHAPTER 4

1. See Stigler, *History of Statistics*, 62ff.

2. See the important works of Ewald, *État providence*, and Link, *Versuch über den Normalismus*.

3. Bernoulli, *Ars conjectandi*, 108.

4. Already in van Schooten's translation of Huygens, *spes* and *exspectatio* (as well as *verisimilius*) occupy the blank space of an interpretation for *kans*. For instance, how much *kans* one allots to oneself and to others for victory or defeat is translated by van Schooten as "winnings to be expected" (*sors et exspectatio*). For this and related examples, see van Schooten's translation in Bernoulli, *Ars conjectandi*, 110–114.

5. See Biermann, "Überblick."

6. "Leibniz on Estimating the Uncertain," trans. Melo, 43.

7. Ibid., 48.

8. Proportions are the point of entry for numerics into the world of alphabetic thinking in general, as Serres, *Système de Leibniz*, chap. 1, shows.

9. De Moivre's classic eighteenth-century formulation reads: "The probability of an event is greater or less according to the number of Chances by which it may happen, compared with the whole number of Chances by which it may either happen or fail. Wherefore, if we constitute a Fraction, whereof the Numerator be the number of Chances whereby an Event may happen, and the Denominator the Number of all chances whereby it may either happen or fail, that Fraction will be a proper designation of the Probability of happening" (Moivre, *Doctrine of Chances*, 1ff.). "[P]robability relates to certainty as the number of cases in which the event in question takes place relates to the number of all possible cases," 's Gravesande writes in *Introduction à la philosophie*, in id., *Œuvres*, 2: 1–154, here 85.

10. This argument requires hope and fear to be analytic, or, in Cartesian terminology, clear concepts. Leibniz does indeed state that number, magnitude, form, and the affects fear and hope are examples of clear concepts (*Meditationes de cognitione, veritate et ideis*, in id., *Kleine Schriften zur Metaphysik*, 34–35).

11. "Leibniz on Estimating the Uncertain," trans. Melo, 48; italicization modified.

12. Michel Serres argues that the—structural—center of the philosophical system in Leibniz presents itself in many equally relevant—thematic—centers. Thus the assumption of an (invisible) center of the system and the effective decentrality of Leibniz's work do not exclude each other (Serres, *Système de Leibniz*).

13. "Leibniz on Estimating the Uncertain," trans. Melo, 45; trans. modified.

14. Leibniz in fact suggests still another way of interpreting the calculation of chances. In this second analysis, Leibniz disregards the soul or interior life of players and instead observes their behavior while gaming. According to Leibniz, the observed choices in a given game can be regarded as representing a particular possibility that, out of all the possibilities of playing the game, is realized. The relation of the realized possibility to all possible cases then allows conclusions to be drawn about the relation between the usual hopes of players and the total winnings. By observing the players, we thus arrive at a proportion of probability at which sides are switched when compared to the proportion described above. The formula for the second solution reads: $F / n = S / R$. It results from a kind of intuitive statistics of the choices of habitual players in the interior of their *tota societas* as observed from the outside. A novel could be written about such probabilities.

15. On the wager text, see esp. Gouhier, *Blaise Pascal*, chap. 5, esp. 247–279; on Pascal's relation to the Port-Royal *La logique, ou l'art de penser*, see Marin, *Critique du discours*.

16. Hacking, *Emergence of Probability*, chap. 8, and Daston, *Classical Probability*, 60–63, do this in exemplary fashion.

17. For an explicit example of negating any relation between the probability of the topoi and mathematical probability theory, see Lübbe, "'Topik," 141. The implication, at least, is the same as in Byrne, *Probability and Opinion*, chap. 1. The present investigation pursues the opposite hypothesis. What is meant, however, is not that the topoi somehow develop into probabilistic thinking. Instead, mathematical probability requires recourse to the topoi in order secure a meaning for itself.

18. This is Hacking's argument, for instance. Hacking goes so far as to question Arnauld's authorship of the last chapter. He introduces instead the manuscript from the Bibliothèque nationale dating from before the publication of *La logique* in which the last four chapters do not appear. Hence Hacking bases his thesis only on these four chapters alone, and although it clearly introduces these, plays down the fifth chapter from the end, calling it "a lame and conventional chapter on what we can know not through demonstration but through faith" (Hacking, *Emergence of Probability*, 75).

19. See Engfer, *Philosophie als Analysis*.

20. This symmetry indeed has a certain point: for the probability that one locates with the help of the topoi is, as Arnauld says, "of . . . little use" (Arnauld and Nicole, *Logic*, 17). In contrast, there can be no doubt of the power and use of the new method of probability, of the calibration of *foi*.

21. The relevance of *fides* comes to the fore right at the beginning of the *Topica*: "Therefore we may define a Place as the location of an argument, and an argument as a reasoning that lends belief to a doubtful issue" (Cicero, *Topica* 2.8, 119). Otherwise, *fides* appears in particular as the term for the reliability of a testimonium, e.g., in chap. 19, 73.

22. Shapiro, *Probability and Certainty*, offers comprehensive information and discussion on the contemporary debate about the reliability of witnesses, showing that the *fides* of testimony plays a great role in historiography as well as in the doctrine of observation in natural sciences of the time.

23. "All methodical treatment of rational discourse involves two skills, invention and judgment" (Cicero, *Topica* 2.6, 119).

24. Arnauld and Nicole, *Logic*, bk. 4, chap. 13. The status of the contingent event—by which the concept of event in the last five chapters is persistently determined—is extended in the course of this argument to singularities in all levels of time. Even past events can thus be analyzed as quasi-future ones with regard to their logical structure, that is, as events that are logically characterized by their ability to occur or not to occur.

25. Arnauld and Nicole, *Logic*, 264.

26. Ibid.

27. Hacking, *Emergence of Probability*, 79ff. Hacking's interpretation, however, takes a wrong turn. Despite the actual wording of *La logique*, he tries to

privilege the "internal circumstances" over the "external" (the testimonies). By reading the "internal circumstances" as "internal evidence," he makes them even into an early precursor of inductive logic to which then probability also belongs (chap. 5). The reading suggested here, however, argues that Arnauld instead neutralizes the differentiation between internal circumstances and testimonies and forges by that token precisely the condition for the empirical sciences and a probabilistic theory of knowledge.

28. This division is also the precondition for Cicero's ability to transfer the credibility (*fides*) of a witness and his statement to the power of a topos to be effective as an unsurpassable argument.

29. See Shapiro, *Probability and Certainty*; Daston, *Classical Probability*, 33–46. For the historiography, see esp. Borghero, *La certezza e la storia* (for the *Logic*, see chap. 3; for *Fides historica*, chap. 5).

30. The concept of data is used in an anticipatory manner here. In part II, chap. 13, it will be introduced in a more systematic way in the context of statistics.

31. Arnauld and Nicole, *Logic*, 273.

32. Cicero, *De oratore* 2.86, 351–360.

33. Arnauld and Nicole, *Logic*, 273–274.

34. Arnauld and Nicole, *La logique*, 428.

35. Arnauld expressly adds this point after concluding the triptych of the princess anecdote, the moralistic of the lottery, and the probabilistic exemplarity of the fair game: "we can show that there is an obvious injustice in the type of games called lotteries because, with the operator of the lottery usually taking a tenth for his share in advance, the whole group of players is duped" (Arnauld and Nicole, *Logic*, 274).

36. Pufendorf, *Le droit de la nature et des gens*, trans. Barbeyrac, 1.5.9, 106–111.

37. The translator Barbeyrac does not hesitate to comment on this. In a footnote he writes, "I do not know what the reason is for our author not to quote the *Art of Thinking* [*La logique, ou l'art de penser*], from which he obviously took this thought" (Pufendorf, *Droit*, 1.5.9, 111).

38. Pufendorf, *Droit*, 1:109, again translated from Barbeyrac's French.

39. Daston, *Classical Probability*, 20–30, discusses the connection of probability to rectificatory justice and the conditioned contracts as one and the same issue. The emphasis in the present study, in contrast, is on the difference. Huygens offers a "systematic" solution under the heading of rectificactory justice, whereas Leibniz gives a rather tentative and partial model for understanding probability through conditioned contracts.

40. Leibniz, *Disputatio juridica posterior de conditionibus*, in *Sämtliche Schriften und Briefe* (cited below as *SSB*), ser. 6, 1: 139.

41. The essential features can also be found again in the *Nova methodus discendae docendaeque jurisprudentiae* (in *SSB*, ser. 6, 1: 258–364), which Leibniz wrote

in 1667 to follow up the two *Disputationes juridicae de conditionibus*. Leibniz began to revise the *Nova methodus* in 1697–1698, at the time of his correspondence with Jacob Bernoulli. This temporal circumstance shows that the restraint from a rigorous mathematization of probability in juristic matters cannot—as Hacking, for example, assumes—be explained by the early date of the two contract treatises. It is rather a systematic element of the juristic logic of probability for Leibniz.

42. Leibniz, *Specimina juris*, in *SSB*, ser. 3, 1: 426.

43. Ibid.

44. The exemplary meaning of jurisprudence is for Leibniz based on the fact that law occupies an intermediate place between rational norm and historical fact. Law is—as natural law—norm, and therefore reasonable and calculable, but also and at the same time—as law in the Roman tradition—a historical and contingent fact and therefore observable and interpretable. Leibniz can hence, on the one hand, contrast the consistency of legal principles with medicine: whatever in law is not prescribed by tradition or written laws is based on bare reason. The "empirical principles, i.e., the observations" in medicine, in contrast, must continuously be renewed (Leibniz, *Nouveux essais*, in *PhS*, 5: 408). On the other hand, jurisprudence should be contrasted with theology: law, like theology, is composed in part of the interpretation of written texts and in part of the exercise of natural reason (Leibniz, *Nova methodus*, 294).

45. See the three fragments from the years 1686–1689, "De veritatibus primis," "De contingentia," and "De dispositionibus internis," in Leibniz, *Kleine Schriften*, 1: 177–189).

CHAPTER 5

1. See Dieudonné, *Abrégé*, 2: 281.

2. Condorcet, *Œuvres*, 1: 496, cited in *Condorcet, mathémathique et société*, ed. Rashed, 104.

3. The reference to the semblance (*Schein*) of truth and its implications in the theory of the novel is taken up in the second part of this book. .

4. Condorcet, *Œuvres*, 11: 496, cited in *Condorcet, mathémathique et société*, ed. Rashed, 104.

5. According to Cournot, producing the rule for fair division provided the occasion for Pascal to invent the probability calculation, though he never suspected "what uses could be made of his geometry of chances in the realm of judgments of possibility and regarding the distribution of natural phenomena." Only Jacob Bernoulli in his *Ars conjectandi* "formally determined the essential purpose and objective value of the theory of chance." Bernoulli's concepts of *probabilitas* and

conjectura, however—and Cournot hereby sets the tone of a critique that leads to the axiomatization of probability—were ambiguous (Cournot, *Exposition*, 86ff.).

6. Nicolaus Bernoulli, preface to *Ars conjectandi*, by Jacob Bernoulli, *Werke*, 3: 108.

7. In his history of probability theory, Condorcet praises Nicolaus Bernoulli for having laid the groundwork for applying calculation to jurisprudence. For Condorect, such application was the liberation from the "ages of ignorance" because it sets a limit to the exclusive competence of specialists and the social ranks they originate from (Condorcet, *Œuvres*, 1: 498, cited in *Condorcet, mathémathique et société*, ed. Rashed, 105).

8. In their views on social statistics as gaining acceptance in the face of the tradition of liberal law (Ewald, *L'État providence*) and winning out over legally decreed norms (Link, *Versuch über den Normalismus*), modern histories of probability still subscribe to this story, first told by Condorcet and Laplace.

9. Hacking was the first to break most clearly with the teleology of the history of probability's application. For him, Bernoulli's *Ars conjectandi* finally realized what he calls the "emergence of probability." In Hacking's story, probability theory appeared suddenly and with a single stroke "[with] its mathematical profundity, its unbounded practical applications, its squirming duality and its constant invitation for philosophizing." This means, in a kind of system-theoretical sense of emergence, that "Probability had fully emerged" (Hacking, *Emergence of Probability*, 143). Despite the elegance of Hacking's work, we might still want to know more about the discursive characteristics that made such an "emergence" possible.

10. For the following details, see Kohli, "Zur Publikationsgeschichte der *Ars conjectandi*," in Jakob Bernoulli, *Werke*, 3: 391–401.

11. Hermann to Leibniz, October 28, 1705, in *PhS*, 4: 285–288.

12. See Fleckenstein, *Johann und Jakob Bernoulli*.

13. Nicolaus Bernoulli, *Dissertatio inauguralis mathematico-juridica, de usu artis conjectandi in jure* (1709), in Jakob Bernoulli, *Werke*, 3: 287–326, esp. *Praefatio*, 289.

14. Bernoulli, *Ars conjectandi*, 239.

15. See chapter 6 below.

16. Bernoulli, *Ars conjectandi*, 250.

17. See Bernoulli to Leibniz, October 3, 1703, in Leibniz, *Mathematische Schriften* (henceforth cited as *MS*), 1: 78.

18. See Pearson, "James Bernoulli's Theorem"; Schneider, "Entwicklung des Wahrscheinlichkeitsbegriffs"; Hacking, *Emergence of Probability*, chap. 17.

19. For the testimonia as *argumenta extrinseca* that do not belong to the technical ("artifical") topoi in the proper sense, see Cicero, *Topica* 19.72ff.; for the *adiuncta* or concomitants, 11.50ff.

20. Although it is generally acknowledged that Bernoulli follows up on Arnauld's concept of probability in bk. 4, this critical attention is limited to passages where Bernoulli mentions Arnauld's name. The much more important but unmarked citation of Arnauld's citation of the *Topica* has been overlooked.

21. See Hacking, "Jacques Bernoulli's 'Art of Conjecturing.'"

22. See the otherwise authoritative essay by Boudot, "Probabilité et logique de l'argumentation selon Jakob Bernoulli."

23. Hess clearly has a connection of this sort in mind when he argues for investigating rhetorical topics in the context of the sciences of the seventeenth century (Hess, "Zum Toposbegriff der Barockzeit," 86).

24. Schmitt-Biggemann, *Topica universalis*.

25. Engfer, *Philosophie als Analysis*.

26. Seifert, *Cognitio historica*.

27. Poovey, *History of the Modern Fact*; Daston and Galison, *Objectivity*.

28. This citation has gone unnoticed in histories of literary verisimilitude as well as in the history of mathematical probability. Byrne's otherwise pioneering book *Probability and Opinion*, which for the first time poses the question of a systematic relation between probability and the topics, also overlooks the layered allusions to Arnauld and Cicero. Although Byrne's first chapter establishes a fundamental gap between mathematical probability and topical probability, he argues in what follows that probability theory occupied the place of the traditional topical notion of probability as acceptance.

29. Aristotle, *Topica* 1.2 (101a35–101b1), *Topics*, trans. Smith, 2–3.

30. Ibid., 1.1 (100b18–20), *Topics*, trans. Smith, 1.

31. Ibid., 1.1 (100b25–101a4), *Topics*, trans. Smith 1.

32. Moraux assumes in "La joute dialectique" that Aristotle wanted to provide himself with technical rules for a downright competitive institution of discussion in the *Topica*. According to Stump, "Dialectic and Aristotle's Topics," however, Aristotle refers at most indirectly to such an institution of discussion, and is concerned primarily with the mnemotechnic use of the topos.

33. In *Zur Struktur der gesellschaftlichen Einbildungskraft*, Bornscheuer, for instance, doubts whether a uniform concept of topos could be reconstructed through historical investigation. He contends that even Aristotle never defined the topos. Though Bornscheuer's methodical concerns are justified, his conclusions are beside the point. The contradiction between the framing conditions and the single move in the game prevents Aristotle from defining the topos otherwise than by its limits in relation to syllogistics, on the one hand, and eristic, on the other.

34. Aristotle, *Topica* 1.4 (101b14–37). Aristotle stresses that *problemata* and *protaseis*—"problems and premises," or pertinent questions and dialectical formulations of questions—are different in terms of their subject ("in form"), but the same in number (*Topics*, trans. Smith, 3–4).

35. See Stump, "Dialectic and Aristotle's *Topics*," 12. Despite all the critique of the assumption of actual dialectical competitions by Moraux, de Pater, and others, Stump insists that Aristotle's topics presuppose a game of speech-delivering partners. The *Topica* is "less a peculiar treatise on logic than a handbook on how to succeed at playing Socrates," she writes (25).

36. Cicero, *Topica*, §8, 119. "Therefore we may define a Place as the location of an argument, and an argument as a reasoning that lends belief to a doubtful issue."

37. Cicero, *Topica* §§19–20, 73–78. For further consequences of this linkage in the early modern history of literature and ideas, see Shapiro, *Probability and Certainty*; and Patey, *Probability and Literary Form*.

38. Cicero, *Topica*, §98, 169.

39. Cicero introduces a division in the topics that is not found in Aristotle: the division, borrowed from rhetoric, between "invention and judgment," *inventio* and *iudicium* (*Topica*, §6). For Cicero, the topics in the narrower sense—the topics that he is writing himself—are a matter of *inventio*. Clearly, however, Cicero is interested only in one side of the division: the *inventio*. The *Topica* do not offer any definition or explication of this issue.

40. "We call dialectic a syllogism that is formed from the arguments of the probable" (Boethius, "Topicorum Aristotelis interpretatio," 1.1, in *Patrologia Latina*, vol. 64, col. 910C).

41. For the insertion of Aristotelian "probability" in the commentary on Cicero's *Topica*, see Boethius, "Commentaria in Topica Ciceronis," in *Patrologia Latina*, vol. 64, col. 1045B–C.

42. Green-Pedersen, *Tradition of the Topics*.

43. See Boethius, "Commentaria in Topica Ciceronis," col. 1048B–C.

44. Ibid., cols. 1048–1054.

45. Boethius retraces the differentiation between a formal and a material aspect of topoi to the Aristotelian relation of premises (*protasis*) and decisive question (*problema*). Ibid., col. 1046ff.

46. For the logical details, see Stump's concise explanation, "Dialectic and Boethius' *De topicis differentiis*."

47. Boethius, "Commentaria in Topica Ciceronis," col. 1052.

48. Boethius, "De differentiis topicis," in *Patrologia Latina*, vol. 64, col. 1185.

49. Agricola, *De inventione dialectica* 1.3.26–27.

50. Ibid., 1.2.12–13.

51. Ibid. The same distinction between the logical-argumentative confirmare and the juridical-argumentative fides can be observed in this passage: "Logical consistency is one thing, and building on trustworthiness [*fides*] is another. Therefore in order that something can be employed to confirm something else [*ad alterius confirmationem*], it must have some connection of a certain kind to

the matter in the proof of which [*cui probando*] it is employed, and must be related in some way to it" (ibid., 1.2.14–15).

52. See Schmidt-Biggemann, *Topica universalis*. Interestingly, the key term "probability" does not play any prominent role in Schmidt-Biggemann's magisterial study of the epistemic function of the topics (see index., ibid.).

53. Agricola, *De inventione dialectica* 1.2.18–19.

54. Ibid., 1.2.14–15.

55. Ibid., 1.10.

56. Ibid., 1.2.18–19.

57. In his *Topica universalis*, Schmidt-Biggemann speaks, with an appeal precisely to this cited passage, of a "substantialization" of the topics. This seems a misunderstanding triggered by the example of the substance category (7 and 12). To speak of "substantialization" in such a manifestly nominalistic context makes little sense. Even more so in reference to Cicero, Schmidt-Biggemann's claim is unjustified that "the logical reference of the categories coincided with the metaphysical ones." Neither for Cicero nor for Agricola would the topoi become "also constituents of the things themselves" (7).

58. Bernoulli, *Ars conjectandi* 4.2.241.

59. Cicero's examples of *adiuncta*: "Matters antecedent to the event which are to be investigated are, for example: preparations, talks, a suitable place, an appointment, a banquet. Contemporaneous matters include: the tapping of feet, people shouting, the shadows of bodies, and what other things of that sort there may be. Subsequent matters include: paleness, a red face, staggering, and if there are any other signs of nervousness and bad consciousness, further an extinguished fire, a sword with blood on it, and the other things which can raise a suspicion that something has been done" (*Topica*, §52, 141).

60. The *adiunctis*, as Cicero notes, is a matter for a trial speaker, and not for the juridical expert or the philosopher. "Now this rhetorical Place is for the most part not only not the province of jurisconsults, but not even of philosophers" (*Topica*, §51, 141). The sphere of the *adiuncta* (concomitants) that, by their ability to organize a story will play an ever greater role in early modernity as the structure of knowledge of singularities, do not have any place in the ancient Aristotelian system of knowledge.

61. Agricola, *De inventione dialectica* 1.10.

62. Ibid., 1.20.120–121.

63. Ibid., 1.21.126–127.

64. Ibid., 1.21.130–131.

65. Thus the epistemological questions that Bernoulli deals with in the first chapter of his fourth book—pace Hacking—arise logically from the topics. They are not introductory problems of a probability void of tradition. This is particularly true of the distinction between subjective and objective certainty (mensura

cognitionis nostrae circa hanc veritatem vs. ipsam veritatem existentiae aut futuritionis illius rei). Hacking takes up the epistemological differentiation with his emergence criterion for probability—the simultaneity of subjective epistemic and objective frequentistic interpretations. But Bernoulli's differentiation does not mean to say anything of the sort, since it does not play a role either within the argumentation theory or the limit theory. In fact, Bernoulli is here concerned exclusively with the theological provisos according to which contingency for us does not mean any contingency of things in creation.

66. Bernoulli, *On the Law of Large Numbers*, 22 (see *Ars conjectandi* 4.3.244).

67. Ibid. (see *Ars conjectandi*, 244ff.).

68. In this regard, see the brilliant chapter 14, "Equipossibility," in Hacking's *The Emergence of Probability*. With good reason, Hacking suspects that an idea of Leibniz can be recognized in Bernoulli's *aeque facile contingere* (125). We should still emphasize that Bernoulli is developing a purely methodical principle. The goal is to depict the topics of singularities in countable contingencies without the interposition of any metaphysical principle of equivalency, as in Leibniz's proposition of equipossibility [Äquipossibilität].

69. The derivation from Huygens's axioms that make up bk. 1 of *Ars conjectandi* is skipped here.

70. Bernoulli, *Ars conjectandi*, 244ff.

71. Namely, the modern terminology according to Boudot, "Probabilité et logique."

72. Bernoulli, *Ars conjectandi*, 246.

73. Hacking champions the idea of considering the partial nonvalidity of the addition rule in Bernoulli (in contrast to Boudot's contention in "Probabilité et logique," esp. 282ff.) not to be a mistake by the author of *Ars conjectandi*. He sees in it even a facet of the discourse-founding power of the *Ars conjectandi* that sets it apart from probability theory in the usual sense (*Emergence of Probability*, 144, 152ff.). In the view proposed here, we are however dealing not with any kind of old remains, but rather with the process of reading the exemplification of the game.

74. When he begins to discuss the combinations of pure, mixed, or pure and mixed probabilities, Bernoulli points one last time to the "existence" of an argument. Since existence and indication represent dependent probabilities, Bernoulli disregards the factor of existence from this point on (*Ars conjectandi*, 245).

75. Hacking, *Emergence of Probability*, 151.

CHAPTER 6

1. See Schneider, ed., *Entwicklung der Wahrscheinlichkeitstheorie* and his discussion of Hacking's claims in "Why Do We Find the Origin of the Calculus of Probabilities in the Seventeenth Century?"

2. See Hartmann, and Klauke, "Anfänge, Formen und Wirkungen der Medizinalstatistik."

3. Seifen provides an overview of the conceptual history in *Der Zufall—eine Chimäre?*

4. Leibniz, *Philosophischen Schriften* (henceforth cited as *PhS*), 3: 193ff.

5. The use of the terms "evidence" and "proof" in this chapter alternates according to the legal context (evidence) or a rather scientific understanding (proof). Yet both words refer to the Latin *demonstratio*, the French *preuve*, and the German *Beweis* as used by Leibniz and his contemporaries. The meaning of "evidence" in English developed under the auspices of common law and especially the "law of evidence" in the Anglo-Saxon world, whereas "proof" became a term reserved for logic, science, and epistemology. In the continental tradition, however, *preuve* and *Beweis* never ceased to function as word and concept in both domains and meanings. The reader might keep in mind that "rules of evidence" and "doctrine of proofs" therefore refer to the same conceptual domain here. See Lesmerle, ed., *Prevue en justice*; Allen, *Law of Evidence*.

6. Leibniz, *MS*, 3/1: 71.

7. The David Speiser's edition of Bernoulli's letters provide helpful information about the biographical connections: Bernoulli, *Briefwechsel*, 104–112.

8. Leibniz, *MS*, 3/1: 71–73n. See also Hofmann, *Entwicklungsgeschichte*.

9. See Jacobi, "Zur Konzeption der praktischen Philosophie bei Leibniz."

10. Leibniz to Burnett, in *PhS*, 3: 194.

11. Serres, *Systéme de Leibniz*.

12. This suggestion follows Koselleck, *Critique and Crisis*, chap. 1: the religious and civil wars of the seventeenth century shook up not only the theological legitimacy of power, but the very possibility of legitimizing power at all.

13. Leibniz, *PhS*, 3: 194.

14. Ibid., 192.

15. Furetière's *Dictionnaire universel* defines *establir* thus: "Poser, asseoir sur quelque chose de stable, & assuré." A long list of figurative uses follows, including those in the sense of reasons and grounds: "Fonder donner commencement à quelque chose" (e.g., "St. Ignace a establi l'ordre des Jesuites") and in the sense of creating a (public) institution ("Establir une manufacture, une imposition, un droit"). In their use "dans les sciences & le raisonnement," the examples reveal a connection between the rhetorical-juridical term "status," the determination of the type of question with which a fact is construed ("Voila un fait bien establi") and the foundational aspect ("Il y a bien des erreurs populaires establies") in legitimizing validities ("Il ne faut pas establir pour principe une chose fausse"). Under the entry for *establissement*, after the general "Action par laquelle on fonde, on establit," we find particularly the foundation of institutions, e.g., a parliament, as well as the establishment of religions, ceremonies, and rituals. The amalgam of

aspects of temporal foundation, legal appointment, and logical validity, finally, sums up: etablissements "des loix, des Magistrats, des imposts, des regles, des principes dans les sciences."

16. "You know that toward the end of his life, Pascal (who died too early) took on the task of establishing the truths of religion [*à établir les vérités de la Religion*]," Leibniz writes Thomas Burnett (*PhS*, 3: 195). Pascal's death too early [*trop tôt*] parallels Leibniz's own situation. Leibniz also has the task still ahead of him, referring the question of *establissement* to theology. Ultimately, the peace and security of the state are at stake for Leibniz in the question of such establishment and foundation and their theological nature.

17. Leibniz, *PhS*, 3: 192.

18. In Leibniz's *Meditationes de cognitione, veritate et ideis*, published in 1684 in the *Acta Eruditorum*, there are allusions to Pascal's text, and to the influence he had on the sequence of chapters on the method and calculation of the probability of events in bk. 4 of the Port-Royal *Logique* (see Leibniz, *Kleine Schriften zur Metaphysik*, 1: 44–45).

19. On members of the London Royal Society influenced by Pascal, see Barker, *Strange Contrarieties*, esp. 47–63, on Locke's, Pepys's, and Petty's reactions to Pascal's wager. An equivalent study of German thinkers has yet to be written, but see Guitton, *Pascal et Leibniz*; Carraud, "Leibniz lecteur des pensées de Pascal."

20. Leibniz, *PhS*, 3: 193.

21. Leibniz, *Nouveaux essais sur l'entendement*, in *PhS*, vol. 5, bk. 4, chap. 2.

22. Shapiro, *Probability and Certainty*, 13, argues that the new science of probability in no way deepened the distance between natural sciences and humanities, but in fact constituted a new point of convergence.

23. On the philosophical aspect, see Cataldi Madonna, *Filosofia della probabilità*; on the historical aspect, see Borghero, *Certezza e la storia*, chaps. 5–7; in general, see Seifert, *Cognitio historica*.

24. Leibniz, *PhS*, 3: 193.

25. "It has been almost thirty years since I made these remarks publicly, and since that time I have done much research in order to lay the foundations for [this]," Leibniz writes at the end of the letter to Burnett. Measuring probative evidence will serve theological truth and the peace of the state, he reiterates in a postscript. His point of departure is the same "establishment" of theological truths to which, he asserts, Pascal had turned toward the end of his life. His own greater expert theological, historical, and legal knowledge may perhaps substitute for Pascal's superior genius in this, Leibniz suggests, but he insists that he would never compare himself to the latter. "Finally, if God grants me continued health and life, I hope he will also give me enough peace and freedom of mind to fulfill the vows I made thirty years ago. Thus I would hope to be able to contribute

to public piety and instruction concerning this matter of utmost significance" (*PhS*, 3: 194, 196).

26. Leibniz, *Nova methodus discendae docendaeque jurisprudentiae* (1667, reworked 1697–1700), in *SSB*, ser. 6, 1: 258–364. This contradicts Hacking's assumption (in id., *Emergence of Probability*, chap. 10) that the nonmathematical nature of Leibniz's juridical logic of probability was due only to his insufficient mathematical education prior to Paris. Even long after his occupation with Pascal, Fermat, and Huygens, however, Leibniz continued to pursue the logic of proofs (probation) in the same way. If at a later stage, Leibniz indeed connected his logic of the probable, which is mathematical only in a metaphorical sense, to games of chance, he did so only at a higher level of conception.

27. Leibniz, *Disputatio posterior de conditionibus* (1665), in *SSB*, ser. 6, 1: 139.

28. See Leibniz, *Specimina juris*, in *SSB*, ser. 6, 1, chaps. 1 and 2, 371–378.

29. Ibid., 420.

30. In his letter of October 3, 1703, Jacob Bernoulli asks Leibniz for suggestions and information in juridical matters in the context of *Ars conjectandi*. In doing so, however, in connection to his series theorem for statistical calculation of a posteriori probabilities, Bernoulli clearly has in mind the data collections for such things as insurance law and annuities. He also seeks information on de Witt's tables (Bernoulli, *Briefwechsel*, 117). Otherwise, the correspondence between Bernoulli and Leibniz about *Ars conjectandi* (April 1703–November 1704) suggests that there had not been any direct exchange between them on this topic previously.

31. Leibniz, *Specimina juris*, in *SSB*, ser. 6, 1: 426.

32. Bernoulli, *Ars conjectandi*, bk. 4, chap. 2, 80.

33. For dates and analysis of the revision, see Leibniz, *SSB*, ser. 6, 1: xviii.

34. Leibniz's arrangement of logical fields largely follows Arnauld and Nicole's *Logique*, where syllogistics and the topics are discussed together in bk. 3 under the heading "Du raisonnement," and probable judgment along with contingent events in bk. 4 under "De la methode."

35. Leibniz, *Nova methodus discendae docendaeque jurisprudentiae* (1667, 1697–1799), in *SSB*, ser. 6, 1: 279–281 (see also the *apparatus criticus*).

36. See Schmidt-Biggemann, *Topica universalis*.

37. Leibniz, *Nova methodus*, *SSB*, ser. 6, 1: 280 (*apparatus criticus*).

38. Leibniz, *New Essays on Human Understanding*, 464–65.

39. For more on the following outline, see Obertello, *John Locke e Port-Royal*; and Cataldi Madonna, *Filosofia della probabilità*.

40. Locke, *Essay Concerning Human Understanding*, bk. 4, chaps. 14–16, 654, and 657.

41. Ibid., 654.

42. Ibid., chaps. 15–16, esp. 655, 657ff. For the clear echo of Pascal and Port-Royal in this passage, see Barker, *Strange Contrarieties*, 50–55.

43. Locke, *Essay Concerning Human Understanding*, 657.

44. See Shapiro, *Probability and Certainty*, 163–167.

45. This parity of our own witnessing with that of others under the title of moral evidence, the type opposed to mathematical evidence, is discussed in 's Gravesande, *Introduction à la philosophie*, in id., *Œuvres*, 2: 1–154, bk. 2, chap. 12, "De l'evidence," esp. 67).

46. It could seem rather misleading to speak of an anti-juridical tendency in view of the many legal examples in Jacob Bernoulli and in particular in view of Nicolaus Bernoulli's *Dissertatio . . . mathematico-juridica, de usu artis conjectandi in jure*. This remark, however, is intended to point to the foundational relation of probability theory to law. Precisely Nicolaus Bernoulli's application of probability to law eventually hides the deeper legal foundation of the theory that he is applying.

47. A key text is the essay "De contingentia," which scholarship has dated between 1686 and 1689, in Leibniz, *Kleine Schriften zur Metaphysik*, 1: 178–187; on contingency in the doctrine of free will, see *Nouveaux essais*, in *PhS*, vol. 5, bk. 2, chap. 21.

48. See Stigler, *History of Statistics*, 3–70.

49. Leibniz to Jacob Bernoulli, December 3, 1703, *MS*, ser. 3, 1: 83ff.

50. Bernoulli to Leibniz, April 20, 1704, ibid., 87ff.

51. Ibid., 83ff.

52. If Hacking in "Equipossibility Theories of Probability" were correct in thinking that Jacob Bernoulli adopted Leibniz's *aequè facile contingere* formula for his recasting of singularities as contingencies it would be profoundly ironic. But in fact the expression can already be found used for the "objective" meaning of *gelijke kans* (equal chance) in van Schooten's translation of Huygens. Bernoulli employs it—in contrast to the equivalency of *possibilitas* and *probabilitas* in Leibniz—in Huygens's purely methodical sense.

53. On the initial meaning of contingency for mathematical probability in and since Leibniz, see Blumenberg, *Paradigms*, chap. 8, esp. 89–92. On the theological and philosophical implications, see Blumenberg, "Kontingenz." See also Estermann, *Individualität und Kontingenz*.

54. Leibniz, *Nova methodus discendae docendaeque jurisprudentiae*, in Leibniz, *SSB*, ser. 6, 1: 281 (*apparatus criticus*).

55. For the translation of *eventus* in this passage, the French *événement* must be considered—the term that dominates the final chapter of Arnauld and Nicole's *Logique*, where for the first time the new field of a logic of judgments is outlined for statements about single events in the past and present, that is, about contingency.

56. The passage in context: "Axiom: If players do similar things in such a way that no distinction can be drawn between them, with the sole exception of the outcome, there is the same proportion of hope to fear. This can be demonstrated by metaphysics: where the appearances are the same, the same judgment can be formed about them, that is, the way of thinking about the future outcome is the same; and the thoughts about the future outcome are hope or fear. If the pool is formed by common, equal contribution of the players, if each one is playing in the same way, and if for the same outcome the same prize or the same penalty is fixed, the game is fair." Leibniz, "Leibniz on Estimating the Uncertain," 43.

57. "If we do not always notice the reason which determines us, or rather by which we determine ourselves, it is because we are as little able to be aware of all the workings of our mind and of its usually confused and imperceptible thoughts as we are to sort out all the mechanisms which nature puts to work in bodies." Leibniz, *New Essays on Human Understanding*, 178.

58. *Inclinare* is used by Leibniz as a pendant to *necessitare*, when he states the unity of the difference between contingency and necessity as regarding the principle of sufficient reason: "In my view it is common to every truth that one can always give a reason for every nonidentical proposition; in necessary propositions, that reason necessitates; in contingent propositions, it inclines." Leibniz, *Philosophical Essays*, 28.

59. Leibniz, *New Essays on Human Understanding*, 183–191.

60. The irrational number in Leibniz is thus not quite yet the operational symbol that Sybille Krämer sees in it in her studies on systems of number writing. Krämer, *Berechenbare Vernuft*, 279–305; pointedly also in id.,"Kalküle als Repräsentation."

61. Leibniz, "Fragment théologique" (VI, 2, f. 11), in id., *Opuscules et fragments*, ed. Couturat, 1–3.

CHAPTER 7

1. Watt, *Rise of the Novel*.
2. For Defoe, see the authoritative study by George Starr, *Defoe and Spiritual Autobiography*, 105–125; for the literary discourse of biographies in general, see Stauffer, *English Biography Before 1700*.
3. Sterne, *Tristram Shandy*, bk. 4, chap. 9.
4. For the background of these formulas, see Starr's *Defoe & Casuistry*. Starr discusses a casuistic approach to decision making in Puritanism at large. Defoe uses forms of casuistry explicitly in his journalistic works; Starr argues that this is indicative for moral decision making in the novels as well. As has been shown,

the word "probabilism," incidentally, derives from (primarily Jesuit) casuistry. The wager or bet is a particularly familiar gesture of such casuistic probabilism.

5. Defoe, *Robinson Crusoe*, in id., *Works* (henceforth *W*), 1: 69, 139, 314.

6. Such concerns resulting from the tradition of realistic interpretation are reflected, e.g., in the subtitle of Hunter, *The Reluctant Pilgrim: Defoe's Emblematic Method and Quest for Form in "Robinson Crusoe"* (1966).

7. The emphasis on culture as a condition of salvation is further strengthened by Defoe in the second volume. The Spaniard, who remains behind on the island after Crusoe's departure for Europe, assures his predecessor on a return visit that only an Englishman could have pulled off the achievement of surviving in bare nature. Crusoe replies that, unlike the Spaniards, he had had the improbable fortune of being able to remove material necessary for survival from the stranded ship. The Spaniard, however, resolves the dispute with a final synthesis: only an Englishman would have been able, in the debacle of the shipwreck with its overwhelming improbability of survival—that is, in nature—to recognize and seize the chance still granted: the survival stock of cultural debris.

8. These are indeed reflections of a legal debate. After Crusoe observes the savages from the neighboring island practicing cannibalism, he considers whether he should treat them as enemies based on this observation, and, consequently, whether he could or should kill them. The arguments of his internal legal debate correspond with objections leveled by champions of natural law against the subjugation and eradication of the native Americans. According to those critics, crime and heathenism, and even cannibalism, are not sufficient grounds for intervention.

9. For an allegorical and military interpretation of the wolves as a Satanic army, see Hunter, *Reluctant Pilgrim*, 196–199.

10. Werner Welzig shows that the term "adventure" split off into two contrary directions after the replacement of the knightly *aventiure*: the adventure of the traveler and wanderer, which amounts to fortunate chance, bad luck, or accident; and the adventure of the businessman, which amounts to risk (Welzig, "Wandel"). As we shall see, Crusoe retrospectively learns to see the probability formulas of the chance adventure as those of a risk adventure or danger. For the erasure of traditional adventure and its reinterpretation in a bourgeois world, see Fohrmann, *Abenteuer und Bürgertum*, 186.

11. See Arbuthnot, "Argument for Divine Providence" and *Of the Laws of Chance.*

12. To this day, interpretations of *Robinson Crusoe* continue to discuss the opposition between circumstantial realism—tending to the modern novel—and allegorical narrative—relying on the traditional genre of conversion biographies. Connected to a first wave of arguments for the developments of circumstantial realism (the author's status as an outsider; the break with tradition around 1700;

the continuation of the picaresque novel; the model of journalistic writing), critics have enlisted probability in the narrative, the details of what is narrated, and comprehensible rationality in the characters' behavior as essential characteristics of novelistic narration. Watt notably brings these observations to bear on the poetological problem of the unconventional, even nonpoetological, form required for the novel. Starr and Hunter in contrast emphasize the allegorical stamp of the Christian conversion familiar from Puritanism in *Robinson Crusoe*. See Watt, *Rise of the Novel*; Starr, *Defoe and Spiritual Autobiography*, chap. 3; Hunter, *Reluctant Pilgrim*.

13. Richetti, *Defoe's Narratives*, promises to analyze a narrative process that undermines such opposition (chap. 1), but he then rather construes an irony of narration (following Paul de Man) that takes a sociohistorical detour via an antagonism between market and individual.

14. "... and I resolved that I would, like a true repenting prodigal, go home to my father" (Defoe, *Robinson Crusoe*, in *W*, 1: 7), as well as other direct and indirect allusions (7–15.) For other instances of the parable of the prodigal son in Puritan conversion literature, see Hunter, *Reluctant Pilgrim*, 26.

15. On these issues, see the commentary of Richetti, *Defoe's Narratives*, 26.

16. For the concept of allegory in this context, see Hunter, *Reluctant Pilgrim*, chap. 5; for Defoe's controversial use of the expression in the preface to the second installment of the Crusoe books, the *Serious Reflections... of Robinson Crusoe*, see Hunter, *Reluctant Pilgrim*, 102–122.

17. Defoe, *Essay upon Projects*, 129, with commentary in id., *Selected Writings*, 23–34. See also Sutherland, *Defoe*, 52–55.

18. A more detailed discussion of this follows in part II, chapters 9–11.

19. The densest and simultaneously most problematic presentation of this figure of thought remains Watt's first chapter in *Rise of the Novel*, "Realism and the Novel Form."

20. The controversy thickens again here between the allegorical reading of *Robinson Crusoe* and that of circumstantial realism. Watt, who bases a "realism" of the narrator in the modern novel first of all on the lack of poetological convention, declares this basis on the other hand to be a "silent agreement." As a validation of its "form," the novel employs, at least by analogy, the juristic conventions of evaluating witness testimony: the (poetological) nonconvention proves to be a (juridical or epistemological) convention (Watt, *Rise of the Novel*, chap. 1, esp. 13ff. and 31ff.). Hunter in contrast considers *Robinson Crusoe* to be steered by a bundle of allegorical indications, one of which is the allegorical structure of the *conversio*. The biographical allegory must, however, offer factual evidence for its formal achievement. Allegory thus becomes a formal technique; and consequently "art begins to masquerade as life." In short, the allegorical form is realized in narration only as nonform (Hunter, *Reluctant Pilgrim*, esp. 207ff.).

21. Defoe, *Robinson Crusoe*, in *W,* 2: 2.
22. Ibid., 196.
23. There is a clear indication of the parallels between his island existence and being marooned in Bengal: ibid., 212.
24. Ibid., 163.
25. Ibid., 197.
26. Ibid., 288.
27. For a virtuoso discussion of Luhmann's observer theory with its application to the modern novel, see Esposito, "Fiktion und Virtualität."
28. Defoe, *Robinson Crusoe*, in *W,* 1: 104. See also Hunter, *Reluctant Pilgrim*, 159ff.
29. Defoe, *Robinson Crusoe*, in *W,* 1: 126–127.
30. Ibid.
31. See Blumenberg, "Concept of Reality" and "Kontingenz." In terms of the theory and history of the novel, this has been discussed most prominently by Frick, *Providenz und Kontingenz*, 8, and Behrens, *Umstrittene Theodizee*, 9ff.

CHAPTER 8

1. "Aesthetic truth (*veritas aesthetica*) . . . therefore is called probability (*verisimilitudo*) for the most part, that is, that degree of truth which, even if it does not arise to perfect certainty, nevertheless might not contain any perceivable falsehood" (Baumgarten, *Aesthetica*, § 483, 1: 309). The fact that the *veritas* of aesthetics is *veri-similitas* has consequences for the epistemic status of the new discipline of aesthetics. According to Baumgarten, aesthetics is *analogon rationis*, an analogy to reason.

2. Baumgarten speaks of probability's gradation and calculation only in a figurative sense. Nevertheless, when Baumgarten, who explicitly professes his allegiance to the Leibniz-Wolffian school, mentions degrees of probability, he is doubtlessly alluding to Leibniz and Wolff's *logica probabilium* (see chapter 6).

3. Krüger, ed., *Probabilistic Revolution*. The most important scholars of the history of probability in North America—Ian Hacking, Lorraine Daston, Theodore M. Porter, and Stephen M. Stigler—all participated in Krüger's research project in Germany.

4. In the German tradition, the so-called academic topical description of the state dating from the seventeenth century kept its influence up through the mid nineteenth century. Rassem and Stagel, eds., *Statistik und Staatsbeschreibung*, is devoted to this rich tradition.

5. See John, *Geschichte der Statistik*; Petty, *Economic Writings*, ed. Hull; Dupâquier, *Histoire de la démographie*; Rassem and Stagel, "Zur Geschichte der Statistik und Staatsbeschreibung," 81–86.

6. In this regard, see principally Dupâquier, *Histoire de la démographie*, 133–137. The text of John Graunt's *Natural and Political Observations upon the Bills of Mortality* (1662) is to be found in Petty, *Economic Writings* 2: 314–343.

7. Sprat, *History of the Royal Society*, 83ff.

8. Graunt, *Natural and Political Observations upon the Bills of Mortality* (1661), in Petty, *Economic Writings*, ed. Hull, 2: 314–343, here 332.

9. Graunt, *Natural and Political Observations*, 2: 320.

10. Petty, *The Political Anatomy of Ireland*, in id., *Economic Writings*, ed. Hull, 1: 9.

11. Behre, *Geschichte der Statistik*, 158–161, 180–193.

12. Pinson de La Martinière, *Estat de la France* (1649); Besongne, *Le parfait Estat de la France* (1656). The English translation, *The Present State of France* (1671), renders the *parfait Estat* back to the present state.

13. See also the abridged Latin version by the jurist Thomas Wood, *Angliae notitia, sive praesens Angliae status succinctè enucleatus* (1686). As with *Estat de la France*, revised editions of *Angliae notitia* appeared at irregular intervals until the second half of the eighteenth century.

14. According to Thomas Sprat, the data of experimental philosophy make up the flesh and muscles of the body of science; dialectics and topical disputation are at most an *exercitium* to keep it limber (Sprat, *History of the Royal Society*, 17ff., cf. 89–91).

15. Regarding Conring, see Dreitzel, "Hermann Conring," and Seifert, "Conring," in *Hermann Conring*, ed. Stolleis. Regarding Conring's and Bose's *Staatsverwaltungswissenschaft* in emulation of Lipsius, see Östreich, "Justus Lipsius," 71.

16. See Seifert, *Cognitio historica*, "Staatenkunde," and "Conring und die Begründung der Staatenkunde."

17. See Conring, *Exercitatio historico-politica de Notitia singularis alicujus Reipublicae*, in *Opera*, 4: 1–47, the lecture transcriptions of the practical statistical exercises with a program that summarizes the theoretical foundations of the *Exercitatio* once again, *Examen rerumpublicarum potiorum totius orbis* (1675), ibid., 48–520; and Bose, *Introductio*.

18. See esp. Conring, *De civili prudentia liber unus* (1672), in *Opera*, 3: 280–421, chap. 12, esp. 350f; and *Exercitatio historica-politica*, chap. 1, § 2.

19. Conring, *Notitia*, in id., *Opera*, 4: 5; Bose, *Introductio*, chap. 8.

20. Seckendorff, *Deutscher Fürstenstaat* (1665), pt. 1.

21. Ibid., chap. 2.

22. In this regard, see Franz, *Dreißigjährige Krieg*, 19–30.

23. On King's manuscript and biography, see Chalmers, *Estimate of the Comparative Strength of the Great Britain* (1804). King, who was in the royal diplomatic service, was an expert in heraldry (that is, *notitia dignitatum*), but his

Natural and Political Observations are political arithmetic entirely untouched by such semantics of the state's organization (for King's dependence on Petty and influence on Davenant, see Dupâquier, *Histoire de la démographique*, 144–152).

24. The shrewd A. F. Lüder was the spokesman in Germany for normative-textual, as opposed to socionumerical, statistics; relevant writings include his *Kritik der Statistik* (1812) and *Kritische Geschichte der Statistik* (1817). Knies, *Statistik als selbstsändige Wissenschaft* (1850), continued this methodological discussion of academic statistics vs. political arithmetic.

25. Leibniz, "Meditatio juridico-mathematica de interusurio simplici," in *Opera omnia*, 3: 151–157.

26. Relevant memoranda, letters, and drafts include Leibniz, "Directiones ad rem medicam pertinentes"; id., "Entwurf gewisser Staatstafeln" and "Quaestiones calculi politici circa hominum vitam," both in *Werke*, 5: 303–314 and 337–340; and "Summarische punctation, die Medicinalische observationes betreffend," in *Werke*, 10: 346–353.

27. Polack, *Mathesis forensis*, 23.

28. Bastineller and Stockmann, *Dissertatio* (1741), 10.

29. Cramer, *Specimen*.

30. This is true not only for the Leibnizian-Wolffian tradition, but also for Thomasius and his followers. Thomasius had limited the use of mathematics to the practical cases of measurement and division, while polemicizing against all speculation, especially about infinity in the manner of Pascal (Thomasius, *Höchstnöthige Cautelen*, 226–260).

31. An example is J. F. Unger, who published the first series of his *Beyträge zur Mathesi forensi* with Vandenhoeck in 1742 and therein established an approach of his own to interest calculation.

32. Polack, *Mathesis forensis*, 4–6.

33. Hacking, "Biopower and the Avalanche of Printed Numbers."

34. Allusions to Polack and the discourse-founding meaning of Leibnizian interest calculation turn up again and again in Kästner's works from 1745 to his move from Leipzig to Göttingen. In Göttingen, too, he announced a course of lectures about Polack's book in 1757. See Kästner, *Pro justitia calculi interusurii Leibnitii*, and *Gradus et mensuram probabilitatis dari*.

35. Kästner, *Abraham Gotthelf Kästner's Selbstbiographie*, 3–7.

36. The following translations by Kästner are examples: William Horsley, *Der allgemeine Kaufmann* (1757); Carl Chassot de Florencourt, *Abhandlungen aus der juristischen und politischen Rechenkunst* (1781); Niels Morville, *Die Lehre von der geometrischen und ökonomischen Verteilung der Felder* (1793).

37. See Kästner, *Anfangsgründe der angewandten Mathematik*.

38. A further link in the academic tradition between *mathesis forensis* and *mathematica applicata* in Halle is the medical doctor and mathematician J. P.

Eberhard, author of the *Beiträge zur Mathesi applicata*, enlarged and revised as *Neue Beiträge zur Mathesi applicata*; cf. also, on Polack, Eberhard's *Gedanken vom Nutzen der Mathematik*, 17–23.

39. Meinert, *Über das Studium der Mathematik*, on Polack, 82; the draft of a discipline of studies of applied mathematics for jurists and cameralists, 132–136.

40. In the section on applied mathematics for jurists the Leibnizian-Huygensian themes of just division—of interest, corporate law, and insurance—are the same as in Polack's *Mathesis forensis*. But they no longer shape the thought of Meinert's enterprise (*Studium der Mathematik*, 29–42).

41. See the instructive excerpts and explanations in Condorcet, *Mathématique et société*, ed. Rashad.

42. The military had immediate practical use for this knowledge. The officer—and not only in the corps of engineers—became a prototype of the universal specialist. He should be able to judge and in emergencies even himself deploy expertise in ballistics, bridge building, mechanics, structural engineering, and, above all, surveying and cartography (Meinert, *Über das Studium der militärisch-mathematischen Wissenschaften auf Universitäten*). Hauptmann in Goethe's *Elective Affinities*, who surveys palace gardens, builds roads, and realizes many of the juridical-economic theories that unify Meinert's *mathesis applicata*, personifies an officer-mathematician of this type (see Kittler, "Ottilie Hauptmann," in id., *Dichter, Mutter, Kind*, 119–148).

43. For both concepts of evidence, see in particular 's Gravesande, "Discourse sur l'Evidence," in id., *Œuvres*, 2: 329–245.

44. Conring, *Notitia* in id., *Opera*, vol. 4; Bose, *Introductio*.

45. Achenwall, *Notitia rerum publicarum*.

46. E.g., John, *Geschichte der Statistik*.

47. This question provides the basis for Ian Hacking's history of statistics, *The Taming of Chance*.

48. Biographical details from Guhrauer, "Leben und Verdienste Caspar Neumanns," in *Schlesische Provinzialblätter*, 7–17, 141–151, 202–210, 263–272.

49. Impressive examples for the polished nature of the tables—for instance for the so-called dying out of a population—can be found in supplements to the later writings of Neumann, which he completed in the course of his cooperation with Edmond Halley. See Grätzer, *Edmund Halley und Caspar Neumann*, 38–48.

50. Neumann's letter to Leibniz, late 1689, quoted in Guhrauer, "Leben und Verdienste," 263ff.

51. Grätzer, *Halley und Neumann*, 49ff.

52. Cited according to Guhrauer, "Leben und Verdienste, " 263ff.

53. Halley, "Estimate of the Degrees of the Mortality. . . at the City of Breslaw," 654–656.

54. See Wilke, "Curriculum vitae: Johann Peter Süssmilch," in Süssmilch, *Die königliche Residenz Berlin und die Mark Brandenburg im 18. Jahrhundert*, 215–259.

55. Nothing is known about a contact with Achenwall's teacher Martin Schmeitzel, a statistician of the old school, during Süssmilch's time in Jena, just as Neumann's relations to his predecessor Bose are not reflected in his numerical-statistical experiments. See Schmeitzel, *Praecognita historiae civilis*.

56. Süssmilch, *Göttliche Ordnung*, 1: 52ff.

57. Ibid., 53–56.

58. Regarding his hardly researched relation to political arithmetic, it is noteworthy that Wolff, who was born in Breslau, received his high school lessons in religion from Neumann. Süssmilch's interest in social welfare associates him with the Pietists in Halle, whereas his geometrical methodology connects him to Wolff.

59. Wolff, preface to Süssmilch, *Göttliche Ordnung*, unnumbered.

60. The wording of the order in cabinet is printed in Behre, *Geschichte der Statistik*, 180; the surviving—but incomplete—population lists from 1748 until the establishment of the statistical bureau in 1805, are also there in appendix 5, 455–462; cf. also 203.

61. See Behre, *Geschichte der Statistik*, 143.

62. The following according to Boeckh, *Geschichtliche Entwicklung*, chap. 1; and Behre, *Geschichte der Statistik*, 131–208.

63. At first, the Great Elector's order of 1683 met with remarkable resistance: the court pastor condemned the census of the population as an infringement on divine providence, and predicted plague as a punishment of God. The Berlin plague of 1682, however, had more likely been what prompted the census in the first place. See Behre, *Geschichte der Statistik*, 133ff.

64. An indication of this switch is the following circumstance: in 1799 the Geistliches Departement ceased compiling its population lists according to the church calendar and began to use the secular one.

65. Leopold Krug was the expert in contemporary economic theory in the first Prussian statistical bureau, established by Frederick Wilhelm III in 1805 on the insistence of Baron Heinrich vom Stein, minister of state for trade; see Boeckh, *Entwicklung der amtlichen Statistik*, chap. 2; Behre, *Geschichte der Statistik*, 378–386.

CHAPTER 9

1. For the iconic status of numbers in contrast to the linguistic element of phonetic script, see Flusser, *Schrift*, chaps. 4 and 17. (A translation of related essays is available in Flusser, *Writings*).

2. See the brief hints in Derrida, *Of Grammatology*, pt. I, chap. 1 (e.g., 9–10, 24–25, 45).

3. *Grand Larousse*, s.v. *tableau*: "Description imagée d'un ensemble, évocation pittoresque par un discours oral ou écrit."

4. For developments characteristic of Leibniz's so-called *Staatstafeln* (state tables), see Vismann, *Files*, 102–109.

5. Leibniz, "Entwurf gewisser Staatstafeln," in id., *Werke*, 5: 303–314, here 308ff.; see also "Von Bestellung eines Registratur Amtes," ibid., 315–320.

6. For the figure of evidence, see Campe, "Shapes and Figures."

7. Quintilian's diction makes an obvious return in Leibniz's language about *in die Enge Treiben* (literally, "driving into a narrow space"; figuratively also "driving into a corner") of the entries in state tables.

8. Quintilian, *Institutio oratoria* 6.1.1: "totam simul causam ponit ante oculos, et, etiam si per singula minus moverat, turba valet"; in *Orator's Education*, trans. Russell, 17).

9. Quintilian, *Institutio oratoria*, 6.2.32–36: "The result will be *enargeia*, what Cicero calls *illustratio* and *evidentia*, a quality which makes us seem not so much to be talking about something as exhibiting it" (*Orator's education*, trans. Russell, 61).

10. Bastian, "Defoe's 'Journal,'" identifies the initials H. F. with an uncle of Defoe's named Henry Foe.

11. See Defoe, *Journal of the Plague Year*, ed. Landa, introduction by David Roberts, vi–xxii, here x–xx.

12. The articles can be found in Lee, *Daniel Defoe*; the first article, from August 12, 1720, appeared in the *Daily Post*, 2: 265, and from then on all the remaining ones were published in *Applebee's Journal*, 2: 277–278, 284–296, 378–379, 399–401, 427–430, 426–438, 449–451, 453–445, 464–465.

13. In this regard, and for a few early, related works of Defoe's, see Bastian, "Defoe's 'Journal,'" and Robert's introduction in Defoe, *Journal of the Plague Year*, ed. Landa, viii–ix. Excerpts from two plague tractates of the time after 1665 that Defoe clearly made use of can be found in Defoe, *Journal of the Plague Year*, ed. Backscheider, 211–225. For the epidemio-political situation of 1722, the time of the *Journal*'s composition, see Novak, "Defoe and the Disordered City."

14. Defoe, *Due Preparations for the Plague, as well for Soul as Body* (1722), in id., *Works*, 15: 1–205, here 3.

15. Starr, *Defoe & Casuistry*, identifies a development of casuistry in the *Journal*'s rhetoric of facts: "a frame of mind which can be brought to bear on perplexities of all kind" (56, 80). Moore, "Governing Discourses," works out an insistence of observation in the journal.

16. John Bender reinterprets transparency and realistic detailing in the novel as corresponding in style to the panoptic architecture of prisons in *Imagining the Penitentiary: Fiction and Architecture of Mind in Eighteenth-Century England*; for transparency and evidence, see chap. 8, for the *Journal*, chap. 3.

17. Defoe, *Due Preparations*, in id., *Works*, vol. 15; for the first part, see 59, 69, 73, 76, 78, 79; for the second part, 99ff., 197ff.

18. Defoe, *Journal of the Plague Year*, ed. Landa, 1.

19. Ibid., 2.

20. Ibid., 3.

21. Ibid., 1.

22. There were certainly periodical publications reporting on the plague even in 1665, such as the *London Gazette*. The *Gazette*, however, was a government newspaper. With this caveat, the remark about a pre-journalistic world around 1665 can well be maintained.

23. See, e.g., Lee, *Defoe*, 285, 291.

24. Defoe had indeed already done so in two or three sketches he contributed to *Applebee's Journal* in 1721. Already there he connects the critique of false or exaggerated stories first with a reminder of the 1665 London plague and second with a sharp critique of the reliability of the *Bills of Mortality* (Lee, *Defoe*, 436–438, 453–455).

25. Several more recent works on the *Journal* look into the differences between official discourses about the plague and private experiences. It is true that the populace's resistance to official measures is an important theme in the *Journal*. But such resistances as H. F.'s doubts about the *Bills of Mortality* are part of the numerical picture of the plague. A dichotomy between the discourse from above and the experience from below, in contrast, is entirely alien to Defoe (see, e.g., Moore, "Governing Discourses").

26. See Mayer, "Reception." In Bastian's words: "What Defoe did in the Journal, and how he did it, is . . . clear enough: and any doubts that remain whether to label it 'fiction' or 'history' arise from the ambiguities inherent in those words" (Bastian, "Defoe's 'Journal,'" 172).

27. The general turn in Defoe scholarship from the "realistic" novel in Ian Watts's sense of the word to the allegorics of conversion has also been attempted for the *Journal*. Identifying patterns of autobiography in the *Journal* meets with difficulties, however; see Zimmermann, "H. F.'s Meditations." On the allegorical technique of the representation of catastrophes in the *Journal* in general, see Rosen, "Plague, Fire, and Typology."

28. Defoe, *Journal of the Plague Year*, ed. Landa, 1.

29. Ibid., 9.

30. Ibid., 12.

31. Ibid., 8.

32. Richetti speaks of an irrational choice as the prerequisite for H.F.'s observer's position. This may however be too hasty a remark. Richetti's understanding of observation as the production of "order from noise" is nevertheless still a pioneering innovation in scholarship on the *Journal* (Richetti, *Defoe's Narratives*, chap. 7).

33. To supply some context for the passage: H.F., observer and reader of tables, has finally begun to keep his own mortality list. In it, he wishes to count "all such, I mean of those Professions and Employments, who thus died, as I call it, in the way of their Duty." From his own private lists, H.F. then turns to reconsidering the *Bills of Mortality* of the City of London (Defoe, *Journal of the Plague Year*, ed. Landa, 237).

34. See Zimmermann, "H.F.'s Meditations"; Rosen, "Plague, Fire, and Typology," 268–273.

35. Defoe, *Journal of the Plague Year*, ed. Landa, 238.

36. Ibid.

37. For the influence of scientific discourses on H. F., see Burke, "Observing the Observer."

38. Defoe, *Journal of the Plague Year*, ed. Landa, 233.

39. This detail sharply expresses the narrator's "dual position of observer and observed," which according to Moore is symptomatic "of a deeper opposition between perspectival unity and informational multiplicity, the two elements fundamental to the Journal's realism" (Moore, "Governing Discourses," 139).

40. The actual title of Schnabel's novel is *Wunderliche Fata einiger See-Fahrer* (Strange Fates of Some Seafarers). However, the Romantic poet Ludwig Tieck published a considerably shortened and revised version in 1828 under the title *Die Insel Felsenburg*, and the latter title has prevailed. For the sake for clarity, I shall stick to the Romantic revision of the early eighteenth-century novel.

41. A fourth installment appeared in 1743. Our reading, however, concentrates on the three first volumes only. Interpretations of *Die Insel Felsenburg* start with deciding on which material to base the reading. Most readers have focused only on the first volume, based on the assumption that the installments make up an open-ended series with the first volume offering all of the most important characteristics (see Brüggemann, *Utopie und Robinsonade*). Grohnert in contrast shows convincingly that at least the first two volumes, probably the first three, follow a coherent plan for a "novel of the state" (*Staatsroman*) (Grohnert, "Schnabels Insel Felsenburg," esp. 606ff.).

42. Fohrmann, *Abenteuer und Bürgertum,* reconstructs this group of novels in the German context.

43. On the transition from limited economic statistics to general administrative statistics in the 1720s in Prussia and the table under twenty-four headings that Frederick Wilhelm instituted in Pomerania in 1722, see Behre, *Geschichte der Statistik*.

44. The public population lists were in fact compiled according to the church year in Protestant states because surveying the data and its first analysis were the responsibility of the spiritual departments in the governments. In Prussia, the

Generaltabelle was not switched to the financial year (beginning March 1) until 1799 (Behre, *Geschichte der Statistik*, 148).

45. Schnabel, *Wunderliche Fata*, vol. 2, 174.

46. Ibid., 78.

47. Behre, *Geschichte der Statistik*, 133ff.

48. The novel can easily be read as an account of colonization in the spirit and style of early modern universal history in Europe.

49. Süssmilch, *Göttliche Ordnung*, introduction, 1: 4–46.

50. Schnabel, *Wunderliche Fata*, 2: 78.

51. Readers so far have shown little interest in the numerical insertions. From the point of view of a dialectic of Enlightenment, however, Mog draws our attention to the fact that the "entire text" is "shot through with numbers." In a similar vein, Fohrmann is primarily interested in the formation of rationalized— "counting"—subjectivity. Mog, *Ratio und Gefühlskultur*, 67ff.; Fohrmann, *Abenteuer und Bürgertum*, 68ff.

52. Schnabel, *Wunderliche Fata*, 2: 174.

53. Ibid., 508.

54. In *Cérémonies de l'information*, Michèle Fogel uses the concept of ritual information in discussing the common criers in the early modern cities of western Europe and the masses of thanks and supplication for the king connected to state events such as military victories and births of royal heirs.

55. The Romantic rediscovery of *Die Insel Felsenburg* by Ludwig Tieck and E. T. A. Hoffmann indulged in the colorful narratives and took the novel in the spirit of a bizarre "old German" or baroque fantasy. Later readers concentrated on finding traces of Enlightenment and early bourgeois culture in German contexts. In both cases the most striking and defining feature went unnoticed: the fact that the narrative is a *Staatsroman*, a "novel of the state."

56. In the 1970s, interest in the politically utopian elements of the novel predominated in scholarship (see Mog, *Ratio und Gefühlskultur*; Jacobs, *Prosa der Aufklärung*, 136–145). Whereas this is a rather forced reading of the novel, earlier critics (Brüggemann's *Utopie und Robinsonade*) convincingly show that *Die Insel Felsenburg* belongs to a literary genre that from early modernity until the mid nineteenth century figured in the German case under the heading of *Staatsroman* (political romance or novel of the state) and, as such, were part of the larger area of *Staatswissenschaft* (science of the state; see Mohl, *Geschichte und Literatur*, vol. 1, chap. 3, "Die Staatsromane"). In this context, Mohl groups, e.g., Plato's *Republic* (whose German title, *Der Staat*, provides Mohl with his terminology) together with Gabriel des Foigny's *Jacques Sadeur* and Xenophon's *Cyropaedia* under the heading of "Staatsroman." For more recent readings in this tradition, see Schings, "Der Staatsroman im Zeitalter der Aufklärung."

57. Botero's *Relationi universali* deals with administration and politics and refers to the reports of a messenger or spy. The German translation *Allgemeine weltbeschreibung* (1596) reminds the reader rather of the scholarly language of historiography as in Sebastian Münster's famous *Cosmographia* (1544).

58. Botero, *Relationi universali* (1595); the 1596 German translation, *Allgemeine weltbeschreibung* (Cologne: J. Gymnicus's Heirs, 1596), "An den Leser" (unnumbered), is cited here.

59. Alpers, *Art of Describing* (1983), reconstructs the aesthetic and epistemological implications of description for Dutch painting in the seventeenth century, drawing on modern theories of the novel. As her earlier essay "Describe or Narrate?" shows, Alpers shares Auerbach's skepticism about the dominance of narration as championed by Lukács, and she follows Barthes's reversion to the *degré zéro* of the *nouveau roman*. Though the Dutch golden age may have little to do with the modern novel, Alpers's discussion of *narratio* and *descriptio* is of highest relevance for the issues at hand.

60. Schnabel, *Wunderliche Fata*, 1: 181ff.

61. Albert Julius is active once more as the founder in the narrower sense when he orders the construction of the church, which is then completed in the second volume. It is in respect to this founding deed, that he declares his desire to hold "a general inspection in this my small kingdom" (ibid., 108).

62. On the early modern "description of people and government," in the German tradition of *Staatsbeschreibung* (description of the state), see the comprehensive discussion in Rassem and Stagel, *Statistik und Staatsbeschreibung*.

63. Schnabel, *Wunderliche Fata*, 3: 230.

64. Ibid., 230–250.

65. Ibid., 7, 10.

66. Ibid., 305, 341.

67. Bose, *Introductio*, 5. See also Conring, *Exercitatio*, in id., *Opera*, 4: 1–4.

68. Bose, *Introductio*, 5.

69. This can be understood as an implementation of a metaphorics of cartography, which was found particularly fascinating in the seventeenth and eighteenth centuries; see Delft, *Littérature et anthropologie*, chaps. 2 and 3.

70. Schnabel, *Wunderliche Fata*, 1: 99.

71. Alpers, *Art of Describing*, 234–264.

72. See, e.g., the map of Amsterdam in Braun and Hogenberg, *Civitates Orbis Terrarum*, 1: 20, and the map in *Die Insel Felsenburg*,.

73. Schnabel, *Wunderliche Fata*, 1: 99.

74. Leibniz, "Entwurf gewisser Staatstafeln," in id., *Werke*, 5: 303–314, here 308ff.

75. Halley, "Historical Account of the Trade Winds." See also Thrower, "Edmond Halley as a Thematic Geo-Cartographer."

76. Tufte, *Visual Display*, esp. 32–34; and *Envisioning Information*.
77. See Funkenhouser, "Historical Development," and Beniger and Robyn, "Quantitative Graphics."
78. See Goody, *Domestication of the Savage Mind*, chaps. 5 and 6.

CHAPTER 10

1. See Voßkamp, *Romantheorie in Deutschland*.
2. "I am surprised at his [Bayle's] not having taken into account that this romance of human life, which makes the history of the human race, lay fully devised in the divine understanding, with innumerable others" (Leibniz, *Theodicy*, § 149; 217).
3. Swift, *Gulliver's Travels*, bk. 1, chaps. 4 (44ff.) and 6 (55); bk. 2, chap. 4 (116ff.); bk. 3, chap. 3 (178–182); bk. 4, chap. 9 (294).
4. Ibid.,138.
5. For Swift's critique of the new science, see Nicolson and Mohler, "Scientific Background."
6. In the original French, the Berlin Academy's question reads: "Les Evénemens de la bonne et de la mauvaise fortune dépendant uniquement de la volonté, ou du moins de la permission de Dieu, on demande si ces événemens obligent les hommes à la pratique de certains devoirs et quelle est la nature et l'étendue de ces devoirs" (Harnack, *Geschichte*, 2: 305).
7. Although the Academy had desired a French version, Kästner entered his essay in Latin. In the preface to the prize publication, which at Kästner's request is in both languages, and in a letter to Maupertuis, Kästner distances himself from the French translation. It is not clear whether, as a follower of Gottsched, he wished to demonstrate against the demand for a French version, or whether on the contrary, he was peeved because his own French had been corrected in Berlin. Prussian Academy of Sciences, *Pièce qui a remporté le prix sur le sujet des événements fortuits* (1751), foreword; Kästner to Maupertuis, October 18, 1751, in *Maupertuis et ses correspondants*, edited by A. Le Sueur (Montreuil-sur-Mer: Notre Dame des Prés, 1897), 304–306.
8. The quotation follows the German version of the prize essay ("Pflichten"), which introduces Kästner's *Vermischte Schriften. Erster Theil* (1755); cited here from Kästner, *Gesammelte poetische und prosaische Schönwissenschaftliche Werke*, pt. 3: 59 (henceforth cited as *GW*).
9. Kästner, "Pflichten," *GW*, 3: 60.
10. Cf. Baasner, *Kästner*, 89–97.
11. For an authoritative discussion of Kästner, see Baasner, *Kästner* (on the prize essay, 449ff. and 454). Baasner tends to overlook the political importance of the competition and Kästner's intervention. Neither does Baasner recognize

Kästner's originality in including Daniel Bernoulli's theory of value in the interpretation of probability.

12. In Göttingen, Kästner wrote the foreword to a German translation of Benjamin Martin's *Philosophia Britannica: oder neuer und fasslicher Lehrbegriff der Newtonschen Weltweisheit, Astronomie und Geographie* (Leipzig, 1778).

13. Battles raged about all three of the questions: between German admirers of Leibniz and sarcastic commentators from the Paris of the *Encyclopédie*; between the Academy and Moses Mendelssohn; and finally between the Academy's president, Maupertuis, and its secretary, Jean-Louis Formey, a Huguenot refugee born in Berlin with the mission of proselytizing his Wolffian agenda (see Harnack, *Geschichte*, 1.1: 401–409).

14. In 1747, Kästner taught Leibniz's calculation of interest, which had been the core of juridical mathematics. In 1749, he lectured on the status of probability theory in Europe. In 1750, he dealt with the juridical-mathematical topic of calculating interest again, this time with a view to working out the value of debt (see *Abraham Gotthelf Kästner's Selbstbiographie*, index of Kästner's works, 16). The titles read: *Pro justitia calculi interusurii Leibnitii*, "Physicae jurisprudentiam illustrantis specimina," and *Gradus et mensuram probabilitatis dari*.

15. Chladenius, *Vernünftige Gedanken*.

16. Rüdiger, *De sensu veri et falsi libri IV* (1709). This work was the twelve-year-old Kästner's first philosophical reading (see *Abraham Gotthelf Kästner's Selbstbiographie*, 3).

17. See Wolff, *Philosophia rationalis sive Logica*, in id., *Gesammelte Werke*, pt. 2, vol. 1, § 539.

18. Baumeister, *Institutiones philosophiae rationalis*; Kahle, *Elementa logicae probabilium*.

19. Süssmilch, *Göttliche Ordnung*; see also Cataldi Madonna, *Filosofia della probabilità*.

20. See Danneberg, "Probabilitas hermeneutica."

21. *Abraham Gotthelf Kästner's Selbstbiographie*, 11.

22. D. Bernoulli, *Werke*, 2: 223–234. Bernoulli's *Specimen theoriae novae* appeared in 1747 in the first edition of the *Hamburgisches Magazin* as a partial German translation under the title: "Versuch einer neuen Lehre, von dem Maaße der Glücksspiele" ("Attempt at a new theory of the measurement of games of chance"), 73–90. The passage Kästner is referring to in Daniel Bernoulli reads as follows: "Ever since mathematicians first began to study the measurement of risk, there has been general agreement on the following proposition: Expected values are computed by multiplying each possible gain by the number of ways in which it can occur, and then dividing the sum of these products by the total number of possible cases where, in this theory, the consideration of cases which are all of the

same probability is insisted upon" (D. Bernoulli, "Exposition of a New Theory on the Measurement of Risk," 22 -23.)

23. Kästner, "Pflichten," *GW,* 3: 71ff.

24. Jorland, "Saint Petersburg Paradox."

25. See the excellent presentation in Daston, *Classical Probability,* 76–108.

26. Lichtenberg, "Betrachtungen über einige Methoden, eine gewisse Schwierigkeit in der Berechnung der Wahrscheinlichkeit beim Spiel zu heben" (Observations on Some Methods to Solve a Particular Difficulty in Calculating Gaming Probability), in id., *Schriften und Briefe,* 3: 9–23.

27. D. Bernoulli, *Exposition of a New Theory on the Measurement of Risk,* 23–24.

28. Daston, *Classical Probability,* 72–76.

29. Because Bernoulli assumes, first of all, that infinitely small winnings correspond to continuously infinitely small increases in advantage, and, second, that the advantages behave in an inverse proportion to the growth of the starting fortune, he comes to a logarithmic curve that has its asymptote in the ordinates of an infinitely large starting fortune. Bernoulli discusses objections that would be brought up again and again in the development of his theorem of econometrics in the nineteenth and early twentieth centuries. First, an individual is capable of simply not adhering to the same model curve as others. The rich miser attributes a greater value to every ducat that he wins or loses than someone who may not be quite so rich, but is also not avaricious. Bernoulli's answer: this objection relates only to the application of the curve to different individuals, not to the construction of the curve. Second, there are cases of higher complexity in which certain assumptions change. According to the modeling curve, to name one example, a given profit has a lesser value for a rich man than for a poor one. The rich man could be taken captive, however, and be obliged to buy his freedom. In this case, the required sum would have a greater value for him than for the poor man. As Daniel Bernoulli remarks, the modeling here proves to be subject precisely to whatever it models: it is thus only "very probable" (Bernoulli, "Specimen theoriae," 224ff.).

30. See, e.g., entry no. 13 in Prussian Academy of Sciences, *Pièce qui a remporté le prix sur le sujet des événements fortuits* (1751), 57–114, esp. 64–92.

31. [J. G. Töllner], "Discours sur le sujet proposé par l'Académie Royale," in Prussian Academy of Sciences, *Pièce qui a remporté le prix sur le sujet des événements fortuits* (1751), 47–56, here 52–56.

32. Johann Gottlieb Töllner, "Gedancken über die Pflichten, zu welchen uns die glücklichen oder unglücklichen Begebenheiten in der Welt verbinden" (Thoughts on the Duties Which Are Imposed Upon Us by the Fortunate and Unfortunate Events in the World), in Prussian Academy of Sciences, *Pièce qui a remporté le prix sur le sujet des événements fortuits* (1751), 114–134, here 121–131. True casuistry is unfolded in "Succincta commentatio Philosophica," ibid., 57–114:

First, there are duties regarding chance in general—we should see the work of God in chance as in everything else; but since we usually do not succeed in doing so, we must humble ourselves before God's omniscience (64–71). Then there are particular duties. The morally good imagine themselves to be in the fortunate state of grace as children of God and behave accordingly to others; the bad should understand fortune as a call to better themselves, but usually only become more foolish. In misfortune, the good should guard against doubting God's purpose; instead, they can see misfortune as a means to bliss if they have assured themselves of its divine origin; the bad should see misfortune as punishment but not despair that they might yet in the end enjoy bliss through the grace of God (71–108).

33. "Succincta commentatio," in Prussian Academy of Sciences, *Pièce qui a remporté le prix sur le sujet des événements fortuits* (1751), 64.

34. Kästner, "Pflichten," *GW,* 3: 61.

35. Ibid., 62.

36. Ibid., 70.

37. Daston, *Classical Probability*. This is how Kästner formulates a context from mathematical theory, which previously had only been accessible in literary adaptations of probability. It is no coincidence that Defoe's *Robinson Crusoe* was Kästner's "bible" (*Leibbuch*) (*Abraham Gotthelf Kästner's Selbstbiographie,* 5).

38. Kästner, "Pflichten," *GW,* 3: 71.

39. Ibid., 75.

40. Aside from God and Moses, Derham is the only author whom Süssmilch quotes in the introduction to *Göttliche Ordnung* (1: 11).

41. See Campe, "How to Use the Future."

42. In an excellent study, *Der andere Roman,* Eckart Meyer-Krentler shows how Gellert's *Schwedische Gräfin* is characterized by differing from the genre expectations of the traditional romance, an operation that was standard in contemporary France and England.

43. Gellert, "Moralische Vorlesungen," in id., *Gesammelte Schriften* (henceforth cited as *GS*), 6: 226. For the pedagogical exercise of the same duty to chance in childhood, see 244ff. For an excellent discussion of Gellert's novel in the context of providence, see Frick, *Providenz und Kontingenz*.

44. Gellert, *GS,* 6: 229ff.

45. Ibid., 193. Gellert works this claim out in his essay "Warum es nicht gut sey, sein Schicksal vorher zu wissen" (Why It is Not Good to Know One's Fate in Advance), in Gellert, *GS,* 5 (pt. 3): 1–20. See also Kästner, "Pflichten," *GW,* 3: 75.

46. Gellert, *Leben der schwedischen Gräfinn von G***, GS,* 4: 36.

47. Gellert, *GS,* 4: 49

48. Ibid., 20 and 14.

49. Ibid., 23.

50. For further commentary on this section of the novel, see Meyer-Krentler, *Der andere Roman*, 67–84.
51. Gellert, *GS*, 4: 19.
52. Ibid., 2, 22.
53. Ibid., 4, 36.
54. Ibid., 23.
55. Ibid., 36.
56. For a comparative discussion of providence with regard to the history of theological and philosophical ideas as well as a concept of narrative form, see Frick, *Providenz und Kontingenz*. For the French tradition, see Behrens, *Umstrittene Theodizee*.
57. Huet, *Traité de l'origine des romans*, 116.
58. Ibid., 114ff.
59. Ibid. For the *translatio* of the novel from the Orient to Greece, see 182; for the Roman reception,183–187.
60. "I could have wished that the novel of these times had a better denouement [*Entknötung*]; but perhaps it has not yet reached its end," Leibniz wrote to the novelist Duke Anton Ulrich von Braunschweig-Wolfenbüttel. "And just as Your Highness is not yet finished with your *Octavia*, so too can our Lord God still have to add a number of volumes to his novel, which thus may finally end up sounding better" (letter of October 3, 1713, cited from Kimpel and Wiedemann, eds.,*Theorie und Technik*, 1: 67).
61. Frick, *Providenz und Kontingenz*, 6.
62. See ibid., 16. For a similar argument, see Behrens, *Umstrittene Theodizee*, 10.

CHAPTER 11

1. For seminal discussions on probability and the rise of modern narration, see Köhler, *Literarische Zufall*; on probability and the modern (French) novel, cf. Kavanagh, *Enlightenment*.
2. Aristotle, *Rhetoric* 1402a (2.24.10).
3. Ibid.
4. The two preceding chapters can easily be recognized as examples for the framing of probabilities. The table is the form in which symbols become data: in the table, data are arranged and transformed according to the technical rules of statistics, and yet we can understand their meaning only from outside of the table's frame. The utility of chance, however, is the precise self-critique of probabilistics by taking into account the improbability of the game of probability. As a result, the view of chance in this case splits the spiritual exercises meant to address the vicissitudes of chance into risk calculation on the one hand and the aesthetics of the game of contingencies on the other.

5. Aristotle, *Poetics* 1456a (18).

6. Formalizing this process of transformation, we may rewrite (1) "It is generally improbable that a" and "It is in particular probable that not a" as (2) "It is specifically improbable (that it is generally probable that not a)." The latter wording pinpoints the structure of peripeteia because, in introducing the moral predicates (evil, unjust), we offer an independent cause for the transformation. The "specific improbability that it should be generally probable for clever men to be mistaken/ heroic men to be defeated" thus receives its own meaning. Such meaning then is the reality of the moral order or poetic justice. This is why I understand peripeteia to be the very transformation of (1) into (2), instead of (2) simply replacing (1).

7. Lugowski, *Form*, esp. 63–78.

8. Shklovsky, *Theory of Prose*. Shklovsky bases the structural analysis of narration in a paradigmatic way on framing. In making this argument, Shklovsky must substantiate at least two claims: (1) we must assume that there are forms of structure on the narrative plane of the text that are comparable with those on the level of the sentence, and (2) we have to accept that devices for arranging elements within a single plot-unity and for joining several plot-unities together are, from a technical point of view, the same. With the first argument, we arrive at the rhetorical plane of the narrative, with the second, the problem of *histoire* is tied to that of *discours* in the narration. Shklovsky develops the first argument in "The Relationship between Devices of Plot Construction and General Devices of Style" (id., *Theory*, 15–51). The second argument is discussed in "The Structure of Fiction" (ibid., 52–71).

9. See Miller, *Empfindsame Erzähler*, 101–105; Davis, *Factual Fictions*, chap. 1.

10. Viktor Shklovsky's 1916 essay, "Iskustvo kak priem," which became famous as a "manifesto of formulism," was translated in into English in 1965 as "Art as Technique," and in 1990 as "Art as Device."

11. Aristotle, *Poetics*, 58 (1460a 24).

12. Dacier, *Poétique d'Aristote*, 430.

13. Bray, *La formation de la doctrine classique en France*, esp. 191–239 . See also Genette, "Vraisemblance et motivation."

14. Breitinger, *Critische Dichtkunst*, 132.

15. Ibid. At this point, Breitinger thus bumps into the pun with *Schein* that has the semblance (*Schein*) of truth, a pun that comes out of the word "probability" (*Wahrscheinlichkeit*). Blumenberg made this wordplay prominent. Breitinger admittedly employs it in a narrower sense by opposing the semblance (*Schein*) of truth to the semblance (*Schein*) of falsehood. As we shall see, both are actually types of the semblance (*Schein*) of truth that emerge from the phenomenologization of probability. See Blumenberg, *Paradigms for a Metaphorology*, "Terminologization of a Metaphor: From 'Verisimilitude' to 'Probability,'" 81–98 (omitting the Breitinger reference).

16. Breitinger, *Critische Dichtkunst*, 130.

17. This is the conclusion of the fifth main section on the miraculous and the sixth section on the probable in Gottsched's *Versuch einer critischen Dichtkunst*, esp. 202–220.

18. Ibid., 150.

19. Ibid.

20. It is not the least of the accomplishments of Wolfgang Kayser's classical study that, right from the beginning, it ties the novelty of the new novel to its dominance for the concept of literature in general: "The role of literature as a field in which the individual feels motivated to view, compare, value, and order the fullness of appearances is now largely carried out by the novel" (Kayser, *Entstehung und Krise des modernen Romans*, 5).

21. Riccoboni picks up again from the passage in the ninth chapter of Aristotle's *Poetics* that claims poetry to be "more philosophical and valuable" than history. Riccoboni explains the greater universality of poetry as an approach to the universality of Platonic "most general forms" (*ultimae formae*), while history deals with "countless singularities" (*singularia innumerabilia*). This approach entails that poetry considers the particular in general (*in universum considerare singularia*). It does so by inventing as many causes and ways it could have happened (*Arten seines Geschehenkönnens*) as possible by the rhetorical topics of circumstances and adding them to a singular event. The poet becomes a model for the historian; he teaches [the historian] probability as an art of conjecture (hence an *Ars conjectandi*). Rhetorical-poetological considerations thus finally return to the relation between poetry and history (and philosophy): "And just as, in singularities, the poet contemplates the things can possibly take place, the historian can also form a judgment in the face of uncertain facts according to the measure of what could possibly have taken place, if he simply treats uncertain facts as uncertain and presents the more possible conjecture" (Riccoboni, *Poetica Aristotelis*, 45).

22. The same relation between the rhetorical *narratio verisimilis* and the probability of poetry is characterized in Dacier's commentary as follows: "the historian rarely explains why what he is narrating has happened. This is because such causes are mostly hidden, and if the historian explicates them, he gives conjectures rather than certainty and true facts. But since the poet is the master of his subject matter, there is not a single incident whose causes and effects he could not explain" (Dacier, *Poétique d'Aristote*, 137).

23. This [claim] applies in particular to Castelvetro's version of poetical probability. His *Poetica d'Aristotele* begins directly with the commentary of the two first sentences of the *Poetics* with the distinction between poetry and history (chap. 9). Castelvetro's basic idea is a parallel introduction, even fusion of the mimesis concept and the poetry/history distinction. The probability of poetry

hence has the same relation to the truth of history as representation to the [thing] represented: just as the represented [object] must be known before its representation, so too must truth (what has happened and the particular) [be known] before probability (the universal). Castelvetro, *Art of Poetry*.

24. Huet, *Traité de l'origine des romans*, 114–116.
25. Davis, *Factual Fictions*, esp. chap. 6.
26. For Germany, see Kimpel and Wiedemann, eds., *Theorie und Technik*; Lämmert et al., eds., *Romantheorie*; Voßkamp, *Romantheorie in Deutschland*. On Leibniz's disputed authorship of the Gotthard Heidegger recension, see Lämmert, *Romantheorie*, 58.
27. Wieland, *Geschichte des Agathon*, 5; the following citations, 5–7.
28. Craig, *Theologiae christianae mathematica christiana*.
29. Battestin, "General Introduction," 1: xvii.
30. Fielding, *Tom Jones*, bk. 8, chap. 1, 346.31. Ibid.
32. Empson, "Tom Jones," 233.
33. Fielding, *Tom Jones*, bk. 8, chap. 1, 348.
34. Ibid., 349.
35. Ibid.
36. Ingarden, *Literary Work of Art*.
37. Fielding, *Tom Jones*, bk. 3, chap. 1, 101.
38. Iser, "Role of the Reader," 51.
39. Fielding, *Tom Jones*, bk. 3, chap. 1, 102.
40. Ibid., bk. 2, chap. 1, 68.
41. See Pufendorf, *Le droit de la nature et des gens*, 1: 110. Alongside and even before the "Genoese lottery," the *bianca* was the prototype for numerical lotteries, and hence a basic form of early modern lottery. The selection took place in two draws for the *bianca*. First, a number of slips of paper, equal to the amount of prizes offered, would be drawn. They were registered with numbers and names that the participants in the lottery had marked with a motto or prayer. Then for every drawn slip, another slip was drawn from a second urn that only contained marked or unmarked (*bianche*: blanks) slips. (Daston, *Classical Probability*, 142ff.).
42. Fielding, *Tom Jones*, bk. 2, chap. 1, 68).
43. Ibid., bk. 7, chap. 2, 321 . Ironically, what amounts to a circumstance to be skipped over is in this case the mention of the historical event. This it is that has remained a vacant space or a blank.
44. For the equivalency between the roles of the narrator and the reader since Fielding, see Lange, "Erzählformen," who traces the nature of narrators' and readers' attention in novelistic writing back to Locke's discussion of linguistic communication. Though Iser ties his investigation of the reader's role to traditional rhetorical and poetological terms, they model the condition of modern aesthetic

experience for him, that is, of the act reading (Iser, "Role of the Reader"). Miller contrasts the aesthetic function of the reader/narrator roles with the investigation of the "ontological conditions of narration," which he sees developed in Henry James. By way of this detour, he manages to distinguish a paradoxical combination of interior and exterior views for the narrator and reader, a characteristic that he specifically connects to sentimentalism (Miller, *Empfindsame Erzähler*, Introduction, esp. 21–27).

45. Wieland, *Agathon*, 33 and 553.

46. Ibid., 17.

47. Ibid.

48. Leibniz, "Über die Methode, reale Phänomene von imaginären zu unterscheiden."

49. Frick, too, reads the gnomons made by the protagonist Agathon and the "narrator" (or, rather, the editor) in context. He recognizes the figure of the probable improbability however only in the case of the "narrator"; for Agathon, Frick speaks in sharp contrast of a mere and, at that, antiquated rhetorical device. By doing so, he fails to recognize the mirror-image reversal between the two positions (Frick, *Providenz und Kontingenz*, 433–436, esp. 460).

50. Wieland, "Theages oder Unterredungen von Schönheit und Liebe: Ein Fragment," in id., *Werke*, 3: 172. The "spirit of the narrative" that the narrator Nicias mentions is an echo of the "poetical spirit," which can, however, only be referred to ironically in the narrative (174).

51. Not even Miller, in his study of the sentimental narrator of the eighteenth-century novel, recognizes the construction of the "author" in Agathon as a third position between the editor-I and the I of the diary-keeping hero. This is all the more surprising because a dual narrator, who switches from distanced overview to instant entanglement, is the exact guiding principle of Miller's own narrative analysis: he speaks of an "attitude of suspense" with regard to Wieland's authorial narrator (Miller, *Empfindsame Erzähler*, 133, 277, 312).

52. For Tom Jones, see, e.g., the thesis of double irony in Empson, "Tom Jones"; for Agathon: Frick, *Providenz und Kontingenz*, 460–463

53. "Unter diesen Umständen wurde Anton geboren, und von ihm kann man mit Wahrheit sagen, daß er von der Wiege an unterdrückt ward" (Anton was born under these circumstances, and it can be truly said that he was oppressed from the cradle on) (Moritz, *Anton Reiser*, *Werke*, 1: 40).

54. Wieland, *Agathon*, 553.

55. Ibid.

56. Ibid.

57. Regarding the circumstances, Frick discusses a tension between psychological realism and fictional description (*Kontingenz und Providenz*, 383–386).

58. Wieland, *Agathon*, XI. 1, 554.

59. In the later version, Wieland deletes the ironic introductory chapter of the final book. In the same stroke, he strengthens the autobiographical grounding of the novel with the "secret story of Danae" and the staging of Agathon as the writer of his own life experiences (*Selbsterlebensschreiber*). But by doing so, he brings the figure of improbable probability out of balance. The first version of Agathon hovered in the middle between a paradox and its resolution: the paradox emerged in the one-dimensional narration and argumentation, whereas the paradox-solving hermeneutics is brought about by the observer who looks from one world into another. This mirrors precisely the fragile balance in the figuration of improbable probability. Since Aristotle this figure, which was emblematically connected to the name of the comedy writer "Agathon," meant both the paradox and the hermeneutics cleared of paradox. Wieland decided in the last version of the novel, however, for dialectical synthesis and the reflective autobiography, and, with it, for a continual reintegration of the paradox in ever new versions from which the paradox has been removed. That synthesis finally becomes a narration negotiated between two parties, the first-person narrator and the editor. In this synthesis, there is no longer any use for the third function.

60. Aristotle, *Rhetoric* 1402a (2.24.10). The line comes from a lost tragedy of Agathon (frg. 8 Snell).

61. For Achenwall's position in *The History of Statistics*, see the striking ambivalence in John, *Geschichte der Statistik*. On the one hand, John assigns Achenwall a place in the early period of a first constitution of academic statistics, 1660–1800, on the other hand, he sees him as already a part of the ensuing period of fully developed statistics up to Quetelet in the nineteenth century.

62. Süssmilch, *Königl[ichen] Residenz Berlin schneller Wachsthum und Erbauung*, 11.

63. Ibid., 25.

64. Assmann, *De rerumpublicarum notitia*, 4ff.

65. Heidegger, *Mythoscopia*. An excerpt can be found in Lämmert, *Romantheorie*, 58–61.

66. Hoffmann, *Entwurff einer Einleitung*; Otto, *Primae lineae*; Gundling, *Ausführlicher Discours*; Schmeitzel, *Praecognita historiae*.

67. The famous sentence Saladin delivers to the wise Nathan reads: "A man like you doesn't stay where the chance of birth threw him" (Lessing, *Nathan der Weise*, act 3, sc. 6, vv. 329–330, in id., *Werke*, 2: 274; *Nathan the Wise* [act 3, sc. 5], 70). Before Saladin even pronounces this formula of the chance of birth, Nathan has already said to the Christian Templar, who rashly agrees: "Neither of us has chosen his people" (act 2, sc. 5, vv. 520–521; *Nathan der Weise*, 253; *Nathan the Wise*, 57). The two fathers, who have either never had any children or have lost them, declare themselves masters of the accident of birth. Recha, the adopted daughter of the one, and in truth the niece of the other, also knows the clause of

the chance of birth. In contrast to the men, she recognizes in contingency a protective moment of security: "And how do you know what clod of earth you are born *for* if it's not for the one that you were born *on*?" (act 3, sc. 1, vv. 42–45; *Nathan der Weise*, 263; *Nathan the Wise*, 63). The structure of *Nathan the Wise* could be developed as a whole from this casuistry of the proposition of the accident of birth. See Schneider, "Zufall der Geburt," and Wellbery, "Zufall der Geburt."

68. Achenwall, *Notitia rerum publicarum*, 8.

CHAPTER 12

1. In his foreword to Süssmilch's *Göttliche Ordnung* (n.p.), Wolff says: "[A]s yet no one has discussed in a comprehensive theory how to use the faculties of reason in order to explore that which is probable. But attempts have been made to test how a theory of probability might be developed for the sake of use in everyday life. The present book is among these."

2. Kahle, *Elementa logicae probabilium*, here esp. Praefatio, xii–xiv, and, in this connection, § III.

3. Ibid., §§ 123 and 192–196.

4. Ibid., § 184; for the a posteriori nature of the observation of event frequencies, see §§ 178 and 181–182.

5. Heidegger, *Essence of Reasons*.

6. Wolff, *Philosophia rationalis sive Logica*, in id., *Gesammelte Werke*, pt. 2, vol. 1, § 578.

7. For the *requisita* analysis of the probabilities of concept and reference in the subject of the proposition, see Kahle, *Elementa logicae probabilium*, §§ 136–140 and § 2, "De logica probabilium."

8. "A singular action together with its outcome is called an event (*eventus*). That which stands in relation to the event is a circumstance (*circumstantia*). The configuration of such relations and of further occurring relations of the event is the occasion (*occasio*), and its cause the occasional cause (*causa occasionalis*)" (Baumgarten, *Metaphysica*, § 323, 98).

9. Seifert, *Cognitio historica*.

10. Patrizi, *Della historia dieci dialoghi*, in Kessler *Theoretiker humanistischer Geschichtsschreibung*, 60.

11. Ibid., 60–63, here 62.

12. Gassendi, *Ad librum D. Edoardi Herberti Angli, De veritate*, in id., *Opera omnia*, 3: 413b.

13. Shapin and Schaffer, *Leviathan and the Air-Pump*, esp. chap. 2; for the concept of literary technology in the ensemble of material and social technologies, see 25ff.

14. Robert Boyle, *New Experiments physico-mechanical*, in id., *Works*, 1: 2.

15. See Solbach, *Evidentia und Erzähltheorie*, as well as Meuthen, *Selbstdarstellung,* specifically "Rhetorische Evidenz, semiotische Transparenz und die Anfänge der Ästhetik," 79–114.

16. Kahle, *Logica probabilium*, § 115 (96).

17. Chladenius, *Nova philosophia definitiva*, 70.

18. Goethe, *Conversations of German Refugees*, 36; *Unterhaltungen deutscher Ausgewanderten*, in *Werke*, HA, 6: 157.

19. "When they pressed him harder he tried to avoid a reply by offering to tell a story himself" (ibid.).

20. Ibid., *Conversations*, 37. *Unterhaltungen*, 159.

21. Ibid., *Conversations*, 20. *Unterhaltungen*, 133.

22. Goethe's *Conversations* are also concerned with material maintenance (as in the Antonelli and Ferdinand novellas) and the upkeep of social forms such as marriage (in the Procurator novella), playing on the homonyms *Unterhaltung* (conversation, amusement) and *Unterhalt* (maintenance).

23. Goethe, *Conversations of German Refugees*, 37; *Unterhaltungen deutscher Ausgewanderten*, in *Werke*, HA, 6: 159.

24. The connection between probability and statistical error theory has not yet been investigated in Lambert scholarship. Histories of philosophy barely acknowledge his mathematical and experiment writings (Arndt, *Möglichkeitsbegriff*). Histories of science, on the other hand, have mostly failed to follow up on indications to Lambert's philosophical writings (Sheynin, "J. H. Lambert's Work on Probability").

25. Lambert, *Cosmologische Briefe*, preface, 89–93.

26. Ibid., 89; Lambert, *Cosmological Letters*, 46.

27. See Lambert, "Anmerkung über die Sterblichkeit, Todtenlisten, Geburten und Ehen," in id., *Beyträge zum Gebrauche der Mathematik*, 3: 476–569.

28. Stigler, *History of Statistics,* shows that construction of "society" as subject to statistical probability derives in a roundabout way from the physical sciences, notably astronomy, rather than directly from Jacob Bernoulli's theory of moral judgment.

29. Hume, *Treatise*, pt. 2, 50–120.

30. Campbell, *Philosophy of Rhetoric*.

31. "This [i.e., the observation of possible results in dice tossing] differs from experience, inasmuch as I reckon the probability here, not from numbering and comparing the events after repeated trials, but without any trial, from balancing the possibilities both sides. But though different from experience, it is so similar, that we cannot wonder that it should produce a similar effect upon the mind" (ibid., 57).

32. Campbell's placement of probability in the new rhetoric follows [that of] Richard Whateley at the beginning of the nineteenth century (Whately, *Elements of Rhetoric*, 76–81).

33. Lambert, *Neues Organon,* in *PS,* vols. 1 and 2, here 2: 318–319.
34. Ibid.
35. Ibid., 318.
36. Ibid., 320.
37. Ibid.
38. On game theory in the narrower sense, see ibid., 320–323. On "a posteriori probability," see the construction of the statistical mean in mortality, a case for which Lambert introduces the principle of indifferent recording of all occurring cases; the organization of regulated and comprehensive observation in the example of meteorology; and the favorite eighteenth-century probability of the "causes" (ibid., 324–328). Lambert deduces the theory of incomplete induction from the difficulties in observations where conclusions are made by reason or consequence of the occurrence or nonoccurrence of an event; he drafts a set of conditions under which conclusions from the part to the whole come about so that the question of more or less complete induction can arise; as Lambert shows moral proofs always belong to this set of inductive conclusions (ibid., 234).
39. Ibid., 234.
40. Ibid., 356.
41. Ibid., 356.
42. Ibid., 358.
43. Ibid.
44. Ibid., 359.
45. At this point in the history of probability, a change from semantic understanding to operative symbolics takes place. Sybille Krämer identifies such tendencies of mathematical notation since modern times (Krämer, "Kalküle als Repräsentation").
46. E.g., Lambert, *Neues Organon,* in *PS,* 2: 360–361, 356, 367.
47. Lambert's concept of *Zeichnung* means semiotics in general, in particular, however, also the diagrammatical representation of logical relations that can be used in the manner of mathematical symbols. Lambert had developed the so-called calculus of lines (*Linienkalkül*) to this end. He offers brief introductions to his calculus of lines in the *Dianoiologie* (The Study of Thought) of the *Neues Organon* in connection with the theory of statements (*PS,* 1: 111–120), the doctrine of conclusions (ibid., 125–140), as well as in the *Semiotik* (*PS,* 2: 5–44). For the history of the calculus of lines, cf. *Anlage zur Architectonic,* Lambert, *PS,* 3: xxi; cf. also the fragment on the algebra of logic calculus (*Logikkalkül*), "Versuch einer Zeichenkunst in der Vernunftlehre," *PS,* 6: 3–14.
48. Lambert, *Anlage zur Architectonic, PS,* 3: 77.
49. "The word 'is' shows a kind of ambivalence . . . , because sometimes it means as much as to exist, and insofar as it is opposed to what is possible or necessary." From this reading of the copula, Lambert derives various modes

of ambiguity: (1) "Is possible or may be"; (2) "Is true or exists"; (3) "Is necessarily there or must be there" (Lambert, *Anlage zur Architectonic*, PS, 3: 196–197).

50. In this regard, see esp. *Anlage zur Architectonic*, chap. 7, "Das Seyn und das Nichtseyn" (Being and Nonbeing) (*PS*, 3: 196), and further §§ 103–104, 163, 245, 265, 314.

51. Lambert, *Anlage zur Architectonic*, PS, 3: 196.

52. For the following account, see *Neues Organon*, in *PS*, 2: 356–357; *Anlage zur Architectonic*, in *PS*, 3: 122–123, 198–199, 283–284.

53. Lambert, *Anlage zur Architectonic*, in *PS*, 3: 123.

54. Ibid.

55. For the historical and systematic introduction of operative terms, see Krämer, *Berechenbare Vernunft*.

56. See Lambert. *Neues Organon* § 152 (*PS*, 2: 321–323) and *Anlage zur Architectonic* § 314 (*PS*, 3: 315–316).

57. For the following, see Lambert, *Neues Organon*, PS, 2: 321–323.

58. For the meaning of this and similar passages, in addition to Sheynin ("Lambert's Work on Probability," 245–246), see Hacking, *Emergence of Probability*, 130. Hacking takes up the passage for the frequency interpretation of probability.

59. In this regard, see Svagelski's exhaustive *L'idée de compensation en France, 1750–1850*.

60. For the following, see Lambert, *Neues Organon*, PS, 2: 323.

61. See Lambert, *Anlage zur Architectonic*, PS, 3, "Hauptstück vom Vorher und Nachher Seyn" (Chapter on Being-Before and Being-After), §§ 313–336.

62. Lambert, *Anlage zur Architectonic*, PS, 4: 464.

63. Ibid., 465.

64. Ibid., 466.

65. Ibid., 469. See also the discussion of error theory in "Anmerkungen und Zusätze zur praktischen Geometrie," "Theorie der Folgen der Fehler," "Folgen der Fehler in zusammen gesetzten Umständen," and "Theorie der Zuverläßigkeit der Beobachtungen und Versuche," in Lambert, *Beyträge zum Gebrauche der Mathematik und deren Anwendung*, 3, pt. 1, 1: 215–274.

66. Lambert, *Anlage zur Architectonic*, PS, 4: 472–473.

67. I. Schneider, "Mathematisierung des Wahrscheinlichen," and Sheynin, "Newton and the Classical Theory of Probability," 217–243.

68. See Sheynin, "J. H. Lambert's Work on Probability," 250–254.

69. "Fragment einer Systematologie" can be found in Lambert, *PS*, 7: 385–413. Cited here from the reprint in Lambert, *Texte zur Systematologie*, 123–144, here §§13–16, 132–133.

70. Lambert, *Texte zur Systematologie*, §§ 17, 133.

71. On Lambert's "systematology" and recent system theory, see Siegwart's introduction in Lambert, *Texte zur Systematologie*, xxxvii–lxxxvii.

72. Lambert, *Neues Organon*, PS, 2: 217–435.

73. Ibid., PS, 2: 218–219, 421–422.

74. Ibid., PS, 2: 421.

75. Ibid., PS, 2: 254–255.

76. The analysis of probability (*Wahr-Scheinlichkeit*) into the appearance (*Schein*) of truth (*Wahrheit*) is characteristic of the entire section on "Phenomenology or the Doctrine of Appearance" in Lambert's *Neues Organon;* see esp. "Von den Arten des Seins" (On the Different Kinds of Being), in PS, 2: 217–237.

CHAPTER 13

1. Lambert to Kant, February 1765, in id., *Briefwechsel*, vol. 1, PS, 9: 342. Lambert's *Neues Organon* had just been published, and his *Anlage zur Architectonic* was in preparation. Lambert had probably heard rumors that Kant was working on a treatise to revise metaphysics. Kant hints at this in his reply to Lambert, see Lambert, *Briefwechsel*, vol. 1, PS 9: 337.

2. Lambert to Kant, November 1765, in id., *Briefwechsel*, vol. 1, PS, 9: 335–336 (Kant, PC, 44).

3. Lambert, *Briefwechsel*, vol. 1, PS, 9: 336. (The passage is translated from the German original; Kant, PC, offers a different wording.)

4. Kant to Lambert, September 2, 1770, in Lambert, *Briefwechsel*, vol. 1, PS, 9: 337 (Kant, PC, 58).

5. Kant in Lambert, *Briefwechsel*, vol. 1, PS, 9: 337 (Kant, PC, 59–60).

6. For the basic argument, see Blumenberg, *Paradigms for a Metaphorology*. Throughout his vast investigations into the history of ideas and the sciences, Blumenberg came back to Lambert again and again. In the above context, we might, with all due caution, think of the fact that Lambert coined the modern notion of phenomenology even before Kant and before Hegel. Blumenberg, for his part, devised "metaphorology" as the historical branch of a Husserlian phenomenology.

7. Lambert to Kant, Berlin, 1770, in id., *Briefwechsel*, vol. 1, PS, 9: 361–362 (Kant, PC, 63–64).

8. Later on, Kant used this formula to reply to Lambert's critique of the transcendental aesthetic: the reality of time (and space) should not be regarded as those of objects, but rather as "the way of representing myself as object" (Kant, *Critique of Pure Reason* [henceforth cited as *CPR*], 182).

9. Lambert, *Briefwechsel*, vol. 1, PS, 9: 362–363. "The whole intellectual world [*Gedankenwelt*] is non-spatial; it does, however, have a counterpart [*Simulachrum*] of space" (Kant, PC, 64).

10. Lambert, *Briefwechsel*, vol. 1, *PS*, 9: 362 (Kant, *PC*, 64).

11. On the status of "simple concepts," see Arndt, *Der Möglichkeitsbegriff bei Christian Wolff und Johann Heinrich Lambert*, 252–277.

12. Lambert, *Briefwechsel*, vol. 1, *PS*, 9: 363 (Kant, *PC*, 65).

13. Ibid. (Kant, *PC*, 66)

14. For this version of metaphorology, which slightly departs from the usual Blumenberg interpretation, see Campe, "From the Theory of Technology to the Technique of Metaphor."

15. Lambert, *Briefwechsel*, vol. 1, *PS*, 9: 365 (Kant, *PC*, 66).

16. The text of Meier's 1752 *Auszug aus der Vernunftlehre*, which Kant relied on, can be found in the *Handschriftlicher Nachlaß, Logik*, in Kant, *Gesammelte Schriften* (henceforth cited as *AA*), 16: esp. 427–443, along with the notes and "reflections" from Kant's own copy of the volume. Of additional interest are the surviving transcriptions of Kant's lectures on logic (see *Vorlesungen zur Logik*, in *AA*, 24: 1–2).

17. Meier, extract §171, cited from Kant, *Handschriftlicher Nachlaß*, in *AA*, 16: 427. Wolff, *Philosophia rationalis sive logica*, in id., *Gesammelte Werke*, pt. 2, vol.1.2, §578: 437).

18. What Kant mostly refers to with "plausibility" vs. "probability" can also be called the "appearance of truth" vs. "probability." See Kant, *Handschriftlicher Nachlaß*, in *AA*, 16, no. 2593: 433.

19. Kant, *Handschriftlicher Nachlaß, Logik*, in *AA*, 16, nos. 2,591, 2,593, 2,595, 2,596, 2,600, 2,601, 2,603, and 2,604); id., *Vorlesungen zu Logik*, in *AA*, 24: Blomberg, 1.144–145, 195–196; Pölitz, 2.555; Busolt, 2.644; Dohna-Wundlacken, 2.742, 882–883; Vienna logic, 2.883–884. See Cataldi Madonna, "Kant und der Probabilismus der Aufklärung."

20. Such genetic understanding is suggested by the following note: "*verisimilitudo* provides the ground of a temporary judgment; *probabilitas*, (the insufficient) ground of a ([practically]) determining judgment to determine the understanding" (Kant, *Nachlaß*, in *AA*, 16, no. 2,595, 434).

21. Kant, *Handschriftlicher Nachlaß*, in *AA*, 16, no. 2,598, 435 .

22. Compare notes mentioning gaming theory with regard to the numbering of homogeneous grounds in Kant's *Handschriftlicher Nachlaß*, nos. 2,605, 2,609, and 2,619 (*AA*, 16: 437–439). On legal proof and the weighing of nonhomogeneous grounds, see Kant, *Logik*, in *AA*, 24.2: 880; id., *Lectures on Logic*, 328: "Here mathematics can provide a certain measure whereby it compares quantities, because quantities contain nothing but what is homogeneous. But in philosophy, this does not work, because here the grounds of the possible winning cases are all non-homogeneous. If I take all possible cases and say that the winning cases must also be contained in these, then the relation of the grounds of holding-to-be-true is non-homogeneous. E.g., there is much testimony for an accused one.

(1.) The deed was done while he was not at home. He actually was not at home. This is a res facti. (2.) Someone else had seen him at the time. Here the credibility of the other man is the ground. The grounds in philosophy are always different as to quality, nonetheless, and they cannot be enumerated, but only weighed."

23. A typical formulation: "All certainty is to be regarded as a unity, and as a complete whole, and thus is the measure of all the rest of our holding-to-be-true, and of each and every one of its degrees. Certainty, however, arises from nothing but the relation of equality between the grounds of cognition that I have and the whole sufficient ground itself" (Kant, *Logik*, in *AA*, 24.1: 144; *Lectures on Logic*, 113–114). For more on the standard, e.g., "With probability there must always exist a standard in accordance with which I can estimate it. This standard is certainty. . . . Such a standard is lacking, however, with mere plausibility" (Kant, *Logik*, in *AA*, 9: 82; *Lectures on Logic*, 583).

24. Kant, *Handschriftlicher Nachlaß*, in *AA*, 16, no. 2,598, 435.

25. For the number as "a representation that summarizes the successive addition of one (homogeneous) unit to another," see Kant, *CPR*, 182). For the principle of reason or, in the usual English rendering of Kant scholars, "grounds," which, in philosophy, in contrast to the pure mathematical construction of concepts, stands under "the condition of time-determination in an experience," see *CPR*, 640.

26. Kant, Logik, *AA*, 9: 82; id., *Lectures on Logic*, 584.

27. Kant, *Handschriftlicher Nachlaß*, in *AA*, 16: 440.

28. Ibid., no. 2,597, 434–435. Jäsche connects the consideration directly to the claim that a *logica probabilium* is impossible. The "universal rules of probability," whose impossibility Kant assumes, thus describe the impossibility of a universal logic of probability. The calculation of the averaged errors appears immediately thereafter as an exception or limitation in relation to the impossibility of the *logica probabilium* (Kant, *Logik*, in *AA*, 9: 82). See Busolt's transcription, Kant, *Vorlesungen zur Logik*, in *AA*, 24.2: 644–645, and the Vienna logic, *AA*, 24.2: 883.

29. See, e.g., Kant, *Logik*, in *AA*, 9: 53–54 (A 77–78).

30. Kant, *Handschriftlicher Nachlaß*, in *AA* 16, 438–439.

31. Kant, *Vorlesungen zur Logik*, in *AA*, 24.2: 644–645.

32. Kant, *Critique of the Power of Judgment* (henceforth cited as *CJ*), §17, 116.

33. Ibid., 118.

34. Ibid.

35. Ibid..

36. Ibid., 119.

37. Manfred Schneider, "Mensch als Quelle," 314.

38. For an important discussion of Michel Foucault's concept of "normalism," see Link, *Versuch über den Normalismus*.

39. Kant, *CJ*,120.

40. Ibid.

41. The normal idea of the human body, although categorically an individual form, in the sense of the aesthetic idea is neither individual nor has any expression. The "idea of the species" has nothing "characteristic"; it lacks "anything specific to a person." In short, "experience also shows that such completely regular faces usually betray an inwardly only average human being." While the "average man" thus betrays his averageness with a glance at his interior, the interior/exterior relation of the aesthetic ideal lives entirely on deviations and idiosyncrasies, in short, from "that which is called genius" (Kant, *CJ*, 119, note).

42. Kant, "On the Amphiboly of Concepts of Reflection," in *CPR*, 369.

43. Kant, *CJ*, 117–18.

44. Ibid., 120.

45. See Kant, *CJ*, Introduction, § 8, 77–80.

46. The reading of the *Critique of the Power of Judgment* as presented here is based on three assumptions: first, the function of the third critique in grounding the unity of all three critiques is to be found in the relation between aesthetic and teleological judgment; second, this relation is asymmetrical; and, finally, Kant provides an irreducible position in this asymmetry to aesthetic presentation. This threefold presupposition can also be found similarly in Jean-Françoise Lyotard's readings of Kant, though admittedly with different justifications, particularly for the last of the three arguments (Lyotard, *Lessons*, chap. 1, "Aesthetic Reflection").

47. "The normal idea must take its elements for the figure of an animal or of a particular species from experience; but the greatest purposiveness in the construction of the figure, . . . the image which has as it were intentionally grounded the technique of nature, . . . lies merely in the idea of the one who does the judging, which, however, . . . can be represented fully *in concreto* as an aesthetic idea in a model image" (Kant, *CJ*, §17, 118).

48. Kant, *CJ*, §59, "On Beauty as a Symbol of Morality." See Gasché's fundamental account, "Some Reflections on the Notion of Hypotyposis in Kant."

49. See Kant, *De principiis formae mundi sensibilis*, in *AA*, 1.2: 403. For the connection between rhetorical and geometrical figures, see Campe, "Shapes and Figures."

50. Hacking, *Taming of Chance*, 35–36; Ewald, *L'État providence*, 108. For a discussion of the "law of large numbers" with respect to the philosophy of history, see Irmscher, "Geschichtsphilosophische Kontroverse," 151.

51. Hertzberg, *Abhandlung über die Bevölkerung der Staaten*.

52. Kant, *Idea for a Universal History with a Cosmopolitan Aim* (1784), in id., *Anthropology, History, and Education*, 108.

53. Ibid.

54. It is just as difficult to derive a law of human history from the "little part" of the course of its history that humanity has thus far completed, as "the course

taken by our sun together with the entire host of its satellites in the great system of fixed stars can be determined from all the observations of the heavens made hitherto" (ibid., 117.)

55. See Stigler, *History of Statistics*.

56. Voltaire, "Nouvelles considérations sur l'histoire" (1744), in id., *Œuvres complètes* (Nendeln ed.), 16: 139ff.

57. This is true from the 1740s up to Voltaire's *Philosophie de l'histoire*, the first paragraphs of which see in id., *Œuvres complètes: Complete Works*, ed. Besterman et al., 59: 89. See also Brumfitt, *Voltaire historian*, chaps. 3 and 4.

58. Willem Kersseboom's tables of average mortality appeared in three installments: the *Eerste verhandeling tot een proeve om te weeten de probable menigte des volks in de Provintie van Hollandt en West-Vriesland* was published at The Hague in 1738, and the second and third followed in 1742. Nicolaas Struyck's mortality tables are found in his *Inleiding tot de algemeene Geographie* (Amsterdam, 1740).

59. Though Voltaire's later *Philosophie de l'histoire* or *Essay sur les mœurs de l'esprit des nations* would in fact hide the numbers in the text.

60. Gatterer, *Ideal einer allgemeinen Weltstatistics*. The tract is closely connected to Gatterer's writings on universal history: *Abriß der Universalhistorie* (1749); and *Einleitung in die synchronistische Universalhistorie* (1771).

61. Gatterer, *Ideal einer allgemeinen Weltstatistik*, 16. See in contrast Büsching's geographical statistics, from which Gatterer distances himself, in Büsching, *Vorbereitung* (1758); a first and still valid discussion of this alternative can be found in Knies, *Statistik* (1850), 73–75.

62. Gatterer, *Ideal einer allgemeinen Weltstatistik*, 20.

63. Ibid., 21–23, cf. 36–40.

64. Ibid., 25–35.

65. See Schlözer, *StatsGelartheit*, pt. 2, vol. 1, *Theorie der Statistik*, 87–88.

66. Schlözer, *StatsGelartheit*, 3.

67. Ibid., pt. 2, vol. 1, *Theorie der Statistik*, 37.

68. In statistics today, this is referred to as the "representation" and "modeling" of data; see, e.g., Schlittgen, *Einführung in die Statistik*.

69. Schlözer, *StatsGelartheit*, pt. 2, vol. 1, *Theorie der Statistik*, 42–46.

70. Ibid. 68–71.

71. Ibid., 61–66, here, 61.

72. Hacking has rediscovered what mid-nineteenth-century German theoreticians of statistics already knew well: "The Prussia that overthrew Napoleon created a conception of society that resolutely resisted statistical generalization. It gathered precise statistics to guide policy and inform opinion, but any regularities they might display fell short of laws of society." Hacking names two conditions for the epistemic and scientific realization of the "statistical law" that he here equates with the "laws of society," or, as Ewald says, with the "sociological

schema," the sociological conception of society: the "avalanche of printed numbers" and "readers of the right kind, honed to find laws of society akin to those laws of nature established by Newton." Though Prussia brought on the avalanche of numbers, it did not, like England and France, produce enough "readers of the right kind" (Hacking, *Taming of Chance*, 35ff.; Ewald, *L'État providence*, 108).

73. See Brumfitt, "La philosophie de l'histoire," in Voltaire, *Œuvres complètes: Complete Works*, ed. Besterman et al., 59, 73.

74. Kant, *Idea for a Universal History with a Cosmopolitan Aim* (1784), in id., *Anthropology, History, and Education*, 108.

75. Ibid., 112.

76. Ibid., 115, 109, 119. A reading similar to that of the "guiding thread" here could be carried out for the word "intent" or "aim" (*Absicht*) from the title (. . . *with a Cosmopolitan Aim*). For one thing, the conventional usage in academic diction means something like "view": the essay takes a closer look at determining "universal history." In the second place, *Absicht* (as a synonym of "plan") is also a word located at the focal point of the categorical considerations of this work on the philosophy of history. One of the text's most important questions repeatedly asks whose intention, and in what sense in general "intention" can be spoken of.

77. For the term "quasi-transcendentality," see Gasché, *Tain*, 293–318. With my use of the term, however, I reverse Gasché's process: while Gasché interprets deconstruction transcendentally, I pick up on the word "quasi transcendental" from Kant's own text and above all in its function between number and word.

78. Kant, *Idea for a Universal History with a Cosmopolitan Aim* (1784), in id., *Anthropology, History, and Education*, 109.

79. For the standard interpretation of the essay in terms of a history of ideas, see Riedel, "Historizismus und Kritizismus."

80. Kant must admit this structural unity, no matter how much the critiques of practical reason and of teleological-aesthetic judgment problematize it. He has at least to imply the idea of it if he wishes to offer a universal and philosophical history. With this in mind, incidentally, Kant's answer to Herder's historical theory should also be reconsidered. The *Idea for a Universal History with a Cosmopolitan Aim* is, as we know, closely related to the first volume of Herder's *Ideen zur Philosophie der Geschichte der Menschheit* (Ideas on the Philosophy of the History of Humanity), which had appeared the preceding year. The two readings of Kant's essay outlined above would make it possible to go beyond the much discussed opposition between Kant and Herder in the theory of history. Instead of juxtaposing critical philosophy on Kant's side and an anthropological theory of history on Herder's, the identical starting situation would become visible for both: namely, the interconnection of history with a basis that Herder conceives of physiologically and Kant conceptualizes in terms of human agency and statistics. Such a revision of the relation

between Kant's and Herder's conception of history would also be important in view of the tradition of the *logica probabilium*. Recent scholarship has related Kant's and Herder's debate of the 1770s on the philosophy of history back to the epistemological turn of the 1760s, when Herder was a student of Kant's in Königsberg, attending and excerpting from his lectures on logic (among other subjects). In brief, Herder's conception of a historical anthropology has been interpreted as a continuation of the epistemological turn, which Kant characterized in his letter to Lambert as a "negative science." Herder's historical anthropology seems, from this perspective, to be a conception that critical philosophy does not ignore but rather (consciously or compelled by misunderstanding) reinterprets as a form of cognition determined in each case historically. In this way, the later confrontation on the field of the theory of history becomes an argument resulting from the different consequences of "negative science." For a groundbreaking study in recent scholarship on this connection, see Irmscher, "Geschichtsphilosophische Kontroverse," esp. 111–116; see also Häfner, *Johann Gottfried Herders Kulturentstehungslehre*.

81. Kant, *Idea for a Universal History with a Cosmopolitan Aim* (1784), in id., *Anthropology, History, and Education*, 109.

82. Ibid.

83. E.g., Kant speaks of "purposelessly playing nature" (ibid.); or of nature that seems to awaken the suspicion "that in the case of the human being alone it is a childish play" (ibid., 110); he also calls for the enlightenment of "such a confused play of things human" (ibid., 119).

84. Süssmilch, *Göttliche Ordnung*, 1: 273–310.

85. Ibid., 280.

86. Ibid., §§ 147–161 (274–299).

87. Ibid., §§ 162–167 (299–319).

88. In this regard, see Paul Hazard's notes on Bossuet's *Discours sur l'histoire universelle* (Hazard, *European Mind*, 210ff.).

89. Süssmilch, *Göttliche Ordnung*, 1: vii.

90. Ibid., 280, 285, 293–297. See the tables corrected by Pasquier in Euler, *Opera omnia*, 1st ser., 7: 507–534, esp. 513, 519–520, 526–531.

91. Euler, "Recherches générales sur la mortalité et la multiplication du genre humain" (1760; 1767), in id., *Opera omnia*, 1st ser., 7: 79–100, here 80ff. and 88–91; quoted 89.

92. Süssmilch, *Göttliche Ordnung*, 1: 291. See Euler, "Sur la multiplication du genre humain," in id., *Opera omnia*, 1st ser., 7: 545, 552.

93. Euler used the tables and formulas he constructed for Süssmilch for an entirely different purpose: for model calculations of actuary mathematics. This is particularly true of the "Recherches générales sur la mortalité et la multiplication du genre humain," which was published in 1767 in the *Mémoires del'Académie des*

sciences in Berlin. A brief "application" under the title "Sur les rentes viagères" followed them there as a second part (Euler, *Opera omnia*, 1st ser., 3: 101–112).

94. Riedel, "Historizismus und Kritizismus."
95. Crome, *Allgemeine* Übersicht, 4ff.
96. Ibid., 3.
97. Crome, *Europens Produkte;* id., *Über die Größe und Bevölkerung der sämtlichen europäischen Staaten* (Dessau: n.p., 1785); id., *Über die Größe und Bevölkerung der europäischen Staaten, als der sicherste Maaßstab ihrer verhältnismäßigen Kultur* (Frankfurt: Jäger, 1793).
98. Playfair, *Statistical Breviary* (1801); the broadly received French translation was made by D. F. Donnant and edited under the title *Élémens de statistique* (1802).
99. Bötticher, *Statistische Übersichtstabellen* (1789). A translation can be found in Bötticher, *Geographical, Historical, and Political Description of the Empire of Germany, Holland, the Netherlands, Switzerland, Prussia, Italy, Sicily, Corsica, and Sardinia* (1800). In the preface to his own work, Playfair mentions that he was commissioned to add supplementary tables to Bötticher's work, which was outdated at the time of his translation: "In the course executing that design, it occurred to me, that tables are by no means a good form of conveying such information.... I have composed the following work upon the principle of which I speak; this, however, I never should have thought of doing, had it not occurred to me, that making an appeal to the eye when proportion and magnitude are concerned is the best and readiest method of conveying a distinct idea" (Playfair, *Statistical Breviary*, 3ff.).
100. See, e.g., Kant to D. Friedländer, November 6, 1787, in id., *AA,* 10: 502, in which Kant supports Bötticher, "who has founded an educational institute with very good progress," in his wish for help in marketing the spinning wheel he had designed.

CHAPTER 14

1. The few documents of Kleist's studies at Frankfurt on the Oder specify that, alongside history lectures and private tutoring in classical languages, he studied natural history, experimental physics, and mathematics under Chr. E. Wünsch and J. S. G. Huth, and took law courses given by the natural law theorist L. G. Mahdin. See Kleist's letter to his sister, October 26, 1800, in id., *Sämtliche Werke und Briefe,* ed. Sembdner (henceforth cited as *SWB-S*), 2: 533; and Sembdner, *Heinrich von Kleists Lebensspuren,* 1: 34.
2. Kleist, *Penthesilea,* in *SWB-S,* 1: 391, vv. 2026–28; Kleist, *Penthesilea,* trans. Agee, 96. Gerhard Fricke has pointed out the fact that the Amazons' "holy law" highlights government and the state in contrast to the legal and contractual arguments of the Greeks (Fricke, *Gefühl und Schicksal,* 102).

3. Behre, *Geschichte der Statistik*, 379–389.

4. For a helpful overview of the "Prussian" orientation of statistics in the nineteenth century, which, in contrast to the sociological model of "western European" statistics since Laplace and Quetelet, juxtaposes society (and statistics) with the state (and its constitutional laws), see Hacking, *Taming of Chance*, chap. 5.

5. Kleist, *Penthesilea*, vv. 83–85, in id., *SWB-S*, 1: 391, 325; Kleist, *Penthesilea*, trans. Agee, 7.

6. Kleist, *SWB-S*, 2: 338.

7. Kleist, letter of January 7, 1805, in *SWB-S*, 2: 750.

8. Müller-Salget has shown that in the first fragment on Tycho and Kepler, Kleist brings together two reminiscences from the *Cosmologische Unterhaltungen* of Chr. E. Wünsch, the professor of natural history whose lectures on experimental physics Kleist had attended. Kleist, *Sämtliche Werke und Briefe*, ed. Barth (henceforth cited as *SWB-B*), 1136.

9. Kleist, "Unwahrscheinliche Wahrhaftigkeiten," in id., *SWB-B*, 3: 376, 379; trans. Carol Jacobs, *Uncontainable Romanticism*, 197, 200.

10. See Kleist, *Michael Kohlhaas*, in id., *SWB-S*, 3: 134.

11. According to Müller-Salget's evidence, "the denial of the so-called gypsy episode (whose function is not understood) has become a nearly obligatory element of the critical literature on Kohlhaas" (*SWB-S*, 3: 715).

12. Müller-Salget omits this subtitle because, though it is listed in the table of contents of the first volume of Kleist's *Erzählungen* in 1810, it does not appear with the story itself (*SWB-S*, 3: 705ff.). It is more likely, however, to assume a vacillation on Kleist's part than a manipulation on the side of the publisher. For this reason, the subtitle should be considered as part of the text.

13. Lugowski, *Wirklichkeit und Dichtung*. For "unmediated reality," see p. 96; for the idea of "overcoming the objective by means of the unmediated," p. 137; for *Improbable Veracities*, pp. 167 and 203. Clemens Lugowski may be called a structuralist avant la lettre, and he was certainly exceptional in German literary studies in the1920s and 1930s. His first book, *Die Form der Individualität im Roman* (1932), only appeared in English translation in 1990 (Lugowksi, *Form, Individuality, and the Novel*). As much as we might be disappointed by discovering what I call the "intellectual fascism" in his book on Kleist, *Wirklichkeit und Dichtung* (1936), comparing the two studies is revealing, even if in frustrating ways. The constructivist, *neusachlich* [new objective] method of the first study, which we can only admire, is uncannily akin to certain ideological assumptions in the book on Kleist. Picking up from where he left things in the first book, and with a thoroughness unmatched in German scholarship, Lugowski analyzes every "substantial" instance of directing or explaining events as rhetorical ("motivating") moves by the narrator. He thus traces the modern novel back to two basic types: the "fairy-tale novel," which attributes events to sheer fortune or misfortune, as

examples of a baroque providence; and the "anti-fairy-tale novel," which, from Madame de La Fayette through Flaubert, replaces external coincidences with psychological or social motivation. After discovering the finally identical structure of narrative figurations (techniques of "motivation") in baroque allegory and realistic probability, Lugowski switches from *neusachlich* constructivism to the ideology of "unmediated reality." This move then results in his view on a—very German—Kleist who functions as a positive counterfigure to the modern realist—and particularly French—novel. A further remark on critical history may be in order here, since it helps us understand why Kleist criticism was an important intermediary between *neusachlich* modernism and intellectual variants of fascist thinking in the 1920s and 1930s in Germany. Lugowski's "unmediated reality" is a direct response to the concept of "absolute feeling" that Fricke had forged in the same context in his own Kleist interpretation from 1929 (Fricke, *Gefühl und Schicksal*): while "absolute feeling" lies beyond all calculation, the position of "unmediated reality," the being-within in the real, comes before any possibility of calculation (Lugowski, *Wahrheit und Dichtung*, 137–160).

14. Jacobs, *Uncontainable Romanticism*, 171–196; on Kleist's "Unwahrscheinliche Wahrhaftigkeiten" (Improbable Veracities), 179–196, esp. 185ff.

15. Chase, "Mechanical Doll, Exploding Machine," in id., *Decomposing Figures*, 141–156, esp. 155.

16. In a fruitful reflection on the Aristotelian distinction between history, poetry, and philosophy, Werner Hamacher assumes the *kath'hekaston* or singularity to withdraw from or to be opposed to contextualization in probable fiction and scientific conceptualization from the beginning. In doing so, he reads the position of the singular characteristic for modern literature back into Aristotle. It is only in modernity, after the probabilistic turn, that literature has become the custodian of singularity (Hamacher, "Über einige Unterschiede," 5–15).

17. For the following, see Kleist, "Unwahrscheinliche Wahrhaftigkeiten," in *SWB-B*, 3: 376–379 (Jacobs, *Uncontainable Romanticism*, 197–200).

18. Sembdner, *Kleist's Lebensspuren*, 1: 34.

19. See Erxleben, *Anfangsgründe der Naturlehre*, §7 "Von der Luft," 173–296, esp. "Künstlich zusammengedrückte Luft," 223–232.

20. For the history of wonder and scientific attentiveness, see Daston and Park, *Wonders and the Order of Nature*.

21. Seifert, *Cognito historica*.

22. See Jacobs, *Uncontainable Romanticism*, and Chase, "Mechanical Doll, Exploding Machine." Hans-Christian von Herrmann also hints at a connection between "Unwahrscheinliche Wahrhaftigkeiten" and probability. A link between Kleist and probability thinkers for him would be the Clausewitzian reform of military strategy. This remark indeed reminds us that nearly all of Kleist's scientific and mathematic considerations occur in the context of the

military—even the narrator of "Unwahrscheinliche Wahrhaftigkeiten" is an officer (Herrmann, "Bewegliche Heere," 227–243). Friedrich Meinert's *Über das Studium der militärisch-mathematischen Wissenschaften auf Universitäten* (1780), arguing for the mathematical and technical education of officers, well exemplifies this military grounding of probability theory.

23. The narrator's irony when he actualizes the "event itself" about which he seems to be narrating in the context of narration should warn us about taking the famous moments of chance in Kleist for instances of a standard "motive" or "situation" in narrative representation. Two important studies can be mentioned here, which, despite their methodical opposition, both conspicuously strive for a typological understanding under the concept of a "situation of chance events": Hans Peter Herrmann's phenomenological study "Zufall und Ich" and Petra Perry's structural study, *Möglichkeit am Rande der Wahrscheinlichkeit*. In his excellent essay "Das Beben der Darstellung," on Kleist's novella *Das Erdbeben in Chili* (The Earthquake in Chile), in contrast, Werner Hamacher emphasizes the fact that the event in its singularity evades classification—even classification as coincidental.

24. Ever since Sembdner, scholars have pointed out that the listener's reference in the third story to the appendix of Schiller's *Abfall der Niederlande* (Revolt of the Netherlands) is deceptive. In fact, the source of the story with the details Kleist reports leads back, not to the poet Schiller, but to Carl Curths, the historian who continued and completed Schiller's history of the Netherlands (see Müller-Salget's commentary, *SWB-B*, 3: 942ff.; Jacobs, *Uncontainable Romanticism*, 191). The story also appears, however, in "Belagerung von Brabant" (The Siege of Brabant), which Schiller published in 1795 in his journal *Die Horen*.

25. Kleist, "Unwahrhaftigkeiten," in *SWB-B*, 3: 376 (Jacobs, *Uncontainable Romanticism*, 197).

26. Kleist, *Michael Kohlhaas*, in *SWB-B*, 3: 134; id., "Michael Kohlhaas" in *The Marquise of O—and Other Stories*, trans. Luke and Reeves, 205; translation modified.

CONCLUSION

1. Poisson, *Recherches*, 31.
2. Ibid., 32.
3. See Hahn, *Pierre Simon Laplace*, 108–112.
4. Laplace, *Philosophical Essay*, 6.
5. Ibid., 11.
6. Hacking is wrong to read the exact equivalency between the "subjective" and "objective" interpretations of probability in Laplace as a mere indecision

between the two concepts (Hacking, *Emergence of Probability*, 131–133; See Daston, *Classical Probability*, 218–221).

7. The trail of this self-reflection of probability theory and its ensuing autonomy as a mathematical theory leads back to Condorcet. We might be reminded above all of the peculiar argumentation in which Condorcet differentiates the value of a particular "probability" from the "motive of belief" in this same probability. The mathematical formula for an event m's probability is the usual quotient of the occurrence of m to all possible events. The "motive of belief" in such probability, in contrast, is expressed by the limit value for $m +/- a$ occurrences of m with variable values for a (Condorcet, *Elémens du calcul des probabilités*, in id., *Sur les elections et autres textes*, art. 4, 537–550).

8. This comparison between Leibniz and Laplace follows Blumenberg, *Lesbarkeit der Welt*, chap. 10.

9. Laplace, *Theorie analytique des probabilités*, 177.

10. Laplace, *Philosophical Essay*, 6; also *Théorie analytique*, 178.

11. Laplace's appearance at the École normale supérieur in 1795 and his epochal *Essai philosophique sur les probabilités* (Philosophical Essay on Probabilities) from 1814 definitely shift the way structure and meaning are related to each other in probability—the theme under scrutiny in this study—to other fields and other methods. Adolphe Quetelet's monumental work on social statistics defines an object and a referent of statistical description, the famous *homme moyen*, the man of sociology. The *homme moyen* is a mathematical construction, but at the same time also the result of a historical tendency, the tendency of realizing the statistical mean or average through the process of socialization (see Quetelet, *Du système social*, 28, 252–256; See Fleming, *Exemplarity and Mediocrity*). By this token, the computation and the meaning of probability enter into a decisively new phase of their relationship. It is from this point that a discussion of statistics and the modern novel would have to make a fresh start for the nineteenth century and, later, through to today.

Bibliography

SOURCE TEXTS

Achenwall, Gottfried (Praes.), and Johann Justus Henno (Resp.). *Notitia rerum publicarum academiis vindicata*. Göttingen: Hager, 1748.
Aglionby, William. *The Present State of the United Provinces of the Low-Countries*. 1669. 2nd ed. London: John Starkey, 1671.
Agricola, Rudolf. *De inventione dialectica libri tres. Drei Bücher über die Inventio dialectica*. Based on the edition of Alardus van Amsterdam, 1539. Translated and edited by Lothar Mundt. Tübingen: Niemeyer, 1992.
d'Alembert, Jean Le Rond. *Œuvres*. 5 vols. Reprint of the 1821–1822 ed. Geneva: Slatkine, 1967.
Aquinas, Thomas, Saint. *Summa theologiae*. Latin text and English trans. New York: McGraw-Hill, 1964–1976.
Arbuthnot, John. *Of the Laws of Chance*. London: Randall Taylor, 1692.
———. "An Argument for Divine Providence, taken from the constant Regularity observ'd in the British Births of Both Sexes." *Philosophical Transactions* 27 (1710): 186–200.
Aristotle. *The "Art" of Rhetoric*. Translated by John Henry Freese. Cambridge, MA: Harvard University Press, 1926.
———. *Topica*. In *Posterior Analytics. Topics*, trans. E. S. Forster, 263–739. Cambridge: Heimann, 1966.
———. *Topics: Books I and VIII*. Translated by Robin Smith. Oxford: Oxford University Press, 1997.
———. *Nicomachean Ethics*. Translated by Joe Sachs. Newburyport, MA: Focus Publishing, 2002.
———. *Poetics*. Translated by Joe Sachs. Newburyport, MA: Focus Publishing, 2005.
Arnauld, Antoine, and Pierre Nicole. *La logique, ou l'art de penser*. 1662. Foreword by Louis Marin. Paris: Gallimard, 1970.
———. *Logic or the Art of Thinking*. Translated by Jill Vance Buroker. Cambridge: Cambridge University Press, 1996.
Assmann, Valentin Jacob. *De rerumpublicarum notitia*. Leipzig: Langenheim, 1735.

Augustine, Saint, bishop of Hippo. *The City of God Against the Pagans.* Translated by Robert W. Dyson, Cambridge: Cambridge University Press, 1998.
Bastineller, Gebhard Christian (Praes.), and Gottfried Wilhelm Stockmann (Resp.). *Dissertatio de praestantia scientiarum mathematicarum in foro juris.* Wittenberg: Eichsfeld, 1741.
Baumgarten, Alexander Gottlieb. *Metaphysica.* 1739. 7th ed., 1779. Reprint, Hildesheim: Olms, 1963.
———. *Aesthetica.* 1750–1758. Facsimile ed., Hildesheim: Olms, 1970.
Barbeyrac, Jean. *Traité du jeu, où l'on examine les principales questions de droit naturel et de morale qui ont du rapport à cette matière.* 2 vols. Amsterdam: Pierre Humbert, 1709.
Baumeister, Friedrich Christian. *Institutiones philosophiae rationalis.* 1735. Reprinted in Christian Wolff, *Gesammelte Werke,* ed. J. École, vol. 24. Hildesheim: Olms, 1989.
Bernoulli, Daniel. "Exposition of a New Theory on the Measurement of Risk." Translated by Louise Sommer. *Econometrica* 22 (1954): 23–36.
———. "Specimen theoriae novae de mensura sortis." In id., *Die Werke,* ed. David Speiser, vol. 2, ed. L. P. Bouckaert and B. L. van der Waerden, 223–234. Basel: Birkhäuser 1982.
Bernoulli, Jakob. *Ars conjectandi.* 1713. In id., *Die Werke,* vol. 3. Basel: Birkhäuser, 1975.
———. *Die Werke.* Vol. 3. Basel: Birkhäuser, 1975.
———. *Der Briefwechsel.* Edited by D. Speiser. Basel: Birkhäuser, 1993.
———. *On the Law of Large Numbers.* Translated by Oscar Sheynin. Berlin: NG Verlag, 2005.
———. *The Art of Conjecturing.* Edited and translated by Edith Dudley Sylla. Baltimore: Johns Hopkins University Press, 2006.
Bernoulli, Nikolaus. *Dissertatio inauguralis mathematico-juridica, de usu artis conjectandi in jure.* Basel: J. C. v. Mechel, 1709. In Jakob Bernoulli, *Werke,* 3: 287–326. Basel: Birkhäuser, 1975.
Besongne, Nicolas. *Le parfait Estat de la France, comme elle est gouvernée à présent, ou il est traitté des principaux points du gouvernement de ce rouyaume.* Paris: Cardin Besongne, 1656.
———. *The Present State of France. Written in French and faithfully Englished.* London: John Starkey, 1671.
Boethius, Manlius Severinus. "Commentaria in Topica Ciceronis"; "Topicorum Aristotelis interpretatio"; "De Differentiis Topicis." In *Opera omnia, Patrologia Latina,* vol. 64. ed. J.-P. Migne, 909–1007; 1039–1172; 173–1216. Paris: Migne, 1860.
———. *In Ciceronis Topica.* Translated by Eleonore Stump. Ithaca, NY: Cornell University Press, 1988.

Bose, Johann Andreas. *Introductio generalis in notitiam rerumpublicarum orbis universi*. Jena: Bielcke, Krebs, 1676.
Botero, Giovanni. *Le relationi universali*. Vicenza: Giorgio Greco, 1595.
———. *Allgemeine Weltbeschreibung*. Cologne: Johann Gymnich Erben, 1596.
Bötticher, Jakob Gottlieb Isaak. *Statistische Übersichtstabellen aller Europäischen Staaten*. 1789. Königsberg: Hartung, 1790.
———. "Statistical Tables: Exhibiting a View of All the States of Europe." In *A Geographical, Historical, and Political Description of the Empire of Germany, Holland, the Netherlands, Switzerland, Prussia, Italy, Sicily, Corsica, and Sardinia*. London: Stockdale, 1800.
Boyle, Robert. *The Works*. 1772. Reprint, ed. Thomas Birch. 6 vols. Hildesheim: Olms, 1966.
Breitinger, Johann Jakob. *Critische Dichtkunst*. 1740. Reprint, Stuttgart: Metzler, 1966.
Braun, Georg, and Franz Hogenberg. *Civitates Orbis Terrarum*. 1572–1618. Reprint, Amsterdam: Theatrum Orbis Terrarum, 1965.
Büsching, Anton Friedrich. *Vorbereitung zur gründlichen und nützlichen Kenntniß der geographischen Beschaffenheit und Staatsverfassung der europäischen Reiche und Republiken*. Hamburg: Bohn, 1784.
Calvin, John. *Institutes of the Christian Religion*. Translated by John Allen. 2 vols. Philadelphia: Presbyterian Board of Christian Education, 1936.
Campbell, George. *The Philosophy of Rhetoric*. 1776. Edited by Lloyd F. Blitzer. Carbondale: Southern Illinois University Press, 1963.
Castelvetro, Lodovico. *Art of Poetry. Slightly Abridged Translation of Lodovico Castelvetro's Poetica d'Aristotele vulgarizzata et sposta*. Binghampton, NY: Medieval and Renaissance Texts and Studies, 1984.
Chamberlayne, Edward. *Angliae notitia, or, The present State of England*. London: N.p., 1669.
———. *Angliae notitia, sive praesens Angliae status succinctè enucleatus*. Abridgment of the preceding work edited by Thomas Wood. Oxford: Henry Clements, 1686.
Chladenius, Johann Martin. *Vernünftige Gedanken von dem Wahrscheinlichen und desselben gefährlichen Mißbrauche*. 1748. Edited by Dirk Fleischer. Waltrop, Germany: Spenner, 1989.
Cicero, Marcus Tullius. *De oratore*. Translated by J. M. May and J. Wisse. New York: Oxford University Press, 2001.
———. *Topica*. Translated by Tobias Reinhardt. Oxford: Oxford University Press, 2003.
Chalmers, George. *An Estimate of the Comparative Strength of the Great Britain: and of the losses of her trade from every war since the Revolution: with an introduction of previous history*. London: John Stockdale, 1804.

468 Bibliography

Condorcet, Jean-Antoine-Nicolas de Caritat, marquis de. *Mathématique et société*. Edited by Roshdi Rashed. Paris: Hermann, 1974.

———. *Sur les élections et autres texts*. Edited by Olivier de Bernon. Paris: Fayard, 1986.

Conring, Hermann. *Opera*. 1730. 7 vols. Edited by Johann Wilhelm Göbel. Reprint, Aalen, Germany: Scientia, 1970–1973.

Corpus Iuris Civilis. Text und Übersetzung. Edited by Okko Behrends, Rolf Knütel, Berthold Kupisch, and Hans Hermann Seiler. Vols. 1–3. Heidelberg: C. F. Müller, 1999.

Cournot, Auguste. *Exposition de la théorie des chances et des probabilités*. Paris: Hachette, 1843.

Craig, Thomas. *Theologiae christianae principia mathematica*. 1699. Edited by Johann Daniel Titius. Leipzig: Lanckisch, 1755.

Cramer, Johann Ulrich. *Specimen novum juris naturalis de aequitate in probabilibus, exemplo emtionis spei illustrata*. Marburg: Müller, 1731.

Crome, August Friedrich. *Europens Produkte: zum Gebrauch der neuen Produkten-Karte von Europa*. Dessau: Crome and Buchhandlung der Gelehrten, 1782.

———. *Allgemeine Übersicht der Staatskräfte von den sämtlichen europäischen Reichen und Ländern*. Leipzig: Fleischer, 1818.

———. *Geographisch-statistische Darstellung der Staatskräfte von den sämmtlichen, zum deutschen Staatenbunde gehörigen Ländern*. 4 vols. Leipzig: Fleischer, 1820–1828.

Dacier, André. *La Poétique d' Aristote: Traduite en françois avec des Remarques Critiques sur tout l'Ouvrage*. 1692. Reprint, Hildesheim: Olms, 1976.

Daneau, Lambert. *Briève remonstrance sur les jeux de sort ou de hazard*. Geneva: Jacques Bourgeois, 1574.

Defoe, Daniel. *The Works*. Edited by Gustavus Howard Maynardier. 16 vols. London: Chesterfield, 1903–1904.

———. *An Essay upon Projects*. 1697. Menston, UK: Scolar Press, 1969.

———. *Selected Writings*. Edited by James T. Boulton. London: Cambridge University Press, 1975.

———. *A Journal of the Plague Year*. 1722. Edited by Louis Landa. Oxford: Oxford University Press, 1990.

———. *A Journal of the Plague Year: Authoritative Text, Backgrounds, Contexts, Criticism*. Edited by Paula R. Backscheider. New York: Norton, 1992.

Eberhard, Johann Peter. *Beiträge zur Mathesi applicata, hauptsächlich zu Mühlenbau- und Bergwerksmaschinen*. Leipzig, 1756.

———. *Gedanken vom Nutzen der Mathematik und ihrem Einfluss in den Staat*. Halle, 1769.

———. *Neue Beiträge zur Mathesi applicata*. Halle: Hemmerde, 1773.

Erxleben, Johann Christian Polycarp. *Anfangsgründe der Naturlehre*. With

improvements and additions by Georg Christoph Lichtenberg. Göttingen: Dieterich, 1794.
Euler, Leonhard. *Commentationes algebraicae ad theoriam combinationum et probabilitatum pertinentes.* In id., *Opera omnia*, 1st ser., vol. 7, ed. Louis Gustave du Pasquier. Leipzig: Teubner, 1923.
Fielding, Henry. *Tom Jones.* 1749. Edited by John Bender and Simon Stern. Oxford: Oxford University Press, 1996.
Frain Du Tremblay, Jean. *Conversations morales sur les jeux et les divertissemens.* Paris: Pralard, 1701.
Furetière, Antoine. *Le Dictionnaire universel.* Facsimile of the edition of 1690. Paris: SNL–Le Robert,1978.
Gassendi, Pierre. *Opera omnia.* 1658. Reprint, Stuttgart: Frommann, 1964.
Gataker, Thomas. *Of the Nature and Use of Lots.* London: John Haviland, 1619.
Gatterer, Johann Christoph. *Ideal einer allgemeinen Weltstatistik.* Göttingen: Vandenhoeck, 1773.
Gellert, Christian Fürchtegott. *Sämtliche Schriften.* 1769–1774. Hildesheim: Olms, 1968.
———. *Gesammelte Schriften.* Edited by Bernd Witte. Berlin: De Gruyter, 1988–2008. Cited as *GS*.
Goethe, Johann Wolfgang von. *Werke.* 14 vols. Edited by Erich Trunz. Hamburg: Christian Wegner, 1948–1960. Cited as HA (Hamburger Ausgabe).
———. *Conversations of German Refugees.* Translated by Krishna Winston. New York: Suhrkamp, 1989.
Gottsched, Johann Christoph. *Versuch einer critischen Dichtkunst.* Darmstadt: Wissenschaftliche Buchgesellschaft, 1962.
's Gravesande, Willem Jacob. *Œuvres philosophiques et mathématiques.* Amsterdam: M. M. Rey, 1774.
Guhrauer, Gottschalk Eduard. "Leben und Verdienste Caspar Neumanns." *Schlesische Provinzialblätter.* Breslau: Doulin, 1863.
Gundling, Nicolaus Hieronymus. *Ausführlicher Discours über den ietzigen Zustand der europäischen Staaten.* 1715. 2 vols. Frankfurt: Spring, 1733–34.
Halley, Edmond. "An Historical Account of the Trade Winds." *Philosophical Transactions* 183 (1686): 153–168.
———. "An Estimate of the Degrees of Mortality of Mankind, drawn from curious Tables of Birth and Funerals at the City of Breslaw." *Philosophical Transactions* 196 (1693): 596–610; 198 (1693): 654–656.
Hertzberg, Ewald Friedrich von. *Abhandlung über die Bevölkerung der Staaten überhaupt und besonders des Preußischen, welche am 27. Januar 1785 in der, wegen des Geburtstages des Königs, gehaltenen öffentlichen Versammlung der Akademie der Wissenschaften zu Berlin vorgelesen ist.* Berlin: n.p., 1785.

Hoffmann, Christian Gottfried. *Entwurff einer Einleitung zu dem Erkänntniß des gegenwärtigen Zustandes von Europa.* Leipzig: Lanckisch, 1720.
Huet, Pierre-Daniel. *Traité de l'origine des romans.* 1670. Edited by Arend Kok. Amsterdam: Swets & Zeitlinger, 1942.
Hume, David. *A Treatise of Human Nature.* 1739–1740. Edited by David Fate Norton and Mary J. Norton. 2000. Oxford: Clarendon Press, 2011.
Huygens, Christiaan. *Œuvres complètes.* 22 vols. The Hague: Nijhoff, 1920.
Huygens, Constantijn. *Fragment eener autobiographie.* Edited by J. A. Worp. The Hague: Nijhoff, 1897.
Isidore of Seville. *Etymologicarum sive originum libri XX.* Oxford: Oxford University Press, 1987.
Kahle, Ludwig Martin. *Elementa logicae probabilium methodo mathematica in usum scientiarum et vitae adornata.* Halle: Renger, 1735
Kant, Immanuel. *Logik, ein Handbuch zu Vorlesungen.* Edited by Gottlob Benjamin Jäsche. Königsberg: Friedrich Nicolovius, 1800.
———. *Gesammelte Schriften.* 29 vols. Berlin: Reimer; subsequently De Gruyter, 1900–. Cited as *AA* (= *Akademieausgabe*).
———. *Philosophical Correspondence, 1759–99.* Translated by Arnulf Zweig. Chicago: University of Chicago Press, 1967. Cited as *PC*.
———. *Lectures on Logic.* Translated by J. Michael Young Cambridge: Cambridge University Press, 1992.
———. *Critique of Pure Reason.* Translated by Paul Guyer and Allen W. Wood. Cambridge: Cambridge University Press, 1998. Cited as *CPR*.
———. *Critique of the Power of Judgment.* Translated by Paul Guyer and Eric Matthews. Cambridge: Cambridge University Press, 2000. Cited as *CJ*.
———. *Anthropology, History, and Education.* Edited by Günter Zöller and Robert B. Louden. Translated by Allen W. Wood. Cambridge: Cambridge University Press, 2007.
Kästner, Abraham Gotthelf. *Pro justitia calculi interusurii Leibnitii.* Leipzig: Langenheim, 1747.
———. *Gradus et mensuram probabilitatis dari.* Leipzig: N.p., 1749.
———. *Dissertation sur les devoirs qui résultent de la conviction, que les événements fortuits dépendent de la volonté de Dieu.* In Prussian Academy of Sciences Berlin, *Pièce qui a remporté le prix sur le sujet des événements fortuits; proposé par l'Académie Royale des Sciences et Belles Lettres de Berlin pour l'année 1751. Avec les pièces qui ont concouru.* Berlin: N.p., 1751.
———. *Anfangsgründe der angewandten Mathematik.* 2 parts. 2nd ed. Göttingen: Vandenhoeck, 1765.
———. *Gesammelte poetische und prosaische Schönwissenschaftliche Werke.* Berlin: Enslin, 1841. Cited as *GW*.
———. *Abraham Gotthelf Kästners Selbstbiographie und Verzeichnis seiner*

Schriften. Edited by Rudolf Eckart. Hannover: Geibel, 1909.
King, Gregory. *Natural and Political Observations and Conclusions upon the State and Conditions of England*. In George Chalmers, *An Estimate of the Comparative Strength of Great Britain*. London: John Stockdale, 1804.
Kleist, Heinrich von. *The Marquise of O—and Other Stories*. Translated by David Luke and Nigel Reeves. New York: Penguin Classics, 1978.
———. *Sämtliche Werke und Briefe*. Edited by Helmut Sembdner. 1961. Munich: Hanser, 1985. Cited as *SWB-S*.
———. *Sämtliche Werke und Briefe*. Edited by Ilse-Marie Barth. Frankfurt: Deutsche Klassiker Verlag, 1990. Cited as *SWB-B*.
———. *Penthesilea: A Tragic Drama*. Translated by Joel Agee. New York: Harper Collins, 1998.
Knies, Carl Gustav Adolph. *Die Statistik als selbständige Wissenschaft*. 1850. Reprint, Frankfurt: Sauer & Auvermann, 1969.
Lambert, Johann Heinrich. *Beyträge zum Gebrauche der Mathematik und deren Anwendung*. Berlin: Im Verlag der Buchhandlung der Realschule, 1772.
———. *Philosophische Schriften*. 10 vols. Hildesheim: Olms, 1965. Cited as Lambert, *PS*.
———. *Cosmological Letters on the Arrangement of the World-Edifice*. Translated by Stanley L. Jaki. Edinburgh: Scottish Academic Press, 1976.
———. *Cosmologische Briefe über die Einrichtung des Weltbaues*. 1761. Edited by Gerhard Jackisch. Berlin: Akademie-Verlag, 1979.
———.*Texte zur Systematologie und zur Theorie der wissenschaftlichen Erkenntnis*. Edited by Geo Siegwart. Hamburg: Meiner, 1988.
Laplace, Pierre-Simon de. *Théorie analytique des probabilités*. Paris: Courcier, 1812.
———. *Essai philosophique sur les probabilités*. Paris: Vve Courcier, 1814.
———. *A Philosophical Essay on Probabilities*. Translated by Frederick Wilson Truscott and Frederick Lincoln Emory. New York: Dover, 1951.
La Placette, Jean. *Divers traités sur des matières de conscience*. Amsterdam: G. Gallet, 1697.
Leibniz, Gottfried Wilhelm. *Essais de Théodicée sur la bonté de Dieu, la liberté de l'homme et l'origine du mal*. Amsterdam: François Changuoin, 1710.
———. *Œuvres philosophiques: Latines et françoises*. Edited by Rudolf Erich Raspe. Leipzig: Schreuder, 1765.
———. *Die Werke von Leibniz gemäß seinem handschriftlichen Nachlasse*. Edited by Onno Klopp. Hannover: Klindworth, 1864–1884.
———. *Philosophische Schriften*. Edited by Carl Gerhardt. Hildesheim: Olms, 1960. Cited as *PhS*.
———. *Sämtliche Schriften und Briefe*. Berlin: Akademie-Verlag, 1960. Cited as *SSB*.

Bibliography

———. *Kleine Schriften zur Metaphysik.* Edited and translated by Hanz-Heinz Holz. Frankfurt: Insel, 1965.
———. *Hauptschriften zur Grundlegung der Philosophie.* Edited by Ernst Cassirer. Hamburg: Meiner, 1966.
———. *Opuscules et fragments inédits de Leibniz.* Edited by Louis Couturat. Hildesheim: Olms, 1966.
———. *Mathematische Schriften.* 7 vols. Edited by Carl Gerhardt. Hildesheim: Olms, 1971. = *Gesammelte Werke*, ed. G. H. Pertz, ser. 3. Cited as *MS*.
———. "Directiones ad rem medicam pertinentes." In *Studia Leibnitiana* 8: 40–68. Wiesbaden: Steiner, 1976.
———. *Theodicy.* Translated by E. M. Huggard. La Salle, IL: Open Court, 1985.
———. *Opera mathematica.* Vol. 3 of *Opera omnia.* 1768–1789. Edited by Louis Dutens. Reprint, Hildesheim: Olms, 1989.
———. *New Essays on Human Understanding.* Translated by Peter Remnant and Jonathan Bennett. Cambridge: Cambridge University Press, 1996.
———. *Philosophical Essays.* Translated by Roger Ariew and Daniel Garber. 1989. Cambridge: Cambridge University Press, 1996.
———. "Leibniz on Estimating the Uncertain: An English Translation of *De incerti aestimatione* with Commentary." Translated by Wolfgang David Cirilo de Melo. *Leibniz Review* 14 (2004): 31–53.
Lessing, Gotthold Ephraim. *Werke.* Edited by Herbert Georg Göpfert. Munich: Hanser, 1970–1979.
———. *Nathan the Wise.* Translated by R. Schlechter. Boston: Bedford, St. Martin's, 2004.
Lichtenberg, Georg Christoph. *Schriften und Briefe.* Edited by Wolfgang Promies. Munich: Hanser, 1967–1992.
Locke, John. *An Essay Concerning Human Understanding.* Oxford: Oxford University Press, 1975.
Lüder, August Ferdinand. *Kritik der Statistik und Politik nebst einer Begründung der politischen Philosophie.* Göttingen: Vandenhoeck & Ruprecht, 1812
———. *Kritische Geschichte der Statistik.* Göttingen: Röwer, 1817.
Martin, Benjamin. *Philosophia Britannica: or, a new and comprehensive system of the Newtownian philosophy, in a course of twelve lectures, with notes.* London, 1747.
Meier, Georg Friedrich. *Auszug aus der Vernunftlehre.* Halle: Gebauer, 1752.
Meinert, Friedrich. *Über das Studium der Mathematik für Juristen, Cammeralisten und Oekonomen auf Universitäten.* Halle: Hendel, 1788
———. *Über das Studium der militärisch-mathematischen Wissenschaften auf Universitäten.* Halle: Hendel, 1788.
Mises, Richard von. "Grundlagen der Wahrscheinlichkeitsrechnung." *Mathematische Zeitschrift* 5 (1919): 52–99.

Moivre, Abraham de. *The Doctrine of Chances: Or, a Method of Calculating the Probabilities of Events in Play*. New York: Chelsea Publishing, 1967.
Moritz, Karl Philipp. *Werke*. 3 vols. Edited by Horst Günther. Frankfurt: Insel, 1981.
Muelen, Johannes Andreas van der. *Forum conscientiae*. Amsterdam: A. van Someren, 1699.
Nietzsche, Friedrich. *Die Geneologie der Moral*. In *Kritische Studienausgabe*, vol. 5. Munich: Deutscher Taschenbuch-Verlag, 1988.
Otto, Eberhard. *Primae lineae notitiae rerumpublicarum*. Jena: Bielck, 1728.
Pascal, Blaise. *Minor Works of Pascal*. Translated by Orlando W. Wright. The Harvard Classics, vol. 48.2. New York: Collier & Son, 1909–1914.
———. *Œuvres complètes*. Edited by Louis Lafuma. Paris: Seuil, 1966.
———. *Pensées*. Translated by A. J. Krailsheimer. London: Penguin Books, 1966.
Patrizi, Francesco. *Della historia dieci dialoghi*. In Eckhard Kessler, *Theoretiker humanistischer Geschichtsschreibung*. Munich: Fink, 1971.
Petty, William. *The Economic Writings of Sir William Petty*. Edited by Charles Henry Hull. Cambridge, 1899.
Pinson de La Martinière, Jean. *Estat de la France comme elle estoit gouvernée en l'an 1648 où sont contenues diverses remarques et particularitez de l'histoire de nostre temps*. 1649. Facsimile, Paris: l'Arche du livre, 1970.
———. *The Present State of France: conteining [sic] the orders, dignities, and charges of that kingdom: newly corrected, and put into a better method than formerly. Written in French, and faithfully Englished*. Translated by Richard Wolley. London: John Starkey, 1671.
Playfair, William. *The Statistical Breviary; shewing, on a principle entirely new, the resources of every state and kingdom in Europe*. London, 1801.
Poisson, Siméon-Denis. *Recherches sur la probabilité des jugements en matière criminelle et en matière civile*. Paris: Bachelier, 1837.
Polack, Johann Friedrich. *Mathesis forensis*. 1734. Leipzig: Lanckisch, 1740.
The Present State of the Affairs Betwixt the Emperor and King of Bohemia. N.p., n.d.
Prussian Academy of Sciences. *Pièce qui a remporté le prix sur le sujet des événements fortuits; proposé par l'Académie Royale des Sciences et Belles Lettres de Berlin pour l'année 1751. Avec les pièces qui ont concouru*. Berlin: N.p., 1751.
Pseudo-Cyprian. *Liber de aleatoribus*. In *Patrologia Latina*, ed. J.-P. Migne, vol. 2, cols. 903–12. Paris: Migne, 1844–1864.
Pufendorf, Samuel von. *De jure naturae et gentium*. 1672. Translated by Jean Barbeyrac as *Le droit de la nature et des gens*. 2 vols. Amsterdam: P. de Coup, 1734.
Quetelet, Adolphe. *Du système social et des lois qui le régissent*. Paris: Guillaumain, 1848.

Quintilian, Marcus Fabius. *Institutio oratoria*. Translated by Donald A. Russell as *The Orator's Education*. Cambridge, MA: Harvard University Press, 2001.
Ramus, Petrus [Pierre de La Ramée]. *Arithmetices libri duo, et Algebra totidem*. Frankfurt: Wechel, 1592.
Riccoboni, Antonio. *Poetica Aristotelis ab Antonio Riccobono latine conversa: eiusdem Riccoboni paraphrasis in poeticam Aristotelis*. Munich: Fink, 1970.
Roccha, Angelo. *Trattato per la salute dell'anime e per la conservatione della robba, e dell' denaro: Contra i giuocchi della carte e dadi*. Rome: Guglielmo Facciotto, 1617.
Rüdiger, Andreas. *De sensu veri et falsi libri IV.* 1709. 2nd ed. Leipzig: Körner, 1722.
Schiller, Friedrich. *On the Aesthetic Education of Man, in a Series of Letters*. Oxford: Clarendon Press, 2005.
Schlözer, August Ludwig. *StatsGelartheit nach ihren Haupttheilen*. Part 2: *Allgemeine Statistik*. Vol. 1: *Theorie der Statistik. Nebst Ideen über das Studium der Politik überhaupt*. Göttingen: Vandenhoeck & Ruprecht, 1804.
Schmeitzel, Martin. *Praecognita historiae civilis universalis*. Jena: Kaltenbrunner, 1730.
Schnabel, Johann Gottfried. *Die Insel Felsenburg oder Wunderliche Fata einiger Seefahrer*. Breslau: Max, 1828.
———. *Wunderliche Fata einiger See-Fahrer*. Frankfurt: Insel, 1973.
Seckendorff, Veit Ludwig von. *Deutscher Fürstenstaat*. 1665. Reprint of 1737 ed. Aalen, Germany: Scientia, 1972.
Spinoza, Baruch de. *Algebraische Berechnung des Regenbogens*. Edited by Hans-Christian Lucas and Michael John Petry. Hamburg: Meiner, 1982.
———. *The Letters*. Translated by Samuel Shirley. Indianapolis: Hackett, 1995.
———.*Theological-Political Treatise*. Translated by Jonathan Israel and Michael Silverthorne. Cambridge: Cambridge University Press, 2007.
Sprat, Thomas. *History of the Royal Society*. Edited by Jackson I. Cope and Harold Whitmore Jones. St. Louis, MO: Washington University, 1958.
Sterne, Laurence. *The Life and Times of Tristram Shandy, Gentleman*. 1759–1767. New York: Norton, 1980.
Stevin, Simon. *L'Arithmétique*. Leiden: C. Plantin, 1585.
———. *De Thiende (Dezimalrechnung)*. Frankfurt: Akademische Verlagsgesellschaft, 1965.
Süssmilch [Süßmilch], Johann Peter. *Die göttliche Ordnung in den Veränderungen des menschlichen Geschlechts, aus der Geburt, dem Tode, und der Fortpflanzung desselben erwiesen*. Reprint of 1765–1766 ed., Augsburg: Cromm, 1988.
———. *Der Königl[ichen] Residenz Berlin schneller Wachsthum und Erbauung in zweyen Abhandlungen*. In id., *Die königliche Residenz Berlin und die Mark*

Brandenburg im 18. Jahrhundert. Edited by Jürgen Wilke. Berlin: Akademie Verlag, 1994.
Swift, Jonathan. *Gulliver's Travels.* 1726. London: Everyman's Library, 1986.
Tertullian. "De spectaculis." Translated by T. R. Glover. In *Minucius Felix, Apology; Tertullian, De spectaculis,* 230–301. Cambridge: Heinemann 1960.
Thiers, Jean-Baptiste. *Traité des jeux et des divertissemens.* Paris, 1686.
Thomasius, Christian. *Höchstnöthige Cautelen Welche ein Studiosus Juris . . . zu beobachten hat.* Halle: Renger, 1713.
Unger, Johann Friedrich. *Beyträge zur Mathesi forensi.* 1742. 2nd ed. Göttingen: Vandenhoeck, 1744.
Vinnius, Arnold. *Tractatus quatuor: Academico-forenses de jurisdictione, pactis transactionibus, collationibus.* Amsterdam: Commelinus, 1651.
Voltaire [François-Marie Arouet]. *Œuvres complètes.* Nendeln, Liechtenstein: Kraus Reprint, 1967.

———. *Œuvres complètes: Complete Works.* Edited by Theodore Besterman et al. Geneva: Institut et Musée Voltaire; Buffalo: University of Toronto Press; Oxford: Voltaire Foundation, 1968.
Whately, Richard. *Elements of Rhetoric.* Edited by Douglas Ehninger. Carbondale: Southern Illinois University Press, 1963.
Wieland, Christoph Martin. *Werke.* Munich: Hanser, 1967.
Wittich, Christoph. *Exercitationes theologicae.* Leiden: Boutesteyn, 1682.
Wolff, Christian. *Gesammelte Werke.* Edited by Jean Ecole et al. Hildesheim: Olms, 1975, 1983.

SECONDARY LITERATURE

Allen, Christopher J. W. *The Law of Evidence in Victorian England.* Cambridge: Cambridge University Press, 1997.
Alpers, Svetlana. "Describe or Narrate? A Problem in Realistic Representation." *New Literary History* 8 (1976–1977): 15–41.
———. *The Art of Describing: Dutch Art in the Seventeenth Century.* Chicago: University of Chicago Press, 1983.
Arndt, Hans Werner. *Der Möglichkeitsbegriff bei Christian Wolff und Johann Heinrich Lambert.* Dissertation. Göttingen: n.p., 1959.
Baasner, Rainer. *Abraham Gotthilf Kästner, Aufklärer(1719–1800).* Tübingen: Niemeyer 1991.
Barker, John C. *Strange Contrarieties: Pascal in England During the Age of Reason.* Montreal: McGill University Press, 1975.
Bastian, F. "Defoe's 'Journal of the Plague Year' Reconsidered." *Review of English Studies* 16 (1965): 151–173.
Battestin, Martin C. "General Introduction." In Henry Fielding. *The History of*

Tom Jones, a Foundling, 2 vols., ed. Fredson Bowers. Middletown, CT: Wesleyan University Press, 1975.
Behre, Otto. *Geschichte der Statistik in Brandenburg-Preußen*. 1905. Reprint, Vaduz, Liechtenstein: Topos, 1979.
Behrens, Rudolf. "Zur anthropologischen Fundierung von Politik und Rhetorik in den *Pensées* Blaise Pascals." *Rhetorik. Ein internationales Jahrbuch* 10 (1991): 16–29.
———. *Umstrittene Theodizee, erzählte Kontingenz*. Tübingen: Niemeyer, 1993.
Bender, John. *Imagining the Penitentiary: Fiction and Architecture of Mind in Eighteenth-Century England*. Chicago: University of Chicago Press, 1987.
Beniger, James R., and Dorothy L. Robyn. "Quantitative Graphics in Statistics." *American Statistician* 32 (1978): 1–11.
Biermann, Kurt-Reinhard. "Überblick über die Studien von G. W. Leibniz zur Wahrscheinlichkeitsrechnung." *Sudhoffs Archiv* 51 (1967): 79–85.
Blumenberg, Hans. "Kontingenz." In *Die Religion in Geschichte und Gegenwart*, vol. 3, cols. 1793ff. Tübingen: Niemeyer, 1959.
———. "The Concept of Reality and the Possibility of the Novel." In *New Perspectives in German Literary Criticism*, ed. Richard E. Amacher and Victor Lange. Princeton, NJ: Princeton University Press, 1979.
———. *Die Lesbarkeit der Welt*. Frankfurt: Suhrkamp, 1981.
———. *Paradigms for a Metaphorology*. Translated by Robert Salvage. Ithaca, NY: Cornell University Press, 2010.
Boeckh, Richard. *Die geschichtliche Entwicklung der amtlichen Statistik des preußischen Staates*. Berlin: Decker, 1863.
Borghero, Carlo. *La certezza e la storia*. Milan: Franco Angeli, 1983.
Bornscheuer, Lothar. *Zur Struktur der gesellschaftlichen Einbildungskraft*. Frankfurt: Suhrkamp, 1976.
Boudot, Maurice. "Probabilité et logique de l'argumentation selon Jakob Bernoulli." *Les études philosophiques*, n.s., 28 (1967): 265–288.
Bray, René. *La formation de la doctrine classique en France*. Paris: Nizet, 1966.
Brüggemann, Fritz. *Utopie und Robinsonade*. Hildesheim: Gerstenberg, 1978.
Brumfitt, John Henry. *Voltaire, Historian*. London: Oxford University Press, 1958.
Brunet, George. *Le pari de Pascal*. Paris: Desclée de Brouwer, 1956.
Burke, John H. "Observing the Observer." *Philological Quarterly* 61 (1982): 13–32.
Byrne, Edmund F. *Probability and Opinion*. The Hague: Nijhoff, 1968.
Campe, Rüdiger. "From the Theory of Technology to the Technique of Metaphor: Blumenberg's Initial Move." *Qui parle* 12 (2000): 105–126.
———. "How to Use the Future: The Old European and the Modern Form of Life." In *Prognosen über Bewegungen*, ed. Gabriele Brandstetter, Sibylle Peters, and Kai van Eikels, 107–120. Berlin: b_Books 2009.

———. "Shapes and Figures: Geometry and Rhetoric in the Age of Evidence." *Monatshefte* 102, no. 3 (2010): 285–299.
Carraud, V. "Leibniz lecteur des pensées de Pascal." *Dix-Septième Siècle* 151 (1986): 107–124.
Cassin, Barbara. *L'effet sophistique.* Paris: Gallimard, 1995.
Cataldi Madonna, Luigi. *La filosofia della probabilità nel pensiero moderno: Dalla Logique di Port-Royal a Kant.* Rome: Cadmo, 1988.
———. "Kant und der Probabilismus der Aufklärung." In *Kant und die Aufklärung*, ed. Norbert Hinske, 25–41. Hamburg: Meiner, 1993.
Chase, Cynthia. "Mechanical Doll, Exploding Machine." In id., *Decomposing Figures: Rhetorical Readings in the Romantic Tradition*, 141–156. Baltimore: Johns Hopkins University Press, 1986.
Colie, Rosalie L. *Some Thankfulnesse to Constantine: A Study of English Influence Upon the Early Works of Constantijn Huygens.* The Hague: Nijhoff, 1956.
Danneberg, Lutz. "Probabilitas hermeneutica. Zu einem Aspekt der Interpretations-Methodologie in der ersten Hälfte des 18. Jahrhunderts." In *Hermeneutik der Aufklärung,* ed. Axel Bühler and Luigi Cataldi Madonna, 27–48. Hamburg: Meiner 1995.
Daston, Lorraine. "Mathematics and the Moral Science (1785–1840)." In *Epistemological and Social Problems of the Sciences in the Early Nineteenth Century,* ed. A. Jahnke and M. Otte, 287–309. Dordrecht: Reidel, 1981.
———. "Rational Individuals Versus Laws of Society." In *The Probabilistic Revolution*, vol. 1, ed. Lorenz Krüger, 295–304. Cambridge, MA: MIT Press, 1987.
———. *Classical Probability in the Enlightenment.* Princeton, NJ: Princeton University Press, 1988.
Daston, Lorraine, and Katharine Park. *Wonders and the Order of Nature.* New York: Zone Books, 1998.
Daston, Lorraine, and Peter Galison. *Objectivity.* New York: Zone Books, 2007.
Davis, Lennard J. *Factual Fictions: The Origins of the English Novel.* New York: Columbia University Press, 1983.
Delft, Louis van. *Littérature et anthropologie: Nature humaine et caractère à l'âge classique.* Paris: Presses universitaires de France, 1993.
Derrida, Jacques. *Of Grammatology.* Baltimore: Johns Hopkins University Press, 1976.
———. *Edmund Husserl's "Origin of Geometry": An Introduction.* Translated by John P. Leavey. Stony Brook, NY: N. Hays, 1978.
Dießelhorst, Malte. *Die Lehre des Hugo Grotius vom Versprechen.* Cologne, Graz: Böhlau, 1959.
Dieudonné, Jean. *Abrégé d'histoire des mathématiques, 1700–1900.* 2 vols. Paris: Hermann, 1978.

Dijksterhuis, Eduard J. *Simon Stevin: Science in the Netherlands around 1600.* The Hague: Nijhoff, 1970.
Dupâquier, Michel. *Histoire de la démographie: La statistique de la population des origines à 1914.* Paris: Perrin, 1985.
Dreitzel, Horst. "Hermann Conring und die politische Wissenschaft seiner Zeit." In *Hermann Conring*, ed. Michael Stolleis. Berlin: Dunker & Humblot, 1983.
Empson, William. *"Tom Jones." Kenyon Review* 20, no. 2 (Spring 1958): 217–249.
Engfer, Hans-Jürgen. *Philosophie als Analysis. Studien zur Entwicklung philosophischer Analysiskonzeptionen unter dem Einfluss mathematischer Methodenmodelle im 17. und frühen 18. Jahrhundert.* Stuttgart: Fromann-Holzboog, 1982.
Esposito, Elena. "Fiktion und Virtualität." In *Medien—Computer—Realität: Wirklichkeitsvorstellungen und neue Medien,* ed. Sybille Krämer, 271–275. Frankfurt: Suhrkamp, 1998.
Estermann, Joseph. *Individualität und Kontingenz: Studie zur Individualitätsproblematik bei Gottfried Wilhelm Leibniz.* Bern: Peter Lang, 1990.
Ewald, François. *L'État providence.* Paris: Grasset, 1986.
Fleckenstein, Joachim Otto. *Johann und Jakob Bernoulli.* Basel: Birkhäuser, 1949.
Fleming, Paul. *Exemplarity and Mediocrity: The Art of the Average from Bourgeois Tragedy to Realism.* Stanford: Stanford University Press, 2009.
Flusser, Vilém. *Die Schrift: Hat Schreiben Zukunft?* Frankfurt: Fischer-Taschenbuch, 1992.
———. *Writings.* Translated by Erik Eisel. Edited by Andreas Strohl. Minneapolis: University of Minnesota Press, 2002.
Fogel, Michèle. *Les cérémonies de l'information dans la France du XVIe au milieu du XVIIIe siècle.* Paris: Fayard, 1989.
Fohrmann, Jürgen. *Abenteuer und Bürgertum: Zur Geschichte der deutschen Robinsonaden im 18. Jahrhundert.* Stuttgart: Metzler, 1981.
Foucault, Michel. *The Order of Things.* New York: Pantheon Books, 1971.
Franz, Günther. *Der dreißigjährige Krieg und das deutsche Volk,* Stuttgart: G. Fischer, 1979.
Freudenthal, Hans. "Huygens' Foundations of Probability." *Historia mathematica* 7 (1980): 113–117.
Frick, Werner. *Providenz und Kontingenz: Untersuchungen zur Schicksalssemantik im deutschen und europäischen Roman des 17. und 18. Jahrhunderts.* Tübingen: Niemeyer, 1988.
Fricke, Gerhard. *Gefühl und Schicksal bei Heinrich von Kleist: Studien über den inneren Vorgang im Leben und Schaffen des Dichters.* Darmstadt: Wissenschaftliche Buchgesellschaft, 1963.
Funkenhouser, H. Gray. "Historical Development of the Graphical Representation of Statistical Data." *Osiris* 3 (1937): 269–404.

Gasché, Rodolphe. *The Tain of the Mirror: Derrida and the Philosophy of Reflection.* Cambridge, MA: Harvard University Press, 1986.
———. "Some Reflections on the Notion of Hypotyposis in Kant." *Argumentation* 4 (1990): 85–100.
Genette, Gérard. "Vraisemblance et motivation," In id., *Figures II,* 71–99. Paris: Seuil, 1969.
Goody, Jack. *The Domestication of the Savage Mind.* Cambridge: Cambridge University Press, 1977.
Gouhier, Henri. *Blaise Pascal.* Paris: Vrin, 1971.
Grätzer, Jonas. *Edmund Halley und Caspar Neumann: Ein Beitrag zur Geschichte der Bevölkerungs-Statistik.* Breslau: Schottlaender, 1883.
Green-Pedersen, Niels Jorgen. *The Tradition of the Topics in the Middle Ages: The Commentaries on Aristotle's and Boethius' 'Topics.'* Munich: Philosophia, 1984.
Grohnert, Dietrich. "Schnabels 'Insel Felsenburg.'" *Weimarer Beiträge* 35 (1989): 602–617.
Guitton, Jean. *Pascal et Leibniz: Étude sur deux types de penseurs.* Paris: Aubier, 1951.
Hacking, Ian. "Equipossibility Theories of Probability." *British Journal for the Philosophy of Science* 22 (1971): 339–355.
———. "Jacques Bernoulli's 'Art of Conjecturing.'" *British Journal for the Philosophy of Science* 22 (1971): 339–355.
———. "The Logic of Pascal's Wager." *American Philosophical Quarterly* 9 (1972): 186–192.
———. *The Emergence of Probability: A Philosophical Study of Early Ideas About Probability, Induction and Statistical Inference.* London: Cambridge University Press, 1975.
———. "Why Do We Find the Origin of the Calculus of Probabilities in the Seventeenth Century?" In *Probabilistic Thinking, Thermodynamics, and the Interaction of the History and Philosophy of Science,* ed. J. Hintikka, 3–24. Dordrecht: Reidel, 1981.
———. "Biopower and the Avalanche of Printed Numbers." *Humanities in Society* 5 (1982): 209–229.
———. *The Taming of Chance.* Cambridge: Cambridge University Press, 1990.
Häfner, Ralph. *Johann Gottfried Herders Kulturentstehungslehre: Studien zu den Quellen und zur Methode seines Geschichtsdenkens.* Hamburg: Meiner, 1995.
Hahn, Roger. *Pierre Simon Laplace, 1749–1827: A Determined Scientist.* Cambridge, MA: Harvard University Press, 2005.
Hamacher, Werner. "Das Beben der Darstellung." In *Positionen der Literaturwissenschaft,* ed. David. E. Wellbery, 149–173. Munich: Beck, 1985.
———. "Über einige Unterschiede zwischen der Geschichte literarischer und

der Geschichte phänomenaler Ereignisse." In *Akten des VII. Internationalen Germanisten-Kongresses,* ed. Albrecht Schöne. Tübingen: Niemeyer, 1986.

Harnack, Adolf. *Geschichte der Königlich Preußischen Akademie der Wissenschaften zu Berlin.* Hildesheim: Olms, 1970.

Hartmann, Fritz, and Hans-Joachim Klauke. "Anfänge, Formen und Wirkungen der Medizinalstatistik." In *Statistik und Staatsbeschreibung in der Neuzeit, vornehmlich im 16.–18. Jahrhundert,* ed. Mohammed Rassem and Justin Stagl, 284–287. Paderborn: Schöningh, 1980.

Hazard, Paul. *The European Mind, 1680–1715.* Translated by J. Lewis May. Cleveland: Word Publishing, 1963.

Heidegger, Martin. *Essence of Reasons.* Translated by Terrence Malick. Evanston, IL: Northwestern University Press, 1969.

Herrmann, Hans Peter. "Zufall und Ich." In *Heinrich von Kleist: Aufsätze und Essays,* ed. Walter Müller-Seidel, 367–411. Darmstadt: Wissenschaftliche Buchgesellschaft, 1967.

Herrmann, Hans-Christian von. "Bewegliche Heere." *Kleist-Jahrbuch* 1998, 227–243.

Hess, Peter. "Zum Topos-Begriff der Barockzeit." *Rhetorik. Ein Internationales Jahrbuch* 10 (1991): 71–88.

Hofmann, Josef Ehrenfried. *Die Entwicklungsgeschichte der Leibnizschen Mathematik während des Aufenthaltes in Paris (1672–1676).* Munich: Leibniz Verlag, 1949.

Hunter, Paul. *The Reluctant Pilgrim: Defoe's Emblematic Method and Quest for Form in "Robinson Crusoe."* Baltimore: Johns Hopkins University Press, 1966.

Husserl, Edmund. *The Crisis of European Sciences and Transcendental Phenomenology.* Translated by David Carr. Evanston, IL: Northwestern University Press, 1970.

Ingarden, Roman. *The Literary Work of Art: An Investigation on the Borderlines of Ontology, Logic, and Theory of Literature.* 1931. Translated by George G. Grabowicz. Evanston, IL: Northwestern University Press, 1973.

Irmscher, Hans Dietrich. "Die geschichtsphilosophische Kontroverse zwischen Kant und Herder." In *Hamann—Kant—Herder,* ed. Bernhard Gajek, 111–192. Frankfurt: Peter Lang, 1987.

Iser, Wolfgang. "The Role of the Reader in Fielding's *Joseph Andrews* and *Tom Jones.*" In id., *The Implied Reader: Patterns of Communication in Prose Fiction from Bunyan to Beckett,* 29–56. Baltimore: Johns Hopkins University Press, 1974.

Jacobi, Klaus. "Zur Konzeption der praktischen Philosophie bei Leibniz." In *Akten des II. Internationalen Leibniz-Kongresses,* 3: 145–173. Wiesbaden: Steiner 1975.

Jacobs, Carol. *Uncontainable Romanticism: Shelley, Brontë, Kleist.* Baltimore: Johns Hopkins University Press, 1989.

Jacobs, Jürgen. *Prosa der Aufklärung: Moralische Wochenschriften, Autobiographie,*

Satire, Roman. Munich: Winkler, 1976.
Jacobs, Paul. *Prädestination und Verantwortlichkeit bei Calvin.* Darmstadt: Wissenschaftliche Buchgesellschaft, 1973.
John, Vincenz. *Geschichte der Statistik: Ein quellenmäßiges Handbuch für den akademischen Gebrauch wie für den Selbstunterricht.* Part 1: *Von dem Ursprung der Statistik bis auf Quetelet.* Stuttgart: Enke, 1884.
Jorland, Gérard. "The Saint Petersburg Paradox, 1738–1937." In *The Probabilistic Revolution,* ed. Lorenz Krüger, 157–190. Cambridge, MA: MIT Press, 1987.
Kavanagh, Thomas M. *Enlightenment and the Shadow of Chance: The Novel and the Culture of Gambling in Eighteenth-Century France.* Baltimore: Johns Hopkins University Press, 1993.
Kayser, Wolfgang. *Entstehung und Krise des modernen Romans.* 1955. Stuttgart: Metzler, 1963.
Kimpel, Dieter, and Conrad Wiedemann, eds. *Theorie und Technik des Romans im 17. und 18. Jahrhundert.* Tübingen: Niemeyer, 1970.
Kirsch, Ulrich. *Blaise Pascals "Pensées" (1656–1662): Systematische "Gedanken" über Tod, Vergänglichkeit und Glück.* Freiburg: Alber, 1989.
Kittler, Friedrich. "Ottilie Hauptmann." In id., *Dichter, Mutter, Kind.* Munich: Fink, 1991.
Köhler, Erich. *Der literarische Zufall, das Mögliche und die Notwendigkeit.* 1973. Frankfurt: Fischer-Taschenbuch, 1993.
Koselleck, Reinhart. *Critique and Crisis: Enlightenment and the Pathogenesis of Modern Society.* Cambridge, MA: MIT Press, 1988.
Krämer, Sybille. *Berechenbare Vernunft: Kalkül und Rationalismus im 17. Jahrhundert.* Berlin: De Gruyter, 1991.
———. "Kalküle als Repräsentation." In *Räume des Wissens: Repräsentation, Codierung, Spur,* ed. Hans-Jörg Rheinberger, Michael Hagner, and Bettina Wahrig-Schmidt, 111–122. Berlin: Akademie-Verlag, 1997.
Krüger, Lorenz, ed. *The Probabilistic Revolution.* 2 vols. Cambridge, MA: MIT Press, 1987.
Lämmert, Eberhard, et al., eds. *Romantheorie: Dokumentation ihrer Geschichte in Deutschland, 1620–1880.* Cologne: Kiepenheuer & Witsch, 1971
Lange, Victor. "Erzählformen im Roman des 18. Jahrhunderts." In *Henry Fielding und der englische Roman,* ed. Wolfgang Iser, 474–489. Darmstadt: Wissenschaftliche Buchgesellschaft, 1972.
Lee, William. *Daniel Defoe: His Life, and Recently Discovered Writings; Extending from 1716 to 1729.* London: J. C. Hotten, 1869.
Leyden, Wolfgang von. *Aristotle on Equality and Justice.* London: Macmillan in Association with the London School of Economics and Political Science, 1985.
Lesmerle, Bruno, ed. *La preuve en justice de l'antiquité à nos jours.* Rennes: Presses universitaires de Rennes, 2003.

Link, Jürgen. *Versuch über den Normalismus: Wie Normalismus produziert wird.* Opladen: Westdeutscher Verlag, 1996.
Lönning, Per. *Cet effrayant pari: Une 'Pensée' Pascalienne et ses critiques.* Paris: J. Vrin, 1980.
Lübbe, Hermann. "'Topik', 'Sinn' und die Geschichte der Zufallstheorie." In *Kontingenz,* ed. Gerhard von Graevenitz and Odo Marquard, 141ff. Munich: Fink, 1998.
Lugowski, Clemens. *Wirklichkeit und Dichtung: Untersuchungen zur Wirklichkeitsauffassung Heinrich von Kleists.* Frankfurt: Diesterweg, 1936.
———. *Form, Individuality, and the Novel.* 1932. Translated by John D. Halliday. Cambridge: Polity Press, 1990.
Luhmann, Niklas. *Risk: A Sociological Theory.* Translated by Rhodes Barrett. New York: De Gruyter, 1993.
———. *Observations on Modernity.* Translated by William Whobrey. Stanford: Stanford University Press, 1998.
———. *Art as a Social System.* Translated by Eva Knodt. Stanford: Stanford University Press, 2000.
Lyotard, Jean-François. *Lessons on the Analytic of the Sublime: Kant's Critique of Judgment.* Translated by Elizabeth Rottenberg. Stanford: Stanford University Press, 1994.
Man, Paul de. "Pascal's Allegory of Persuasion." In id., *Aesthetic Ideology,* ed. Andrzej Warminksi, 51–69. Minneapolis: University of Minnesota Press, 1996.
Marin, Louis. *La critique du discours: Sur la "Logique de Port-Royal" et les "Pensées" de Pascal.* Paris: Minuit, 1975.
———. "Die Fragmente Pascals." In *Fragment und Totalität,* ed. Lucien Dällenbach and Christian L. Hart Nibbrig, 160–181. Frankfurt: Suhrkamp, 1984.
Mayer, Robert. "The Reception of a 'Journal of the Plague Year.'" In *English Literary History* 57 (1990): 529–599.
McKenna, Antony. "L'argument 'Infini–rien.'" In *Méthodes chez Pascal: Actes du Colloque à Clermont-Ferrand 1976,* 497–508. Paris: Presses universitaires de France, 1979.
Meijer, Reinder P. *Literature of the Low Countries: A Short History of Literature in the Netherlands and Belgium.* Assen: Van Gorcum, 1971,
Melzer, Sarah: *Discourses of the Fall: A Study of Pascal's "Pensées."* Berkeley: University of California Press, 1986.
Mesnard, Jean. *Pascal et les Roannez.* Bruges: Desclée de Brouwer, 1965.
———. *Les "Pensées" de Pascal.* Paris: Société d'édition d'enseignement supérieur, 1976.
Meuthen, Erich. *Selbstüberredung: Rhetorik und Roman im 18. Jahrhundert.* Freiburg: Rombach, 1994.

Meyer-Krentler, Eckardt. *Der andere Roman. Gellerts "Schwedische Gräfin."* Göppingen: Kümmerle, 1971.
Miller, Norbert. *Der empfindsame Erzähler. Untersuchungen an Romananfängen des 18. Jahrhunderts.* Munich: Hanser, 1968.
Mog, Paul. *Ratio und Gefühlskultur: Studien zu Psychogenese und Literatur im 18. Jahrhundert.* Tübingen: Niemeyer, 1976.
Mohl, Robert von. *Die Geschichte und Literatur der Staatswissenschaften.* Erlangen: F. Enke, 1855.
Moore, Benjamin. "Governing Discourses: The Problem of Narrative Authority in 'A Journal of the Plague Year." *The Eighteenth Century* 33 (1992): 133–147.
Moraux, Paul. "La joute dialectique d'après le huitième livre des *Topiques*." In *Aristotle on Dialectics: Proceedings of the Third Symposium Aristotelicum*, ed. G. E. L. Owen, 277–311. Oxford: Oxford University Press, 1968.
Nicolson, Marjorie, and Nora M. Mohler. "The Scientific Background of 'Voyage to Laputa.'" In *Jonathan Swift*, ed. Alexander Norman Jeffares. Harlow, Essex: Longman, 1976.
Novak, Maximilian E. "Defoe and the Disordered City." *PMLA* 92 (1977): 241–252.
Obertello, Luca. *John Locke e Port-Royal. Il problema della probabilità.* Trieste: Istituto di filosofia, 1964.
Orcibal, Jean. "Le fragment *Infini–rien* et ses sources." In *Blaise Pascal: L'homme et l'œuvre: Cahiers de Royaumont*, 158–185. Paris: Minuit, 1956.
Östreich, Gerhard. "Justus Lipsius als Theoretiker des neuzeitlichen Machtstaates." In id., *Geist und Gestalt des frühmodernen Staates*, 35–79. Berlin: Dunker & Humblot, 1969.
Othmer, Sieglinde C. *Berlin und die Verbreitung des Naturrechts in Europa. Kultur- und sozialgeschichtliche Studien zu Jean Barbeyracs Pufendorf-Übersetzungen und eine Analyse seiner Leserschaft.* Berlin: De Gruyter, 1970.
Patey, Douglas Lane. *Probability and Literary Form: Philosophic Theory and Literary Practice in the Augustan Age.* Cambridge: Cambridge University Press, 1984.
Pearson, K. "James Bernoulli's Theorem." *Biometrika* 17 (1925): 201–210.
Perry, Petra. *Möglichkeit am Rande der Wahrscheinlichkeit: Die "phantastische Situation" in der Kleistschen Novellistik.* Cologne: Böhlau, 1989.
Poovey, Mary. *A History of the Modern Fact: Problems of Knowledge in the Sciences of Wealth and Society.* Chicago: University of Chicago Press, 1998.
Rassem, Mohammed, and Justin Stagl. "Zur Geschichte der Statistik und Staatsbeschreibung." *Zeitschrift für Politik* 24 (1977): 81–86.
———, eds. *Statistik und Staatsbeschreibung in der Neuzeit, vornehmlich im 16.–18. Jahrhundert.* Paderborn: Schöningh, 1980.
Richetti, John. *Defoe's Narratives: Situations and Structures.* Oxford: Clarendon Press, 1975.

Riedel, Manfred. "Historizismus und Kritizismus: Ihre Entwicklung und ihr Verhältnis zur Aufklärung. Kants Streit mit G. Forster und J. G. Herder." In *Deutschlands kulturelle Entfaltung: Die Neubestimmung des Menschen*, ed. Bernhard Fabian, Wilhelm Schmidt-Biggemann, and Rudolph Vierhaus, 31–48. Munich: Kraus International Publications, 1980.

Riffaterre, Michael. *Semiotics of Poetry*. Bloomington: Indiana University Press, 1978.

Rosen, Alan. "Plague, Fire, and Typology in Defoe's *A Journal of the Plague Year*." *Connotations* 1 (1993): 258–282.

Schings, Hans-Jürgen. "Der Staatsroman im Zeitalter der Aufklärung." In *Handbuch des deutschen Romans*, ed. Helmut Koopmann, 161–169. Düsseldorf: Bagel, 1983.

Schlittgen, Rainer. *Einführung in die Statistik: Analyse und Modellierung von Daten*. Munich: Oldenbourg, 1990.

Schmidt-Biggemann, Wilhelm. *Topica universalis: Eine Modellgeschichte humanistischer und barocker Wissenschaft*. Hamburg: Meiner, 1983.

Schneider, Helmut J. "Der Zufall der Geburt: Lessings *Nathan der Weise*." In *Körper/Kultur: Kalifornische Studien zur deutschen Moderne*, ed. Thomas W. Kniesche, 100–118. Würzburg: Königshausen & Neumann, 1995.

Schneider, Ivo. "Die Entwicklung des Wahrscheinlichkeitsbegriffs in der Mathemathik von Pascal bis Laplace." Diss., Munich, 1972.

———. "Mathematisierung des Wahrscheinlichen und Anwendung auf Massenphänomene im 17. und 18. Jahrundert." In *Statistik und Staatsbeschreibung in der Neuzeit, vornehmlich im 16.–18. Jahrhundert*, ed. Mohammed Rassem and Justin Stagl. Paderborn: Schöningh, 1980.

———, ed. *Die Entwicklung der Wahrscheinlichkeitstheorie von den Anfängen bis 1933. Einführungen und Texte*. Darmstadt: Wissenschaftliche Buchgesellschaft, 1988.

Schneider, Manfred. "Der Mensch als Quelle." In *Der Mensch—das Medium der Gesellschaft?* ed. Peter Fuchs and Andreas Göbel, 297–322. Frankfurt: Suhkamp, 1994.

Schobinger, Jean-Pierre. *Blaise Pascals Reflexionen über die Geometrie im allgemeinen*. Stuttgart: Schwabe, 1974.

Seifen, Johannes. *Der Zufall—eine Chimäre? Untersuchungen zum Zufallsbegriff in der philosophischen Tradition und bei Gottfried Wilhelm Leibniz*. Sankt Augustin, Germany: Academia, 1992.

Seifert, Arno. *Cognitio historica: Die Geschichte als Namengeberin der frühneuzeitlichen Empirie*. Berlin: Dunker & Humblot, 1976.

———. "Staatenkunde." In *Statistik und Staatsbeschreibung in der Neuzeit, vornehmlich im 16.–18. Jahrhundert*, ed. Mohammed Rassem and Justin Stagl, 217–244. Paderborn: Schöningh, 1980.

———. "Conring und die Begründung der Staatenkunde." In *Hermann Conring (1606–1681): Beiträge zu Leben und Werk*, ed. Michael Stolleis. Berlin: Dunker & Humblot, 1983.
Sembdner, Helmut, ed. *Heinrich von Kleists Lebensspuren: Dokumente und Berichte der Zeitgenossen*. 1957. Frankfurt: Insel,1984.
Serres, Michel. *Le système de Leibniz et ses modèles mathématiques: Étoiles—Schémas—Points*. Paris: Presses universitaires de France, 1968.
Shapin, Steven, and Simon Schaffer. *Leviathan and the Air-Pump: Hobbes, Boyle, and the Experimental Life*. Princeton, NJ: Princeton University Press, 1985.
Shapiro, Barbara J. *Probability and Certainty in Seventeenth-Century England*. Princeton, NJ: Princeton University Press, 1983.
Sheynin, Oscar B. "J. H. Lambert's Work on Probability." *Archive for History of Exact Sciences* 7 (1970–1971): 244–256.
———. "Newton and the Classical Theory of Probability." *Archive for History of Exact Sciences* 7 (1970–1971): 217–243.
Shklovsky, Viktor. "Art as Technique." In *Russian Formalist Criticism: Four Essays,* ed. Lee T. Lemon and Marion J. Reiss, 3–24. Lincoln: University of Nebraska Press, 1965.
———. *Theory of Prose*. Translated by Benjamin Sher. Elmwood Park, IL: Dalkey Archive Press, 1990.
Solbach, Andreas. *Evidentia und Erzähltheorie: Die Rhetorik anschaulichen Erzählens und ihre antiken Quellen*. Munich: Fink, 1994.
Starr, George A. *Defoe & Casuistry*. Princeton, NJ: Princeton University Press, 1971.
———. *Defoe and Spiritual Autobiography*. 2nd ed. New York: Gordian Press, 1971.
Stauffer, Donald A. *English Biography Before 1700*. Cambridge, MA: Harvard University Press, 1930.
Stigler, Stephen M. *The History of Statistics: The Measurement of Uncertainty Before 1900*. Cambridge MA: Harvard University Press, 1986.
Struik, Dirk J. *The Land of Stevin and Huygens: A Sketch of Science and Technology in the Dutch Republic During the Golden Century*. Dordrecht: Reidel, 1981.
Stump, Eleonore. "Dialectic and Aristotle's *Topics*." In id., *Dialectic and Its Place in Medieval Logic,* 11–30. Ithaca, NY: Cornell University Press, 1989.
———. "Dialectic and Boethius' *De topicis differentiis*." In id., *Dialectic and Its Place in the Development of Medieval Logic,* 31–56. Ithaca, NY: Cornell University Press, 1989.
Sutherland, James. *Defoe*. London: Methuen, 1950.
Svagelski, Jean. *L'idée de compensation en France, 1750–1850*. Lyon: Éditions l'Hermès, 1981.

Thirouin, Laurent. *Le hasard et les règles: Le modèle du jeu dans la pensée de Pascal.* Paris: J. Vrin, 1991.
Thrower, N. J. W. "Edmond Halley as a Thematic Geo-Cartographer." *Annals of the Association of American Geographers* 59 (1969): 652–657.
Topliss, Patricia. *The Rhetoric of Pascal: A Study of His Art of Persuasion in the "Provinciales" and the "Pensées."* Leicester: Leicester University Press, 1966.
Tufte, Edward Rolfe. *The Visual Display of Quantitative Information.* Cheshire, CT: Graphics Press, 1982.
———. *Envisioning Information.* Cheshire, CT: Graphics Press, 1990.
Vismann, Cornelia. *Files: Law and Media Technology.* Stanford: Stanford University Press, 2008.
Voßkamp, Wilhelm. *Romantheorie in Deutschland: Von Martin Opitz bis Friedrich von Blankenburg.* Stuttgart: Metzler, 1973.
———. "Die Macht der Tugend. Zur Poetik des utopischen Romans am Beispiel von Schnabels 'Insel Felsenburg' und von Loens' 'Der redliche Mann am Hofe.'" In *Dichtungstheorien der deutschen Frühaufklärung*, ed. Theodor Verweyen, 176–186. Tübingen: Niemeyer, 1995.
Vries, Katja de. "De calculus der billijkheid: Over de *juridische* oorsprong van de kansrekening." PhD diss., Leiden University, 2006. www.vub.ac.be/LSTS/pub/Devries/005.pdf (accessed June 9, 2012).
Watt, Ian. *The Rise of the Novel: Studies in Defoe, Richardson and Fielding.* London: Chatto & Windus, 1967.
Wellbery, David E. "Der Zufall der Geburt." In *Kontingenz*, ed. Gerhard von Graevenitz and Odo Marquard, 291–318. Munich: Fink, 1998.
Welzig, Werner. "Der Wandel des Abenteuertums." In *Pikarische Welt: Schriften zum europäischen Schelmenroman*, ed. H. Heidenreich, 438–454. Darmstadt: Wissenschaftliche Buchgesellschaft, 1969.
Wolters, Gereon. *Basis und Deduktion: Studien zur Entstehung und Bedeutung der Theorie der axiomatischen Methode bei J. H. Lambert (1728–1777).* Berlin: De Gruyter, 1980.
Zimmermann, Everett. "H. F.'s Meditations: *A Journal of the Plague Year.*" *PMLA* 87 (1972): 417–423.

Cultural Memory | in the Present

Jean-Luc Marion, *In the Self's Place: The Approach of Saint Augustine*
Rodolphe Gasché, *Georges Bataille: Phenomenology and Phantasmatology*
Niklas Luhmann, *Theory of Society, Volume 1*
Alessia Ricciardi, *After La Dolce Vita: A Cultural Prehistory of Berlusconi's Italy*
Daniel Innerarity, *The Future and Its Enemies: In Defense of Political Hope*
Patricia Pisters, *The Neuro-Image: A Deleuzian Film-Philosophy of Digital Screen Culture*
François-David Sebbah, *Testing the Limit: Derrida, Henry, Levinas, and the Phenomenological Tradition*
Erik Peterson, *Theological Tractates*, edited by Michael J. Hollerich
Feisal G. Mohamed, *Milton and the Post-Secular Present: Ethics, Politics, Terrorism*
Pierre Hadot, *The Present Alone Is Our Happiness, Second Edition: Conversations with Jeannie Carlier and Arnold I. Davidson*
Yasco Horsman, *Theaters of Justice: Judging, Staging, and Working Through in Arendt, Brecht, and Delbo*
Jacques Derrida, *Parages*, edited by John P. Leavey
Henri Atlan, *The Sparks of Randomness, Volume 1: Spermatic Knowledge*
Rebecca Comay, *Mourning Sickness: Hegel and the French Revolution*
Djelal Kadir, *Memos from the Besieged City: Lifelines for Cultural Sustainability*
Stanley Cavell, *Little Did I Know: Excerpts from Memory*
Jeffrey Mehlman, *Adventures in the French Trade: Fragments Toward a Life*
Jacob Rogozinski, *The Ego and the Flesh: An Introduction to Egoanalysis*
Marcel Hénaff, *The Price of Truth: Gift, Money, and Philosophy*
Paul Patton, *Deleuzian Concepts: Philosophy, Colonialization, Politics*
Michael Fagenblat, *A Covenant of Creatures: Levinas's Philosophy of Judaism*
Stefanos Geroulanos, *An Atheism that Is Not Humanist Emerges in French Thought*

Andrew Herscher, *Violence Taking Place: The Architecture of the Kosovo Conflict*

Hans-Jörg Rheinberger, *On Historicizing Epistemology: An Essay*

Jacob Taubes, *From Cult to Culture*, edited by Charlotte Fonrobert and Amir Engel

Peter Hitchcock, *The Long Space: Transnationalism and Postcolonial Form*

Lambert Wiesing, *Artificial Presence: Philosophical Studies in Image Theory*

Jacob Taubes, *Occidental Eschatology*

Freddie Rokem, *Philosophers and Thespians: Thinking Performance*

Roberto Esposito, *Communitas: The Origin and Destiny of Community*

Vilashini Cooppan, *Worlds Within: National Narratives and Global Connections in Postcolonial Writing*

Josef Früchtl, *The Impertinent Self: A Heroic History of Modernity*

Frank Ankersmit, Ewa Domanska, and Hans Kellner, eds., *Re-Figuring Hayden White*

Michael Rothberg, *Multidirectional Memory: Remembering the Holocaust in the Age of Decolonization*

Jean-François Lyotard, *Enthusiasm: The Kantian Critique of History*

Ernst van Alphen, Mieke Bal, and Carel Smith, eds., *The Rhetoric of Sincerity*

Stéphane Mosès, *The Angel of History: Rosenzweig, Benjamin, Scholem*

Pierre Hadot, *The Present Alone Is Our Happiness: Conversations with Jeannie Carlier and Arnold I. Davidson*

Alexandre Lefebvre, *The Image of the Law: Deleuze, Bergson, Spinoza*

Samira Haj, *Reconfiguring Islamic Tradition: Reform, Rationality, and Modernity*

Diane Perpich, *The Ethics of Emmanuel Levinas*

Marcel Detienne, *Comparing the Incomparable*

François Delaporte, *Anatomy of the Passions*

René Girard, *Mimesis and Theory: Essays on Literature and Criticism, 1959-2005*

Richard Baxstrom, *Houses in Motion: The Experience of Place and the Problem of Belief in Urban Malaysia*

Jennifer L. Culbert, *Dead Certainty: The Death Penalty and the Problem of Judgment*

Samantha Frost, *Lessons from a Materialist Thinker: Hobbesian Reflections on Ethics and Politics*

Regina Mara Schwartz, *Sacramental Poetics at the Dawn of Secularism: When God Left the World*

Gil Anidjar, *Semites: Race, Religion, Literature*

Ranjana Khanna, *Algeria Cuts: Women and Representation, 1830 to the Present*

Esther Peeren, *Intersubjectivities and Popular Culture: Bakhtin and Beyond*

Eyal Peretz, *Becoming Visionary: Brian De Palma's Cinematic Education of the Senses*

Diana Sorensen, *A Turbulent Decade Remembered: Scenes from the Latin American Sixties*

Hubert Damisch, *A Childhood Memory by Piero della Francesca*

José van Dijck, *Mediated Memories in the Digital Age*

Dana Hollander, *Exemplarity and Chosenness: Rosenzweig and Derrida on the Nation of Philosophy*

Asja Szafraniec, *Beckett, Derrida, and the Event of Literature*

Sara Guyer, *Romanticism After Auschwitz*

Alison Ross, *The Aesthetic Paths of Philosophy: Presentation in Kant, Heidegger, Lacoue-Labarthe, and Nancy*

Gerhard Richter, *Thought-Images: Frankfurt School Writers' Reflections from Damaged Life*

Bella Brodzki, *Can These Bones Live? Translation, Survival, and Cultural Memory*

Rodolphe Gasché, *The Honor of Thinking: Critique, Theory, Philosophy*

Brigitte Peucker, *The Material Image: Art and the Real in Film*

Natalie Melas, *All the Difference in the World: Postcoloniality and the Ends of Comparison*

Jonathan Culler, *The Literary in Theory*

Michael G. Levine, *The Belated Witness: Literature, Testimony, and the Question of Holocaust Survival*

Jennifer A. Jordan, *Structures of Memory: Understanding German Change in Berlin and Beyond*

Christoph Menke, *Reflections of Equality*

Marlène Zarader, *The Unthought Debt: Heidegger and the Hebraic Heritage*

Jan Assmann, *Religion and Cultural Memory: Ten Studies*

David Scott and Charles Hirschkind, *Powers of the Secular Modern: Talal Asad and His Interlocutors*

Gyanendra Pandey, *Routine Violence: Nations, Fragments, Histories*

James Siegel, *Naming the Witch*

J. M. Bernstein, *Against Voluptuous Bodies: Late Modernism and the Meaning of Painting*

Theodore W. Jennings, Jr., *Reading Derrida / Thinking Paul: On Justice*

Richard Rorty and Eduardo Mendieta, *Take Care of Freedom and Truth Will Take Care of Itself: Interviews with Richard Rorty*

Jacques Derrida, *Paper Machine*

Renaud Barbaras, *Desire and Distance: Introduction to a Phenomenology of Perception*

Jill Bennett, *Empathic Vision: Affect, Trauma, and Contemporary Art*

Ban Wang, *Illuminations from the Past: Trauma, Memory, and History in Modern China*

James Phillips, *Heidegger's Volk: Between National Socialism and Poetry*

Frank Ankersmit, *Sublime Historical Experience*

István Rév, *Retroactive Justice: Prehistory of Post-Communism*

Paola Marrati, *Genesis and Trace: Derrida Reading Husserl and Heidegger*

Krzysztof Ziarek, *The Force of Art*

Marie-José Mondzain, *Image, Icon, Economy: The Byzantine Origins of the Contemporary Imaginary*

Cecilia Sjöholm, *The Antigone Complex: Ethics and the Invention of Feminine Desire*

Jacques Derrida and Elisabeth Roudinesco, *For What Tomorrow . . . : A Dialogue*

Elisabeth Weber, *Questioning Judaism: Interviews by Elisabeth Weber*

Jacques Derrida and Catherine Malabou, *Counterpath: Traveling with Jacques Derrida*

Martin Seel, *Aesthetics of Appearing*

Nanette Salomon, *Shifting Priorities: Gender and Genre in Seventeenth-Century Dutch Painting*

Jacob Taubes, *The Political Theology of Paul*

Jean-Luc Marion, *The Crossing of the Visible*

Eric Michaud, *The Cult of Art in Nazi Germany*

Anne Freadman, *The Machinery of Talk: Charles Peirce and the Sign Hypothesis*

Stanley Cavell, *Emerson's Transcendental Etudes*

Stuart McLean, *The Event and Its Terrors: Ireland, Famine, Modernity*

Beate Rössler, ed., *Privacies: Philosophical Evaluations*

Bernard Faure, *Double Exposure: Cutting Across Buddhist and Western Discourses*

Alessia Ricciardi, *The Ends of Mourning: Psychoanalysis, Literature, Film*

Alain Badiou, *Saint Paul: The Foundation of Universalism*

Gil Anidjar, *The Jew, the Arab: A History of the Enemy*

Jonathan Culler and Kevin Lamb, eds., *Just Being Difficult? Academic Writing in the Public Arena*

Jean-Luc Nancy, *A Finite Thinking*, edited by Simon Sparks

Theodor W. Adorno, *Can One Live after Auschwitz? A Philosophical Reader*, edited by Rolf Tiedemann

Patricia Pisters, *The Matrix of Visual Culture: Working with Deleuze in Film Theory*

Andreas Huyssen, *Present Pasts: Urban Palimpsests and the Politics of Memory*

Talal Asad, *Formations of the Secular: Christianity, Islam, Modernity*

Dorothea von Mücke, *The Rise of the Fantastic Tale*

Marc Redfield, *The Politics of Aesthetics: Nationalism, Gender, Romanticism*

Emmanuel Levinas, *On Escape*

Dan Zahavi, *Husserl's Phenomenology*

Rodolphe Gasché, *The Idea of Form: Rethinking Kant's Aesthetics*

Michael Naas, *Taking on the Tradition: Jacques Derrida and the Legacies of Deconstruction*

Herlinde Pauer-Studer, ed., *Constructions of Practical Reason: Interviews on Moral and Political Philosophy*

Jean-Luc Marion, *Being Given That: Toward a Phenomenology of Givenness*

Theodor W. Adorno and Max Horkheimer, *Dialectic of Enlightenment*

Ian Balfour, *The Rhetoric of Romantic Prophecy*

Martin Stokhof, *World and Life as One: Ethics and Ontology in Wittgenstein's Early Thought*

Gianni Vattimo, *Nietzsche: An Introduction*

Jacques Derrida, *Negotiations: Interventions and Interviews, 1971–1998*, ed. Elizabeth Rottenberg

Brett Levinson, *The Ends of Literature: The Latin American "Boom" in the Neoliberal Marketplace*

Timothy J. Reiss, *Against Autonomy: Cultural Instruments, Mutualities, and the Fictive Imagination*

Hent de Vries and Samuel Weber, eds., *Religion and Media*

Niklas Luhmann, *Theories of Distinction: Re-Describing the Descriptions of Modernity*, ed. and introd. William Rasch

Johannes Fabian, *Anthropology with an Attitude: Critical Essays*

Michel Henry, *I Am the Truth: Toward a Philosophy of Christianity*

Gil Anidjar, *"Our Place in Al-Andalus": Kabbalah, Philosophy, Literature in Arab-Jewish Letters*

Hélène Cixous and Jacques Derrida, *Veils*

F. R. Ankersmit, *Historical Representation*

F. R. Ankersmit, *Political Representation*

Elissa Marder, *Dead Time: Temporal Disorders in the Wake of Modernity (Baudelaire and Flaubert)*

Reinhart Koselleck, *The Practice of Conceptual History: Timing History, Spacing Concepts*

Niklas Luhmann, *The Reality of the Mass Media*

Hubert Damisch, *A Theory of /Cloud/: Toward a History of Painting*

Jean-Luc Nancy, *The Speculative Remark: (One of Hegel's bon mots)*

Jean-François Lyotard, *Soundproof Room: Malraux's Anti-Aesthetics*

Jan Patočka, *Plato and Europe*

Hubert Damisch, *Skyline: The Narcissistic City*

Isabel Hoving, *In Praise of New Travelers: Reading Caribbean Migrant Women Writers*

Richard Rand, ed., *Futures: Of Jacques Derrida*

William Rasch, *Niklas Luhmann's Modernity: The Paradoxes of Differentiation*

Jacques Derrida and Anne Dufourmantelle, *Of Hospitality*

Jean-François Lyotard, *The Confession of Augustine*

Kaja Silverman, *World Spectators*

Samuel Weber, *Institution and Interpretation: Expanded Edition*

Jeffrey S. Librett, *The Rhetoric of Cultural Dialogue: Jews and Germans in the Epoch of Emancipation*

Ulrich Baer, *Remnants of Song: Trauma and the Experience of Modernity in Charles Baudelaire and Paul Celan*

Samuel C. Wheeler III, *Deconstruction as Analytic Philosophy*

David S. Ferris, *Silent Urns: Romanticism, Hellenism, Modernity*

Rodolphe Gasché, *Of Minimal Things: Studies on the Notion of Relation*

Sarah Winter, *Freud and the Institution of Psychoanalytic Knowledge*

Samuel Weber, *The Legend of Freud: Expanded Edition*

Aris Fioretos, ed., *The Solid Letter: Readings of Friedrich Hölderlin*

J. Hillis Miller / Manuel Asensi, *Black Holes / J. Hillis Miller; or, Boustrophedonic Reading*

Miryam Sas, *Fault Lines: Cultural Memory and Japanese Surrealism*

Peter Schwenger, *Fantasm and Fiction: On Textual Envisioning*

Didier Maleuvre, *Museum Memories: History, Technology, Art*

Jacques Derrida, *Monolingualism of the Other; or, The Prosthesis of Origin*

Andrew Baruch Wachtel, *Making a Nation, Breaking a Nation: Literature and Cultural Politics in Yugoslavia*

Niklas Luhmann, *Love as Passion: The Codification of Intimacy*

Mieke Bal, ed., *The Practice of Cultural Analysis: Exposing Interdisciplinary Interpretation*

Jacques Derrida and Gianni Vattimo, eds., *Religion*

The authorized representative in the EU for product safety and compliance is:
Mare Nostrum Group
B.V Doelen 72
4831 GR Breda
The Netherlands

www.ingramcontent.com/pod-product-compliance
Lightning Source LLC
Chambersburg PA
CBHW021813300426
44114CB00009BA/151